T0312606

Transportation Project Management

Rob Tieman

This edition first published 2023
© 2023 Rob Tieman

All rights reserved. No part of this publication may be reproduced, stored in a retrieval system, or transmitted, in any form or by any means, electronic, mechanical, photocopying, recording or otherwise, except as permitted by law. Advice on how to obtain permission to reuse material from this title is available at http://www.wiley.com/go/permissions.

The right of Rob Tieman to be identified as the author of this work has been asserted in accordance with law.

Registered Office(s)
John Wiley & Sons, Inc., 111 River Street, Hoboken, NJ 07030, USA
John Wiley & Sons Ltd, The Atrium, Southern Gate, Chichester, West Sussex, PO19 8SQ, UK

For details of our global editorial offices, customer services, and more information about Wiley products visit us at www.wiley.com.

Wiley also publishes its books in a variety of electronic formats and by print-on-demand. Some content that appears in standard print versions of this book may not be available in other formats.

Trademarks: Wiley and the Wiley logo are trademarks or registered trademarks of John Wiley & Sons, Inc. and/or its affiliates in the United States and other countries and may not be used without written permission. All other trademarks are the property of their respective owners. John Wiley & Sons, Inc. is not associated with any product or vendor mentioned in this book.

Limit of Liability/Disclaimer of Warranty
While the publisher and authors have used their best efforts in preparing this work, they make no representations or warranties with respect to the accuracy or completeness of the contents of this work and specifically disclaim all warranties, including without limitation any implied warranties of merchantability or fitness for a particular purpose. No warranty may be created or extended by sales representatives, written sales materials or promotional statements for this work. This work is sold with the understanding that the publisher is not engaged in rendering professional services. The advice and strategies contained herein may not be suitable for your situation. You should consult with a specialist where appropriate. The fact that an organization, website, or product is referred to in this work as a citation and/or potential source of further information does not mean that the publisher and authors endorse the information or services the organization, website, or product may provide or recommendations it may make. Further, readers should be aware that websites listed in this work may have changed or disappeared between when this work was written and when it is read. Neither the publisher nor authors shall be liable for any loss of profit or any other commercial damages, including but not limited to special, incidental, consequential, or other damages.

A catalogue record for this book is available from the Library of Congress

Hardback ISBN: 9781394185474; ePub ISBN: 9781394185498; ePDF ISBN: 9781394185481
Obook: 9781394185504

Cover image: © Veni/E+/Getty Images
Cover design by Wiley

Set in 9.5/12.5pt STIXTwoText by Integra Software Services Pvt. Ltd, Pondicherry, India

SKY10048045_051723

Contents

1

Project Management 101

"Leaders think and talk about the solutions. Followers think and talk about the problems."
– Brian Tracy

1.1 Introduction

"Expect the best, plan for the worst, and prepare to be surprised."
– Denis Waitley

What is a Project Manager?

A) Someone convinced they can successfully leverage nine women to deliver a baby in one month.
B) An organizational leader dedicated to the imposition of order upon chaos, even if chaos is perfectly happy with the status quo.
C) The person assigned by the performing organization to achieve the project objectives.

While there may be some truth in each of these options, the sixth edition of the *Project Management Institute (PMI) Guide to the Project Management Body of Knowledge* (*PMBOK*) defines a Project Manager (PM) as choice C. After twenty-five years of transportation project, program, and portfolio management, I like choice B.

The mental image of boldly facing the raging tempest and taming the chaos seems especially relevant to transportation projects. Most other engineering disciplines operate in closed systems where you can quantitatively define, and directly control, most project risks. Transportation projects are decidedly different.

Project boundaries and scopes can be expanded with citizen input. Project objectives and success criteria can unpredictably change with the political tide. Legislative action and agency promulgation can dramatically change the rules of engagement mid-project by modifying processes, procedures, standards, and grandfathering new regulations. Weather can directly impact your schedule. Market conditions may increase your estimates even as your disconnected budget is declining. Other private entities are also constantly playing in your sandbox and often do not coordinate their efforts with your project. You work closely

Transportation Project Management, First Edition. Rob Tieman.
© 2023 John Wiley & Sons Ltd. Published 2023 by John Wiley & Sons Ltd.

with a myriad of local, state, and federal organizations that often have very different definitions of "a helpful and timely response." Bureaucratic inertia, indecision, and personal and professional agendas are commonplace. And so much more.

Transportation Project Management is a fascinatingly complex game. It will challenge your patience, intelligence, and determination. And it will fill you with pride as you develop and deliver tangible and enduring products that improve the safety and quality of life for the community.

This book will paint a broad-stroke, contextual picture of foundational transportation knowledge, provide a comprehensive examination of project management fundamentals, and offer practical guidance on how to efficiently plan, develop, and deliver public transportation projects.

1.1.1 The Big Picture

At their core, most transportation owner-operators are funding organizations. This is increasingly the case as more design, development, construction, maintenance, and operations activities are being contracted out and not performed with in-house staff. Public agencies operate under completely different paradigms than private companies. They are not a for-profit enterprise. They are also not a bank. Their continued existence and success is not dependent upon the size of their cash reserves or its ability to optimize a positive financial advantage in the transaction of goods or services. Their directive is to spend the money in a reasonable and responsible way that directly and significantly improves the public's safety, mobility, and quality of life. While the public entity may own and operate the transportation asset, it is important to understand that the money that flows through its various projects, programs, and portfolios supports and enables a host of other public agencies, consultants, contractors, manufacturers, suppliers, and services that drive an entire industry with far-reaching economic impacts.

Most state Departments of Transportation (DOTs) have remarkably similar mission statements. They typically include planning, delivering, operating, and maintaining safe, accessible, and reliable multimodal transportation systems that enable efficient movement of people and goods, enhance the economy, and improve the overall quality of life. Most DOTs also cite shared values that typically include safety, trust, honesty, integrity, accountability, transparency, fiscal responsibility, innovation, sustainability, and responsive service. These are ambitious and noble goals. The PM plays a critical role in the execution of this mission.

To meet these challenges, DOTs and other transportation system owners and operators must work collaboratively and strategically to ensure they are planning, developing, and delivering the right projects. Significant time and effort is spent evaluating and prioritizing the projects that make cost-effective and significant impacts to the transportation systems and the traveling public. Once a project is selected, funded, and initiated, it becomes the PM's responsibility to develop and deliver the project.

I have yet to meet an engineer who went through the gauntlet of an engineering education so they could live their dream of filling out forms and shuffling papers. Engineers solve problems. Engineers create and innovate. Engineers build things. That is why we are engineers. A transportation PM has to juggle many factors and risks, work with a wide range of stakeholders, all while strategically and stubbornly navigating their project to satisfy sometimes complex and conflicting success criteria. At times this can be exhausting and overwhelming. Resist surrendering that sense of wonder and excitement you had building Lego creations as a kid. Remember, the goal is to build a solution that solves a problem and improves the community's quality of life. That is a noble goal.

Life is short. Harvey Mackay said, "*Time is free, but it's priceless. You can't own it, but you can use it. You can't keep it, but you can spend it. Once you've lost it you can never get it back.*" In your professional career you only get so many projects to manage and lead. Seize the limited opportunities to make a tangible and significant difference in the world.

1.2 Triple Constraint

> "*Nothing is so simple that it cannot be misunderstood.*"
>
> – Freeman Teague, Jr.

In 2016, the worldwide management consulting firm McKinsey & Company estimated large transportation projects typically finish 80% over budget and 20% beyond the original schedule. Pending the sample set of projects, the results can be even more bleak. Bent Flyvbjerg, of Oxford Business School, estimated 90% of mega projects go over budget. Meanwhile, on the other end of the spectrum, small local transportation projects may be no better. While the lack of holistic and consistent performance metrics over the life of the project and across state and locality lines can make it difficult to determine if transportation organizations are developing and delivering projects on-time and on-budget compared against a common schedule and budget baseline, there is no denying transportation projects have a history of being chronically late and over budget.

Why is this? In a recent Gallup poll of most trusted and ethical professions, engineers rank second behind nurses, ahead of medical doctors and pharmacists. Top college engineering programs are among the most competitive for entrance, and most rigorous to complete. Transportation professionals are typically well-intentioned, bright engineers who are driven to solve problems and build solutions. Where is the disconnect?

Many of the reasons for this uninspiring performance will be discussed and addressed in this book. Some of these reasons are outside of our control, but most are not. An organization or individual cannot expect success if their level of transportation project management expertise and experience is low. This book will equip you to succeed.

We start at the very beginning. Project Management 101 is the Triple Constraint. It is so simple. Most every engineer, manager, or administrator can describe it. Yet project performance indicates most don't understand it, properly utilize it, or grasp its power.

As PM, the Triple Constraint is fundamental to all you do. It is graphically shown as a triangle, with the three legs being labelled for Budget, Scope, and Schedule.

This triangle, illustrated in Figure 1.1, represents your project. Your project's budget, scope, and schedule are individual constraints that are inherently interconnected.

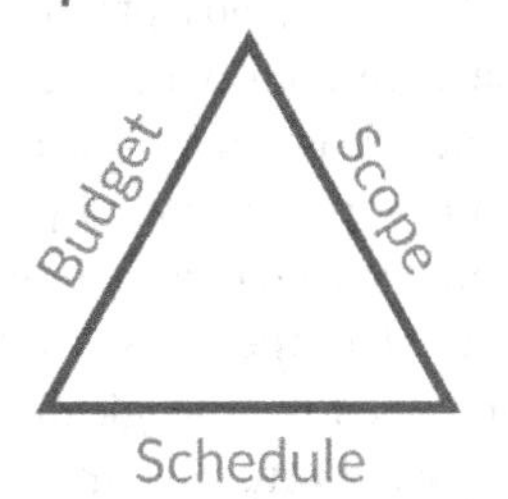

Figure 1.1 Triple constraint.

Once set, you cannot change one without impacting the other two. Think of the area within the triangle as fixed. The option of simply enlarging the triangle is rarely realistic without compromising the project's success. So, if you extend the schedule line, the budget and/or scope must decrease in order for the area within the triangle to remain the same. This results in the common quip that you can have your project fast, cheap, or good – pick two of the three.

It is important to recognize that you don't have a project unless that project has a budget, scope, and schedule. You may have an idea, a hope, a plan, or even a commitment, but you do not have a project until you have a budget, scope, and schedule. Once the project is active, the triple constraint is established. Once established, the PM lives within this triangle. All project work should be conducted within this space. To be successful, you must fiercely monitor and guard these three critical constraints. As PM, this is your primary job.

George Box said, "*All models are wrong, but some are useful.*" This is true of the Triple Constraint. The three constraints have very different units of money (budget), time (schedule), and content (scope). As such, there may be other constraints of similar units. For instance, a decrease in funding, an anticipated funding shortfall due to a revised construction cost estimate, or higher than anticipated project expenses may all fall under the budget of the Triple Constraint. It can be tempting to try to read too much into this model. However, while the concept is an oversimplification, it does effectively illustrate a project's fundamental competing and interdependent constraints. This understanding and mindset correctly frames the context for successful Project Management and Change Management. The PM must constantly search for, and find, balance between these three constraints.

When considering the Triple Constraint, it is important to realize it provides different insights in different scenarios. When two of the three constraints are fixed, it can be problematic to adjust the third. If your budget and schedule are fixed, it is difficult to increase scope by adding design features. If your budget and scope are fixed, the project cannot extend its schedule without complications. If the schedule and scope are fixed, a budget decrease may jeopardize success.

The situation is different when only one constraint is fixed. In these situations, the remaining two constraints must be adjusted to balance the project. If the scope is fixed and your schedule is extended, your budget may need to be increased. If your scope is fixed and your budget is decreased, you will need to compress your schedule. If your schedule is fixed and your scope is increased, you may need to increase your budget. If your schedule is fixed and your budget is decreased, you may need to reduce scope. If your budget is fixed and your schedule is accelerated, you may need to reduce scope. If your budget is fixed and your scope is increased, you may need to fast-track your schedule.

In reality, most projects experience some variance in all three constraints. The scope may be increasing, which necessitates extending the schedule and increasing the budget. Or the schedule is suddenly accelerated, which necessitates increasing the budget to add resources and reducing scope by eliminating nonessential design elements. Or the budget is cut, which necessitates reducing the scope and modifying the schedule.

In all of these cases, effective Change Management is critical to project success. As PM, you need to strategically be preparing and moving the game pieces so that you are positioned to win, meaning that you develop and deliver the project on-time and on-budget in a way that satisfies the Project Management Plan (PMP) and achieves established success criteria.

Effectively managing a project's Triple Constraint requires purposeful action. In the life of most projects, the Triple constraints will be strained. Past performance shows us these situations are not always handled well. Successful PMs are flexible in approach and adaptable to change while maintaining a laser-like focus on objectives, risks, and constraints. Planned change is one of the most powerful tools a PM can use to nimbly develop and deliver a successful project. Conversely, nothing will jeopardize a project faster than unplanned or unmanaged change. The PM is fully responsible to monitor and manage any and all changes to the Triple Constraint. Change Management is discussed at length in Chapter 10, *Controlling the Project*, Section 10.3, *Change Management*.

Many mature organizations have a project-centric approach. For our purposes, this means that every project has a schedule. Every schedule has essential tasks. Every task has a responsible resource. Every schedule has a defined critical path. Every project has a PM. Every project has a budget. Every dollar has a project. Every scope is defined. Every change is processed per approved change management procedures. Underlying all of these tenets is a fundamental reliance upon the triple constraint, which is the unyielding foundation for all successful projects.

While your primary responsibility as PM is the Triple Constraint, there are other factors you must also consider. Underlying the Triple Constraint of Budget, Scope, and Schedule, are the constant undercurrents of Risk, Resources, and Quality. None of these are inherently project constraints; although changes to budget, scope, and schedule may be dependent upon, or directly impact, Risks, Resources, and Quality. As such, any of these may become constraints that become entangled with the Triple Constraints. Identifying, managing, and responding to Risks can directly impact any or all of the triple constraints. The same can be true for Resources and Quality. Intentionally and proactively managing Risks, Resources, and Quality will position you for project success. Conversely, ignoring or mismanaging Risks, Resources, and Quality will almost certainly lead to failure. Each of these will be discussed at length in subsequent chapters.

1.3 The Project Management Plan

"An hour of planning can save you ten hours of doing."

– Dale Carnegie

A project's Purpose and Need defines the *Why* of the project. The project scope details the *What* to satisfy the *Why*. The Project Management Plan (PMP) details the *How*, *Who*, and *When*. This living document provides guidance and clarification on how the final product will be managed during development. It provides clarity and consistency, and serves as a tremendous resource during unexpected times or in the midst of critical staff transitions. The PMP is the responsibility of the PM.

When you see "*Plan*," think "*How*." The project schedule is not the PMP. At its core, the PMP is a collection of other plans, each one detailing how a critical area of the project will be managed. Much of this may already be defined in your organizational guidance. The intention of the PMP is not to recreate it, but to highlight key components. The final PMP should not be a physical or virtual binder that sits proudly on a shelf. The PMP should be a living document and a dynamic tool that is used and leveraged throughout the project.

All projects should have a PMP. Larger and more complex projects can have massive PMPs. However, even smaller projects can benefit from a very quick and simple PMP. Many organizations have PMP guidance or templates that should be used and followed. While each project is unique, the format can remain consistent. The content and level of detail should be tailored to the size and complexity of the project, per your organization's guidance. If your organization does not have PMP guidance or template, you may use the following format illustrated in Figure 1.2.

Figure 1.2 Project management plan template.

1.3.1 Project Details

Every PMP should start with a section on Project Details. This should be tailored to the protocol and expectations of your organization. Typically, this would detail the project's Purpose and Need, objectives, and goals. This section should also include a project narrative. High-level summaries of project phase details and the selected project delivery method should also be incorporated. If your organization requires a Project Charter or a Consultant's Scope of Work, it should be included in this section.

1.3.2 Resource Management Plan

The Resource Management Plan details how the project team will interact throughout the project. An organizational chart and list of Roles and Responsibilities are critical components of this plan. Detailed information on developing and managing project teams is included in Chapter 7, *Managing Resources*. Additional information on the Resource Management Plan is included in Section 7.3, *Resource Planning*.

1.3.3 Scope Management Plan

The Scope Management Plan details how the project scope is defined, validated, and controlled throughout the project. This should describe the project development approach. The project's Issue Log is a critical work product within this plan. Detailed information on Scopes is included in Chapter 4, *Scope*. Additional information about the Scope Management Plan is included in Section 4.4, *Scope Management*.

1.3.4 Schedule Management Plan

The Schedule Management Plan details how the schedule will be developed, monitored, and controlled throughout the project. The Project Schedule and Baseline Schedule are critical work products under this plan. Detailed information on schedules is included in Chapter 5, *Schedule*, with Section 5.2, *Schedule Management Plan*, focusing specifically on the Schedule Management Plan.

1.3.5 Cost Management Plan

The Cost Management Plan details how the project budget and estimates will be developed, monitored, and controlled throughout the project. The project's budget and estimates, along with the baseline estimate, are critical work products under this plan. Detailed information on these financial issues included in Chapter 3, *Budgets and Estimates*. Additional information on the Cost Management Plan is included in Section 3.4, *Managing the Budget*.

1.3.6 Risk Management Plan

The Risk Management Plan details how risk will be identified, analyzed, and managed throughout the project. This includes the processes that will be followed, and the tools that will be utilized. The Risk Register and Assumption Log are key documents that should be part of this plan, and regularly updated. Detailed information on risk management and the Risk Management Plan are included in Chapter 6, *Managing Risk*, and Section 6.2, *Risk Management Plan*.

1.3.7 Communications Plan

The Communications Plan details how information will be collected and disseminated to the project team, leadership, and stakeholders throughout the project. The Stakeholder Register and Stakeholder Engagement Plan will be large components of the Communications

Plan. Detailed information is included in Chapter 9, *Communications*, particularly in Section 9.1, *Project Communications*, and Section 9.2, *Stakeholder Involvement and the Stakeholder Engagement Plan.*

1.3.8 Quality Management Plan

The Quality Management Plan details how quality will be defined and measured throughout the project. This should address both process (via Quality Assurance) and the product (via Quality Control). Detailed information is included in Chapter 8, *Managing Quality.* Information specific to the Quality Management Plan is included in Section 8.2, *Monitoring and Controlling Quality.*

1.3.9 Change Management Plan

Change is inevitable. The Change Management Plan details how change will be managed and processed throughout the project. Established Change Management Procedures are critical to project success. The Change Log documents all project changes throughout the project. Detailed information on Change Management Plan is included in Chapter 10, *Controlling the Project*, within Section 10.3, *Change Management.*

1.4 Organizational Basics

> "*The war is not won with bayonets, but with effective organization.*"
>
> – Unknown

1.4.1 Organizational Structures

Organizations develop and deliver projects under different frameworks. These frameworks have far-reaching implications for how your organization operates, including who makes what decisions. Generally speaking, there are three different types of organizational structures for project management: Functional, Matrix, and Projectized. Matrix organizations can be subdivided into Weak Matrix, Balanced Matrix, and Strong Matrix organizations. Figure 1.3 shows the full range from Functional on one end to Weak Matrix, to Balanced Matrix, to Strong Matrix, with Projectized on the other end. This spectrum represents the scale of an organization from being Department Oriented to Project Oriented. The balance of power and authority between the Functional Manager and Project Manager transition along this range.

A Functional Organization is divided into discipline-specific departments. Resources are controlled within their functional group, which is controlled by the functional manager. Employees are specialized by area. Rigid, disconnected organizational hierarchies can create "silos" that forces team members to first go up their own vertical chain of command for information outside of their silo. While this structure can concentrate expertise within efficient departments, it can also isolate resources and introduce turf battles. Decision making can be slow, and loyalty can first be to the silo and not the organization or project. In this arrangement, PMs have little-to-no power or authority over the project or team resources.

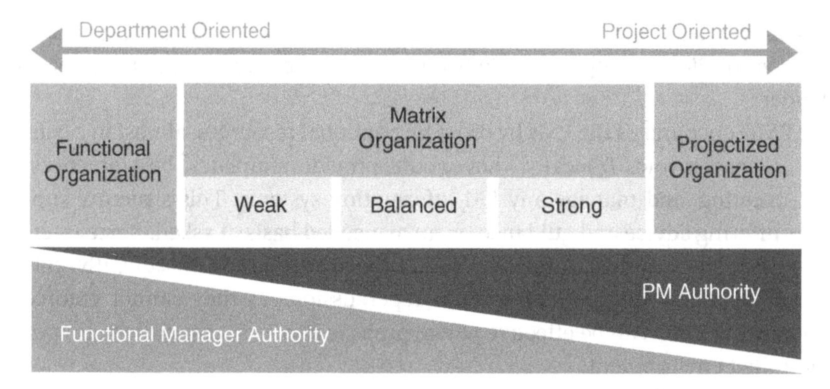

Figure 1.3 Organizational structures for project management.

On the other end of the spectrum are Projectized Organizations. These organizations are highly project focused, and arrange their staffs accordingly as all activities are intentionally managed through projects. In this framework, the PM has tremendous power and authority. Resources are completely controlled by the PM within this dynamic and adaptive environment. This structure can enable nimble progress as decisions can be quickly made for a team that is first loyal to the project success. However, the inherent temporary nature of this structure can challenge continuity and introduce employee uncertainty as they may be reshuffled after each project. Additionally, while encouraging collaboration, it can limit discipline-specific expertise that is essential in technical industries.

Most transportation organizations are Matrix Organizations. A Matrix Organization sits comfortably in the middle, offering the best, and worst, of both worlds. An organization can tailor their resources and strengths to enable flexibility and encourage efficiencies across multiple objectives. Here resources answer to both the Functional Managers and PM. Which manager has more power and authority determines if the organization is Weak, Balanced, or Strong. In a Weak Matrix, the budget and schedule are controlled by the Functional Managers. The PM may be part-time and is more of a coordinator. In a Strong Matrix, the budget and schedule are predominantly controlled by the PM. Functional Managers fill a valuable support role. In a Balanced Matrix, they share control of the budget and schedule. Typically, the PM will lead project-related tasks while the Functional Manager is responsible for resource-related issues. This dynamic structure enables concentration of expertise and successful collaboration across department lines. However, it does necessitate effective communication to clarify potentially competing roles, responsibilities, and objectives.

1.4.2 Project Management Office (PMO)

PMI's *PMBOK* defines a Project Management Office (PMO) as, "*A management structure that standardizes the project-related governance processes and facilitates the sharing of resources, methodologies, tools, and techniques.*" This typically means the PMO defines and maintains project management standards, while realizing economies of scale in project development. They should be the organization's experts in project management practices and methodologies.

Not all PMOs are created equal. They too can have different structures and focus. There are three basic types of PMOs: Supporting, Controlling, and Directive. Think of them as a range, in this order.

A Supporting PMO encourages the PMs by providing optional resources, almost in a consultant, advisor, or coach role. Typically, they would provide templates, best practices, lessons learned, training, and manage any PM information systems. This structure supports the PMs by offering advice and guidance on an as-needed basis, if asked. Since interactions are voluntary, their control, influence, and authority are low. Consequently, this type of PMO can be summarily ignored without repercussions as they cannot enforce anything. A Supporting PMO can be effective where projects can be successfully managed in a loosely controlled environment.

The most common type of PMO is a Controlling PMO, which strikes the balance between being hands-off and a taskmaster. In this sweet spot, they emphasize support, governance, and conformance. Typically, they would also provide project management policies, procedures, forms, reporting, and other tools. This type of PMO works to establish the project management framework, and then enforces compliance with organizational practices and methodologies. A Controlling PMO can be extremely effective in complex environments where compliance with established practices is critical to success, such as in the transportation industry.

A Directive PMO directly manages the projects. In these instances, all PMs work directly for the PMO. Success requires strong governance framework as their goal is to become efficient project execution factories. A Directive PMO can be very effective in rigid, highly regulated, high-risk environments, or in larger matrix organizations.

Assuming your transportation organization has a PMO, it is likely some version of a Controlling PMO. As such, it will likely have some common objectives regarding project management, including but not limited to: Governance, Standards, Policies, Processes, Resources, Forms, Templates, Documentation, and Training. This all enables the organization to provide consistency in project development and delivery between PMs and across district, divisional, or department lines. This includes establishing and maintaining common processes and performance metrics. It also often includes administrating the organization's project management software, tracking, and management tools.

The PMO should promote a pervasive project management culture that provides consistency and harmony while increasing productivity and consistent results. The influence footprint of a successful PMO is much larger than might be expected. It can be the tiny rudder that turns the entire ship. Often one of the most important responsibilities of a PMO is to minimize conflict and improve efficiencies at the critical links between people, processes, and technology. As such, the PMO may be in a unique position to break down organizational silos and build bridges of innovation. As a Transportation PM, understand the expertise within your PMO and leverage it to your advantage.

1.4.3 Projects, Programs, and Portfolios

Projects, programs, and portfolios are three distinct entities. Understanding the implications of their fundamental differences will better enable you to successfully manage your projects within your organization's structure.

A Project is a temporary effort to develop and deliver a product within the established budget, scope, and schedule. A Program is a collection of projects that share a meaningful relationship. A Portfolio is the full collection of projects and programs that share the same strategic objectives. To use a music analogy, consider each individual song to be a project. An artist's individual albums would be separate programs. The artist's full catalog of albums and stray songs would be their portfolio. In transportation, every project is a project. Projects may then be grouped into Programs by some meaningful relationship between the projects. This may be a geographic district, a specific funding source, a defining discipline (e.g., bridges), who is administering the project (e.g., DOT managed or Locality managed), or some other useful distinction. For the purpose of performance metrics, projects may be analyzed in more than one program. For instance, you may first look at a geographic district's entire program, and then look at all bridge projects across district lines. Every project should have a PM. Pending the size and structure of your organization, you may also have Program Managers. The portfolio would then be the entirety of programs and projects that share a strategic objective. This will typically be all construction projects within your organization's multiyear funding plan. Figure 1.4 illustrates the relationship between these three entities.

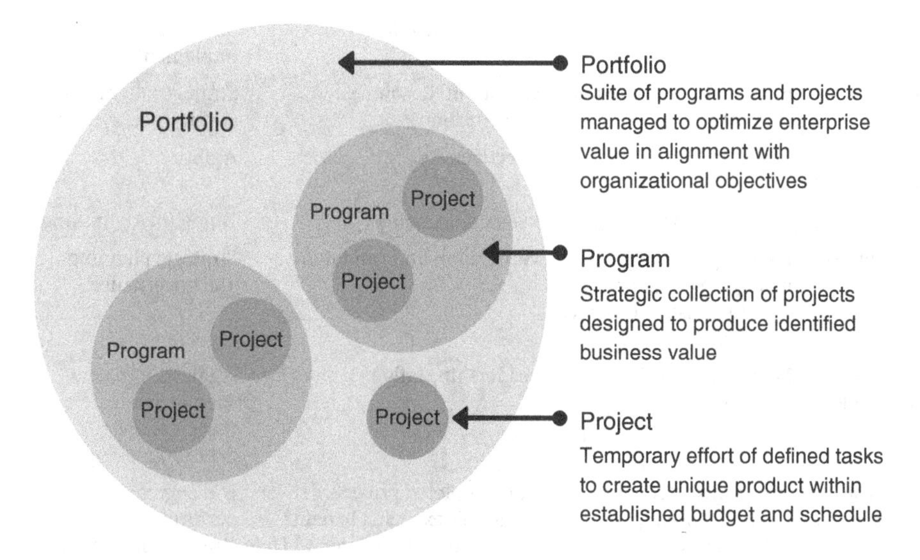

Figure 1.4 Projects, programs, and portfolios.

Projects, programs, and portfolios operate at different levels that all need to work together to ensure organizational success. Those responsible at each level operate at very different altitudes and levels of expertise. It is helpful to realize each may have different focus and success criteria. Pending the size and maturity of your organization, these different levels may be clearly delineated or mostly blurred. The same person may fill one or more of these roles in smaller organizations. Larger transportation organizations may have different managerial and leadership individuals or entities at each level. For instance, there may be a hierarchy of PMs that are under District program PMOs, that in turn are under the Portfolio (or Enterprise) PMO in Central Office.

It is worth discussing the differences in approaches. On one side of the spectrum is Project Management, which is uniquely tactical in nature. This concentrates on doing, resolving the how, and focusing on day-to-day execution. Project Management drives the deliverables, and wants to improve "what is." On the other side of the spectrum is Portfolio Management, which is uniquely strategic in nature. This concentrates on planning, determines the *what* and *why*, and focuses on defining the future. Portfolio Management drives the outcomes, and considers "what can be." Program Management hovers in the middle, balancing strategic direction with tactical realities. Figure 1.5 summarizes some of the major differences between each level.

Comparison	Project	Program	Portfolio
Purpose	Develop and deliver the scoped project on-time and on-budget	Satisfy program's objectives by executing projects	Support organization's mission by guiding and administering strategy
Goal	Output	Capacity and effect	Strategic implementation
Priorities	Budget, Scope, Schedule, Quality	Validation, Resources, Prioritization	Strategic alignment, governance, synergy
What is it about?	Efficiency	Effectiveness	Agility
Key Imperative	Doing things right	Realizing benefits	Doing the right things
Key Activities	Planning, scoping, managing budget, scope, schedule, risks, resources, and quality	Verification and validation	Strategic planning and governance
Key Competencies	Management skills	Leadership skills	Analysis, problem solving, strategic thinking, and decision making
How do you define success?	Was project delivered on-time, on-budget, within scope, and with acceptable customer satisfaction?	Did completed projects deliver program's expected benefit? Did you spend the money? Did you meet target performance metrics?	Did aggregate performance satisfy strategic objectives and defined priorities in business plan?
Success Measures	Budget, Scope, Schedule, and Quality	Efficient utilization of resources to meet business objectives	Strategic objectives and enterprise performance
Leadership Focus	Task delivery within defined success criteria	Managing relations and conflict resolution	Adding value and improving processes
Management Focus	Task performance	Project performance and program's efficacy	Enterprise performance and achieving business objectives

(Continued)

(Continued)

Comparison	Project	Program	Portfolio
Key Management Skills	Technical knowledge and interpersonal skills	Leadership, support, and organization	Insight, innovation, and synthesis
Scope	Well-defined (Budget, Scope, and Schedule)	Larger objectives to satisfy expected program benefits	Business scope to accomplish defined strategic objectives
Change	Fully managed with established change management processes	Manage change within program (trade-offs) to maintain program's expected benefits	Monitor and react in broad environment to maintain strategic objectives
Risks typically viewed as...	Threats	Opportunities, balanced against threats	Deviations from leadership expectations
Tries to answer questions to...	Eliminate uncertainty and forward project progress	Reduce ambiguity and focus on program benefits	Bring clarity of vision and eliminate systemic obstacles
Monitoring	Work packages that control and produce the project deliverables	Process of program, and project progress to meet expected benefits	Aggregate performance, resource allocations, risks, and changes to strategic objectives
Guidance, Policies, and Procedures	Follow it	Implement and enforce it	Establish it

Figure 1.5 Comparison of projects, programs, and portfolios.

1.5 A Look Ahead

"Man's mind, once stretched by a new idea, never regains its original dimensions."
– Oliver Wendell Holmes, Jr.

When a PM traverses their completed transportation projects, they see the improvement in ways no one else does. They see the property with the belligerent owner who tried to kill the project. They see the now grassed area of the test hole that confirmed they had a massive utility conflict. They see the pipe outfall that needed to be extended and required additional survey that stressed the schedule. They see the storm inlets in the roundabout that were nearly impossible to adequately drain on the challenging incline. But that is not all they see. They also see the median break strategically positioned to accommodate the church's masterplan for expansion. They see the pedestrian path that was extended to connect two neighbourhoods so residents could safely walk and bike to the nearby swim club. They see the improved roadway geometry and safety features that dramatically reduced accidents. They see the innovative bridge abutment design that saved that massive old oak tree.

When a PM looks at their completed project, they see so much more than just the completed assets. They see how the finished product came to be. They see the journey. Some of these memories are wildly successful tales of innovation and persistence. Others are pride-swallowing lessons learned. But they are all part of the story. They are all part of your journey.

Whether each memory and issue went your way or not, they are all testimonials to successful resolutions. PMs solve problems and make things happen, from megaprojects that change the landscape and character of cities and corridors, to small projects that improve safety at a pedestrian crossing. There are very few systems that directly impact our lives as much as the transportation network. Be it our individual travel for work or pleasure, or the transport of goods and services that enables our economy to survive and thrive. The economic and time benefits of transportation improvements are real.

It is said PMs find a way, or an excuse. As PMs, you drive this ship. You have the power to push development of an effective and efficient design. The larger your organization, perhaps the easier it is to lose the urgency of your mission. If the project is delivered next May or next September, does it really matter? If you are improving traffic efficiency, think of all the time you are saving and giving back to each person who drives that road. If you are improving a bike or pedestrian facility, think of the added joy families have using it, and of those who now run on it every day. If you are improving safety, think of those who may have been in serious or fatal accidents if the project were delayed. I guarantee that their loved ones would say your efforts mattered a great deal. This is exciting! You are making a difference. This should motivate you to wake up each day and drive your projects forward. You are making a very real and tangible difference in the lives of the community in which you and many, many others live, work, and play. As Ralph Waldo Emerson said, "*The reward of a thing well done is to have done it.*"

2

Project Development Process

"If you can't describe what you are doing as a process, you don't know what you are doing."

– William Edwards Denning

The PM is very similar to the conductor of an orchestra. They don't need to know how to play the viola, oboe, or timpani, but they do need to know the instruments' capabilities and constraints. They acknowledge the musicians are the experts, and work to position them to perform their best. The conductor approaches the performance from a different perspective. They see the entirety of the score, understanding how and when which instruments can be combined to enhance the results. They make adjustments to the plan to emphasize key musical components. They guide and shape the performance in a way that enables the full orchestra to reach their collective potential. And while they may speak for the full orchestra, the conductor graciously thanks all those involved, highlighting key musicians and soloists.

2.1 Safety First

"Safety isn't expensive, its priceless."

– Jerry Smith

2.1.1 A Matter of Life and Death

In the transportation industry, "Safety First" is more than a slogan. It is a mantra. Safety is a foundational tenet of transportation policy and practice, upon which we should be ever vigilant.

The National Society of Professional Engineers (NSPE) clearly advocates that a licensed Professional Engineer (PE) should hold paramount their obligation to protect the health, safety, and welfare of the public. PEs bear an ethical responsibility to do so, and can lose their licensure if it is determined they were negligent in this regard.

Roadway safety is a global issue. In 2021, the World Health Organization (WHO) estimates traffic accidents are responsible for approximately 1.3 million fatalities each year,

Transportation Project Management, First Edition. Rob Tieman.
© 2023 John Wiley & Sons Ltd. Published 2023 by John Wiley & Sons Ltd.

with an additional 20 to 50 million annually suffering nonfatal injuries. More than half of all traffic road deaths are pedestrians, cyclists, and motorcyclists. Over 90% of the world's traffic fatalities occur in less developed countries. These deaths come with considerable economic impact, costing most countries 3% of their gross domestic product.

Developed countries are not immune from these tragedies. Traffic deaths are rising in the United States. The National Highway Safety Traffic Administration (NHSTA) estimated there were 42,915 traffic fatalities in 2021. This represents a 10.5% increase over 2020. It should be noted that estimated vehicle miles traveled in 2021 increased by about 325 billion miles, representing an 11.2% increase over 2020. Normalized for miles traveled, 2021 saw a decrease to 1.33 fatalities per 100 million vehicle miles traveled, down from 1.34 in 2020. While this may seem like good news, perhaps we should think about it this way – 33 states experience more than one traffic fatality each day, with California, Florida, and Texas experiencing over 10 traffic fatalities each day. Each death is a tragic loss that shatters the world of their family and loved ones. This is unacceptable.

2.1.2 The Usual Suspects

The Insurance Institute for Highway Safety (IIHS) and Highway Loss Data Institute (HLDI) thoroughly examine fatality crash data. There are some interesting and unsurprising themes.

Human error causes most fatalities. Distracted driving, impaired driving, speeding, and other unwise behavior decisions are major causes of fatalities. Younger drivers are more likely to have a fatal crash by lane mile than experienced drivers. States that have legalized marijuana are seeing an increase in traffic fatality numbers and rates. About half of all highway fatalities are unbuckled. This percentage is even higher in pickup trucks. And so forth.

Other behavior themes can be more directly addressed with proactive, redundant safety designs. Such as about half of all fatalities, including about ¾ of bike and pedestrian deaths, occur at night. About ¼ of all traffic fatalities and about ½ of severe accidents occur at intersections. About half of all highway fatalities are roadway departures. And so forth.

The root causes of traffic fatalities have largely remained unchanged; however, what is changing is how we approach addressing this situation.

2.1.3 Safe System

Traffic safety is a complicated problem. It extends beyond any one engineering discipline and lends itself to broader systems thinking. Peter Senge said, *"Systems thinking is a discipline for seeing wholes. It is a framework for seeing interrelationships rather than things, for seeing 'patterns of change' rather than static 'snapshots.' It is a set of general principles spanning fields as diverse as physical and social sciences, engineering and management."* This seems especially applicable to traffic safety.

Safe System is an intentional engineering and system approach that aims to eliminate fatal and serious injuries for all road users. This philosophy has been embraced and adopted by a number of countries throughout the world which involve some significant mindset shifts from traditional transportation methodologies, some of which are detailed in Figure 2.1.

Traditional Approach	Safe System Approach
Prevent crashes	Prevent deaths and serious injuries
Improve human behavior	Design for human mistakes and limitations
Control speeding	Reduce system kinetic energy
Individuals are responsible	Share responsibility
React based on crash history	Proactively identify and address risk

Figure 2.1 Traditional vs. Safe system.

In America, the Federal Highway Association (FHWA) has embraced this safe system approach. This important initiative is built upon the following six guiding principles:

1) Deaths and serious injuries are unacceptable
2) Humans make mistakes
3) Humans are vulnerable
4) Responsibility is shared
5) Safety is proactive
6) Redundancy is crucial

These six principles should be advocated and applied to all five elements of a safe system:

1) Safe road users
2) Safe vehicles
3) Post-crash care
4) Safe roads
5) Safe speeds

It is worth noting that transportation system stewards and designers can only directly impact these last two elements: safe roads and safe speeds. Implementation of this philosophy involves anticipating human mistakes, and designing transportation infrastructure to reduce and mitigate risks of fatal or serious accidents in a holistic and overlapping manner.

As transportation system stewards and designers, we need to recognize that users make mistakes and bad decisions. It is incumbent that we take measures to keep those impacts at tolerable levels. At its core, injury severity is all about kinetic energy (Kinetic Energy = ½ mass × velocity squared). All auto safety features being relatively equal, in a head-on collision you can't avoid the law of conservation of momentum (momentum = mass × velocity). In both these simple equations, velocity is critical. Figure 2.2 is a FHWA prepared graphic that shows anticipated fatality risks based on impact speeds.

These sobering realities should motivate designers to encourage safe speeds and manipulate crash angles to reduce injury severity. This understanding has accelerated efforts like traffic calming, speed management, and focusing on target speeds. Similarly, innovative intersections are being embraced and advanced, including roundabouts, Diverging Diamond Interchanges (DDI), Restricted Crossing U-Turns (RCUT), Continuous Flow Intersections (CFI), and the like.

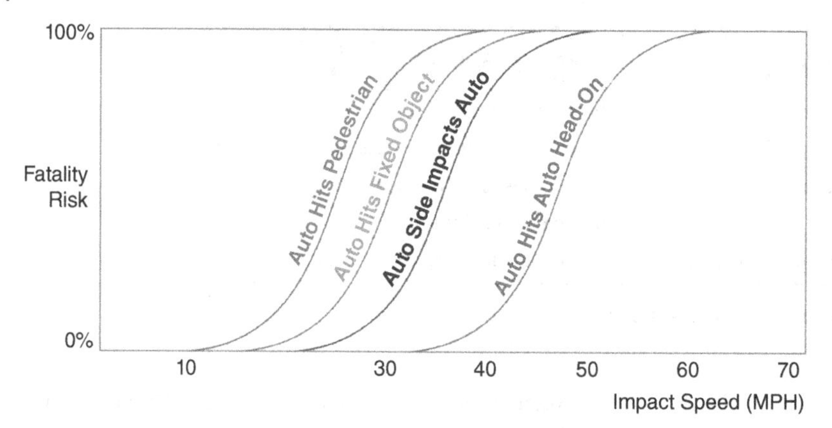

Figure 2.2 Fatality risks based on impact speeds.

2.1.4 Standards Do Not Equal Performance

It is critical to recognize and acknowledge that designing and building a project to standards does not directly equal safety performance. Figure 2.3 shows how historically most transportation owner operators have behaved as if meeting or exceeding design standards will inherently ensure acceptable levels of safety, while failure to meet the standards may seriously jeopardize safety. Roadways that meet the minimum standards are deemed to be nominally safe, by the fact that they meet the established standards.

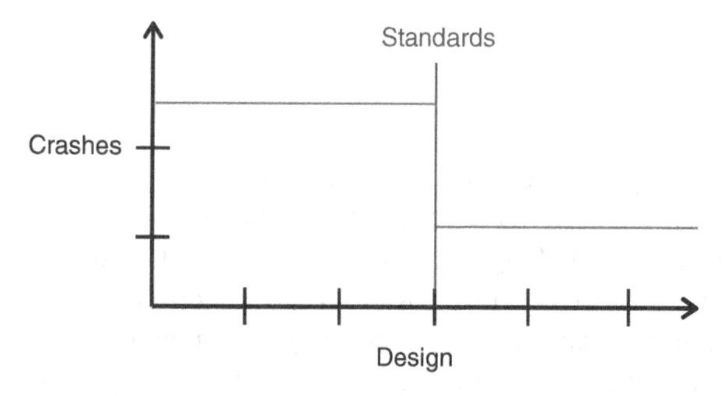

Figure 2.3 Safety standards – historic approach.

In some situations, there is an instantaneous increase in realized safety benefits at certain design decision points or detailed demarcations; however, Figure 2.4 shows how it is far more common that most safety benefits are gradual in nature. This more progressive approach shifts away from nominal safety toward substantive safety, which focuses on the expected or actual long-term safety performance of the road.

This approach does not diminish the importance and significance of safety and design standards. They remain the north star of design, but it shouldn't end there. In certain situations, design standards may not provide the best achievable level of safety. The transportation industry is steadily moving toward Performance-Based Design (PBD), which relies on standards but grants flexibility to more easily use engineering judgment and context-sensitive

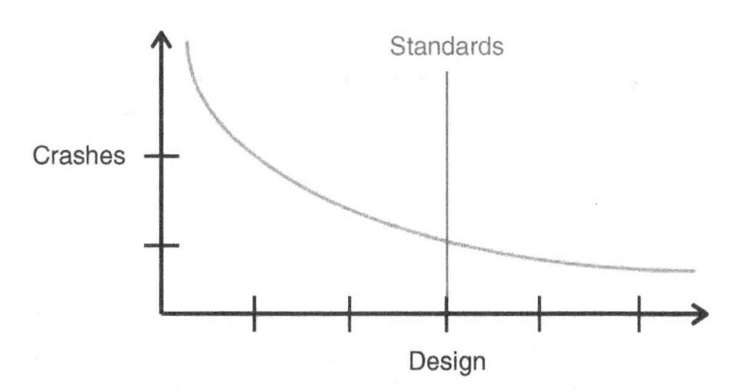

Figure 2.4 Safety standards – progressive approach.

solutions that tailor solutions to each specific project. This mindset is perfectly suited toward identifying and implementing reasonable and effective safety measures.

Every transportation project presents an opportunity to increase safety. As transportation stewards and designers, we need to ensure the design is about decisions, not just dimensions. A safe system culture places safety first and foremost in the system investment decisions.

2.2 Overview

"Slow down. Calm down. Don't Worry. Don't Hurry. Trust the process."

– Alexandra Stoddard

2.2.1 Five Project Stages

A public transportation project has five stages during its development and delivery, before it is transferred to operations. Different states or localities may vary in how they delineate these phases, but key elements of the workflow remain, as shown in Figure 2.5.

Initiation is the stage where the idea evolves and becomes a project. Before you manage a project, there has to be a project to manage. This first stage involves project planning, prioritization, selection, and initiation. Every project begins with an idea that originates from somewhere. Perhaps it is a corridor study, citizen input, economic development, political desires, accident data, structurally deficient ratings, or an obvious need; regardless, this idea somehow gains momentum and support over other ideas. This can be inclusion in strategic long-term plans, a political champion, or a connection to economic development. Once the idea is properly prioritized, adequate funding must be identified. Our transportation systems have many more identified needs and wants than available resources. As such, each idea is inherently competing against other ideas for limited dollars. Be it through a criteria-based application processes, strategic processes, or political decisions; some prioritized ideas are selected for funding and become projects. Project initiation includes vetting the idea, identifying funding, setting up the project in all applicable systems, and including the project in all applicable multi-year transportation plans. This stage typically ends around PE Authorization when the project is assigned to a PM to develop. One enduring challenge in this stage is building and establishing the preliminary budget, scope, and schedule with very little information.

Figure 2.5 Five project stages.

Preliminary Engineering is the process by which the idea of the initiated project is designed to advance the project to construction. This is the Project Development Process (PDP) that will be discussed at length in this chapter. The PDP is the responsibility of the PM. While you should understand the other stages, you need to know your agency's PDP.

Advertise is the process that incorporates advertisement and award of the construction contract. These activities are unique in that they have one foot in project development and the other in delivery. While it involves both design and construction, it is neither. More accurately, it is the agency's transition between development and delivery. At its heart, these are procurement tasks aimed at the agency eventually making the business decision whether to award the contract or not. Since many PMs guide their projects through this stage, we will discuss this in more detail later in this chapter.

Construction is project delivery. This is the actual construction of the project. While its official start and stop may vary by agency, it generally starts when the owner agency issues the Notice To Proceed (NTP) to the contractor, and ends with Final Acceptance. Start triggers may vary slightly, such as the contract execution date, or the actual date the contractor begins working the site. NTP is most typical as the Contractor will often order supplies and submit required paperwork submissions prior to mobilization, which is still part of the delivery effort. Prior to Final Acceptance, there may be Conditional Acceptance and Punch List resolution. Final Acceptance may also incorporate resolution of all construction change orders, payment of Final Invoice, and release of any retainage. Delivery activities are typically not overseen by the Development PM, but rather the owner agency's Construction staff, or their appointed representatives.

Closeout is the backend office work to balance and reconcile all accounting and programming records. This involves reconciling all outstanding bills, settling claims, finalizing paperwork, closing federal phase agreements, and archiving project records. The ultimate closeout goal is to close the project in a timely way that maintains systemic records and financial integrity, while identifying surplus funds that can be transferred and put to use on another project. Some organizations require project expenditure certifications throughout the life of the project (e.g., quarterly) which can tremendously expedite administrative closeouts. A common standard for PM expense certification approvals is "reasonable" and "expected."

2.2.2 The Project Development Process (PDP)

The federal government and state Departments of Transportation (DOTs) have their Project Development Process (PDP) well documented in manuals, policies, procedures, interactive webpages, and color-coded workflow schematics. Many of these show process granularities down to the task level with dependencies. Many localities have similar resources. Devour these resources.

While the PDP is well documented, it is often not well understood by those who rely on its information. As Charles Kettering said, "*Knowing is not understanding. There is a great difference between knowing and understanding; you can know a lot about something and not really understand it.*"

The PDP are the rules by which the transportation development game is played. Imagine gathering with family or friends to play a new, very complicated board or card game, but choosing not to read the rules or instructions because you kind of have an idea how it should generally work. When confusion and frustration inevitably ensue, someone looks for a specific answer to the specific question at hand, only to realize they have been doing some things wrong that now need to be redone. So, the group takes multiple steps backward to take one step forward, only to repeat this same predictable cycle again and again. Likewise, many transportation projects follow this example because the PM doesn't understand the rules of the game.

Discipline-specific engineers are that discipline's Subject Matter Experts (SME) on the PDP. By contrast, the PM needs to know the entire PDP, understanding the purpose of each milestone and the relationships between tasks.

There are five questions every PM should be able to answer at any time on their projects:

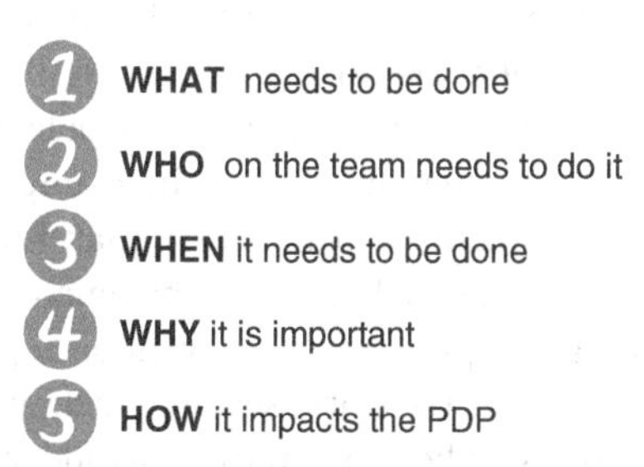

Figure 2.6 Five questions every PM should be able to answer.

Within accepted organizational limitations, the PM may be allowed to expand or compress the standard PDP based upon the specifics and complexity of the project. Exceptional PMs are effective decision makers. This is only possible if they understand the PDP, and can answer the five questions in Figure 2.6. PMs should only adjust the PDP when they fully grasp the significance and interconnectivity of tasks between disciplines, and can articulate its implications to the project's budget, scope, and schedule. Additionally, at the federal and DOT level, most milestones and tasks are directly or indirectly tied to federal and state requirements. So, while there can be flexibility to modify some of these processes, proceed with caution. Be sure your workflow changes do not unknowingly compromise your project, or conflict with governing requirements.

Established PDPs can vary for a variety of reasons, including specific policies, laws, agreements, funding source requirements, asset ownership, maintenance responsibilities, and historic precedents. In spite of that, much of the transportation Design-Bid-Build PDP remains the same.

Most of the guidance displays PDP workflow in a waterfall model that clearly delineates a linear path of sequential steps. Federal authorizations are for separate and sequential phases; first Preliminary Engineering (PE), then Right-of-Way (RW), then Construction (CN). The 30% design meeting proceeds the 60% Design meeting, which proceeds the 90%

design meeting. Citizen Information Meetings proceed the Public Hearing which must happen before you secure Design Approval. Wetlands are flagged, then confirmed, then you move forward with applicable permits. Traffic counts are collected, then they are integrated into the traffic models which drives the traffic study that recommends intersection design improvements. And so forth. These linear relationships are easy to define, and powerful to display, particularly for tasks within each engineering discipline.

Mature PDPs use these linear relationships for higher level milestones as their framework, while recognizing and optimizing the concurrent opportunities within different engineering disciplines. The many separate, but related, engineering discipline trains should be running down their own tracks, strategically positioned and timed to converge at key milestones. The PM needs to keep everyone focused and on schedule, ensuring all are doing what needs to be done, when it needs to be done, so the project will progress as planned. The only way for this to happen is if the PM actually knows and understands the entire PDP workflow.

The power of concurrent engineering is realized only when the interconnectivity between disciplines is well understood. The water resources train of tasks may appear rather straightforward until you realize the hydraulic analysis can be an iterative process with the structural engineer as they collectively optimize the bridge design to pass the required design storm with the most cost-effective structure. The Traffic Signal design may appear straightforward until a utility test hole dictates the traffic pole placement must be moved, which impacts right-of-way and easement lines, which impacts limits of disturbance, which impacts Erosion and Sediment Control (ESC) quantities and requires additional measures that must be accommodated. The noise abatement train of tasks may appear straightforward until the designers place the noise wall on top of the proposed drainage pipes which necessitates redesigning the storm sewer, which was already redesigned once because of a sanitary sewer extension betterment that was added to the scope. And so forth.

Exceptional PMs study and know the PDP. They have a breadth of experience that grants them insight into the unique and sometimes complex interconnectivity of engineering discipline tasks. They have the knowledge to prioritize, and the wisdom of perspective. They are humble. They ask questions. They strategically push the envelope a bit. They build a strong team, and then get out of the way.

2.2.3 Five PDP Phases

While much of the workflow is the same, different DOTs and organizations may divide their PDP into different numbers of phases. For our purposes, we will discuss project development in the five phases shown in Figure 2.7.

2.3 Scoping Phase

"Deciding what not to do is as important as deciding what to do."

– Steve Jobs

Scoping generally begins after PE is authorized and ends after the 30% design meeting. This critical phase positions the project for success. When PE is authorized, you have an idea. This

1	2	3	4	5
Scoping	Preliminary Design	Detailed Design	Final Design	Advertise
% Plans				
0 % 30 %	50 %	60 %	90 %	100 %
Key Milestones				
• Assemble project team • Scoping • 30% Plans	• VE • Public Hearing • Design Approval	• RW authorization • UFI • 60% Plans	• RW acquisition • URL • 90% Plans	• PS&E • Advertise • Award

Figure 2.7 Five PDP phases.

may be a mature idea that was thoroughly researched, planned, and vetted. Or perhaps it was not. At project initiation, you should have the prescoping level budget, scope, and schedule. The Scoping Phase further develops these preliminary ideas and transitions them into a project.

As PM, the Scoping Phase oscillates between hurry-up-and-wait and herding cats. It can feel like you are captain of a ship that has just left to cross the ocean. However, as you set sail you are still gathering your crew, the ship itself is still under construction, and your destination is still being finalized.

The PM's primary responsibility is to develop and deliver the project within the established triple constraint of budget, scope, and schedule. The challenge in this first PDP phase is that these constraints are not yet settled. Preliminary expectations have likely been established that can guide you. But the whole point of the Scoping Phase is to clearly define the triple constraint of budget, scope, and schedule in which you will operate for the remainder of the project. Meanwhile, you are assembling your team, identifying stakeholders, and building the Project Management Plan.

This phase has three overarching goals: Data Collection, Scoping, and 30% plans. Each of the activities within this phase should forward one or more of these objectives.

During the first part of the Scoping Phase, imagine a giant funnel, as shown in Figure 2.8. A wide variety of inputs go in the top which are sifted and combined and evaluated and eventually filtered to output the Triple Constraint of the Scoped Budget, Scope, and Schedule.

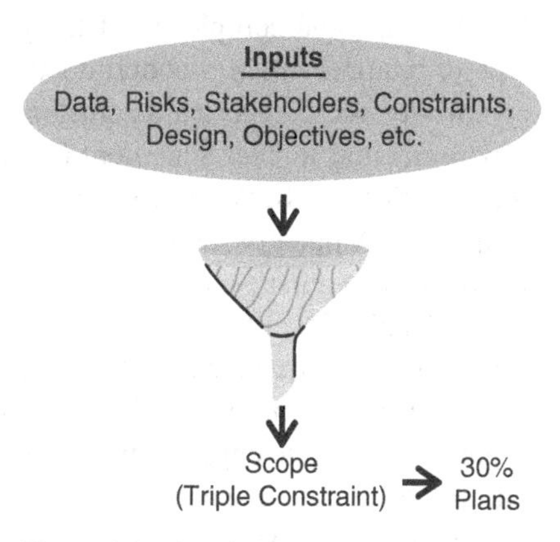

Figure 2.8 Scoping inputs.

As PM, you control and guide the inputs. You are the project's primary gatekeeper and facilitator. It is important to make sure all that should be considered is, and that which should not is not. This begins with any information that was prepared during project initiation. Field data from all relevant engineering disciplines should then be added. At this stage, data collection is primarily from noninvasive investigations (e.g., ground and aerial survey, wetland flagging, traffic counts, etc.). Invasive investigations (e.g., geotechnical borings, test holes, historic resource digs, etc.) are typically performed in the next PDP phase. Other critical inputs may include internal and external stakeholder expectations, financial realities, political constraints, and all other nontechnical considerations.

As information is input into the top of the funnel, the PM is responsible to navigate the scoping process. This can be a bit iterative as all relevant information will not be available at the same time. The challenge is to leverage this information and work with the applicable SMEs to qualitatively identify project risks. Each engineering discipline will identify its own risks. The PM must find the interdependence between the disciplines, evaluate the identified threats and opportunities, and synthesize these with the project's objectives and resources. Risks will likely not change the project's Purpose and Need, but they may significantly impact design, resources, constraints, and securing stakeholder buy-in. These qualitatively identified risks must then be quantitatively converted to project impacts that are reflected in the Triple Constraint of budget, scope, and schedule. Effectively identifying, analyzing, and responding to risk is the very essence of scoping. These processes are further described in Chapter 6, *Risk Management*.

Scoping clarifies the project expectations, constraints, and resources, while setting up the schedule and budget baselines against which you will be measured for the remainder of the project. These scoping baselines are essential in order to correctly position the right amounts of the right types of money at the right times in the multi-year improvement plan so the projects and programs can be successfully developed and delivered. The Scoping Process is further described in Chapter 4, *Scope*.

Most every project should have a kick-off and scoping meeting. Often these are combined. The purpose is to gather the team together. This may be the first time many members are interacting. As such, a meet-and-greet may be important, as well as discussing expected roles, responsibilities, and ground rules for the project and future team interactions. This is a great chance to share initial thoughts and set the tone for project expectations, while laying the foundation of the desired team culture. Many of these points should be included in the Project Management Plan.

NEPA is the National Environmental Protection Act. This is a fundamental federal law that directs federal agencies to conduct environmental reviews to consider a project's potential environmental impacts. The range of NEPA's scope is vast, and expanding. The size and complexity of the proposed transportation project will determine the type of required environmental document. The three main types, in ascending order of complexity, are Categorical Exclusion (CE or CX), Environmental Assessment (EA), and Environmental Impact Statement (EIS). The time and effort associated with these documents varies widely. Where a CE can be relatively painless, an EIS can drive the multi-year project development schedule. If a project represents little to no impact on the environment, one can pursue a Findings of No Significant Impact (FONSI).

The details of NEPA, and its associated workflows, are complicated, nuanced, and evolving. This can be true for both the regulations and their interpretations, at the federal and state levels. The more complex the project and required environmental document, the more critical it is to engage environmental professionals who live and breathe these regulations. Beyond their specialized knowledge, exceptional environmental experts also bring with them positive and productive working relationships with the various decision-makers and entities that hold some influence or jurisdiction for this process.

Environmental professionals will swim in their lane, and shepherd the NEPA process throughout the life of the project. From a PM perspective, it is imperative that you understand your project's environmental commitments and risks. It is crucial that you connect the environmental professional to the rest of the team, particularly the designer and hydraulics SMEs, to ensure each discipline is aware of existing and changing constraints and commitments throughout the development process. During the scoping phase, it is essential that the PM works with the environmental professionals to determine the environmental document type and help navigate this restrictive and prescriptive workflow. Once the environmental document type is determined, the budget, scope, and schedule should be adjusted to accommodate it, if needed.

Pending the type of environmental document, there may be associated stakeholder engagement requirements. The PM should work closely with the environmental professionals to integrate NEPA-related public outreach into the project's broader stakeholder engagement plan, if and when it makes sense.

Generally speaking, environmental activities should never be critical path tasks. The PM should endeavor to ensure concurrent engineering is utilized such that environmental tasks are pursued at the proper time, but such that they never become critical path tasks. This can require attention at both ends of the schedule spectrum. First, you must confirm all necessary and required paperwork is prepared, submitted, and approved at the proper time such that it does not delay progress. Conversely, a PM should ensure environmental permits and the like are not secured too soon. Environmental activities become troublesome when they are completed either too late or too early.

The Scoping Phase can produce a number of important deliverables. Each engineering discipline may produce a preliminary work product that includes data collection and preliminary thoughts, recommendations, or design solutions. The Project Management Plan should be assembled, as further described in Chapter 4, *Scope*. Risks should be appropriately documented, as further described in Chapter 6, *Risk Management*. Stakeholders should be engaged, as further described in Chapter 9, *Communications*. Scoping is closed with the formal approval of a Scoping document, as per your organizational requirements.

The different discipline building blocks all come together in the 30% plans. Your DOT or organization may refer to the 30% design meeting by another name, such as Preliminary Field Inspection (PFI), Preliminary Design Meeting, or Staff Review Meeting. The design percentage may also vary slightly, but is typically between 20% and 30%. While the exact name and percentage is subjective, the intent and purpose of this first plan submission should be the same. The goals are to pull together all the initial thoughts, identify and answer big questions, and chart the course for a design solution.

The first challenge in data collection is to ensure you have all the information you need. This discovery process can seem relatively straightforward. Were the wetlands flagged and confirmed? Are there Threatened and Endangered species in your corridor? Where are the underground utilities? Did Traffic collect all the needed counts for the intersection analysis? Did the survey collect all the data points to evaluate sight distance on that complicated curve? And so forth. The second challenge is to absorb and synthesize this information across disciplines. Do you need to collect additional survey since the Hydraulics Engineer had to chase a creek farther than expected to demonstrate an adequate outfall? Do you need to expand your wetland delineation footprint since it appears you may need more temporary pavement for Maintenance of Traffic (MOT) since you cannot close the road during construction? The PM should work closely with the lead designer and team to ensure the right engineering disciplines, SMEs, and decision makers are involved in the right conversations to pull all these thoughts and issues together in a way that paints a more complete project picture. The power and expertise of the whole project team is greater than the sum of its parts.

The 30% plans provide an opportunity to ask and answer big project questions. What are the high-risk issues that could significantly impact the project if not resolved? These are the proverbial forks in the road whose chosen course of action will directly and indirectly impact the design solution. Can you close the road during construction, or do you need to keep it open to traffic? How many left-turn lanes will be at the signalized intersection? Is there space to stage a crane for the bridge construction? Do you have an adequate outfall? Because if you don't have an outfall, you don't have a project. The PM and lead designer should concentrate on questions like these with focused intensity as they need to be aggressively addressed and resolved at this stage in the project, before it is formally scoped.

While the 30% plans paint the design solution with broad strokes, some details are starting to emerge. There may not be too much meat on the bones, but the bones should be there. For roadway projects, the goal is to define the horizontal alignment. You should have all field survey and located subsurface utilities. Any required Interchange Access Report (IAR) should be secured. The environmental document type should be settled. Horizontal and vertical roadway alignments are proposed, along with typical sections. It is understood that vertical details may need to be adjusted as the storm sewer design or related utility work advances. You should demonstrate an adequate outfall. Stormwater management approaches are reflected, along with conceptual MOT strategies. The design approach should be logical and consistent with the existing corridor characteristics, the project stakeholder expectations, and the project's purpose and need. The PM should work closely with the team and all stakeholders to facilitate the close cooperation that is needed to define and address all risks and issues that may impact the road's horizontal alignment, across all engineering disciplines.

The 30% plans should be routed for review by all applicable engineering disciplines, SMEs, and other appropriate reviewing entities. Comments should be received and evaluated to discern if other major qualitative risks or conflicts were identified that should be quantitatively converted to time and money in the scoping schedule and budget before it is officially submitted, reviewed, and approved.

The PM should then assemble the formal scoping documents, as required by your organization. This should detail, at a minimum, the project's budget, scope, and schedule. This may be a form, a report, or be presented in another format. Regardless, this documentation should then be reviewed and approved in a manner consistent with your organization's applicable procedures. The approved Scoping document should be retained by the PM, and routinely used as the guiding reference for the budget, scope, and schedule constraints within which they are to operate for the remainder of the project.

During the scoping phase the team is forming, as further described in Chapter 7, *Managing Resources*, and each engineering discipline is running full steam ahead down their own critical path. The PM must know the pulse of their progress, and be able to quickly identify when to pump the brakes and bring other disciplines in on specific issues. Otherwise, David G. Farragut's "*Damn the torpedoes, full steam ahead*" approach will necessitate rework down the line. The PM doesn't need to know all the answers, but they should know the right questions to ask at the right time. Team communication, as further described in Chapter 9, *Communications*, is critical to Scoping success.

It is difficult to overemphasize the importance of field visits for the PM and the entire team. Walk the site. Despite increasingly detailed and accurate remote tools and applications, there is no substitute for walking the site. It can provide invaluable context. If possible, walk the site with key team SMEs from each engineering discipline. You might be surprised how quickly and efficiently risks can be identified and potential solutions flushed out at a high level if you have the right people on site at the same time.

Figure 2.9 provides general guidance on key activities, milestones, and deliverables during the Scoping Phase, divided by engineering discipline. While each discipline has its own

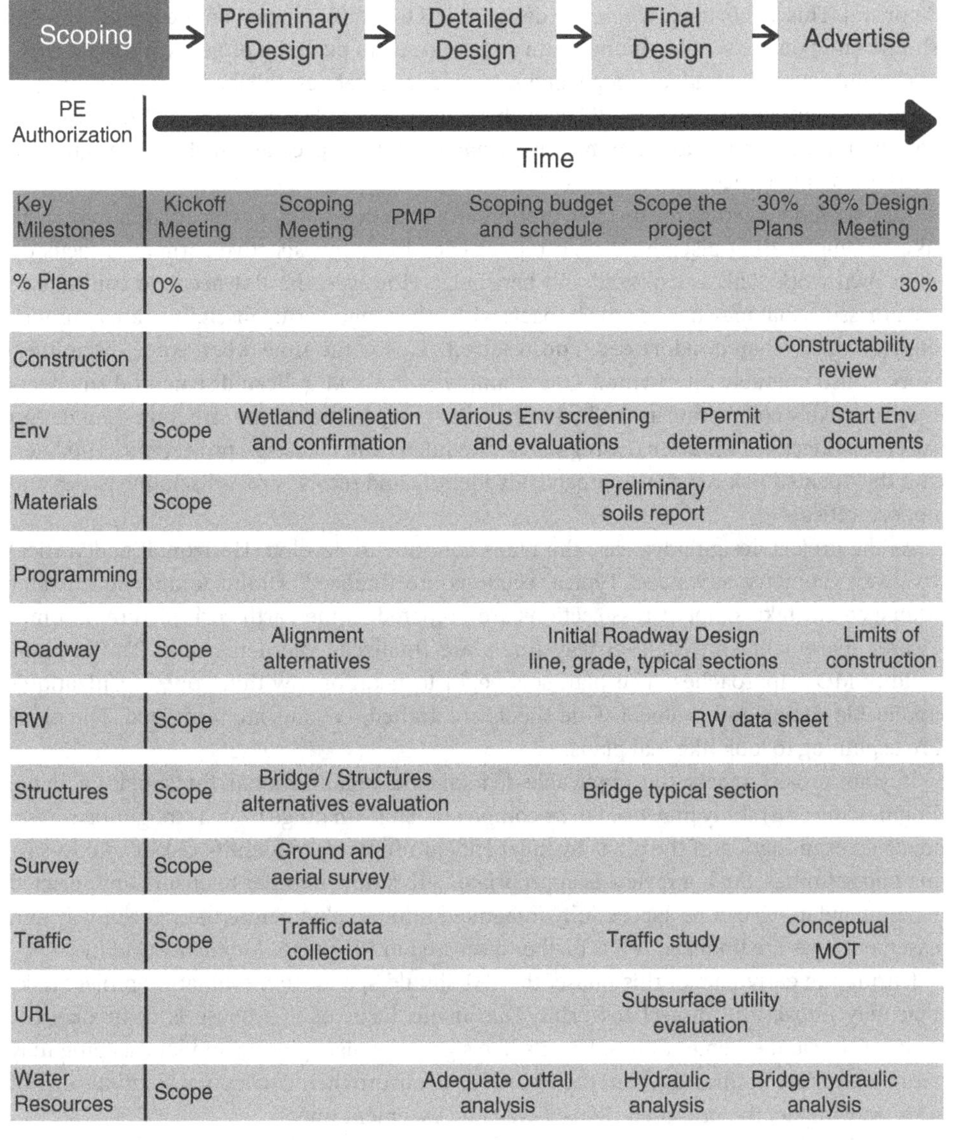

Figure 2.9 Scoping phase.

critical path, their work is remarkably interconnected. Exceptional PMs understand and leverage these interdependencies to increase efficiency and drive productivity.

2.4 Preliminary Design Phase

"Design is thinking made visual."

– Saul Bass

Preliminary Design generally begins after the 30% design meeting and ends with Design Approval. This is the phase where the design starts to put some meat on the bones. The PM should build on the scoping momentum and continue to push the project within the triple constraint criteria of budget, scope, and schedule that were established during Scoping. It is especially important that the PM actively manage the triple constraint as Scope Creep is common in this phase due to public involvement and the pace at which the design solutions are being refined.

After coming together for the 30% preliminary plan design review meeting, it may seem like the engineering disciplines then retreat back to their respective corners to focus on their own work. This is expected and beneficial. However, the PM needs to continue to ensure issues that may have ripple impacts with other engineering disciplines are appropriately elevated, shared, addressed, and resolved. This is the time when some disciplines' work is also uniquely intertwined. One example is how hydraulic and structural engineers may iteratively collaborate as the hydraulic report may influence the structure span design specifics. Effective PMs often use regularly scheduled team meetings, project status updates, and the updated risk register to proactively identify and resolve cross-discipline issues and opportunities.

As the project design advances, the plans continue to develop. Horizontal and vertical roadway geometry advances. Typical sections are finalized. Drainage and Stormwater Management take shape. Cross-sections are prepared as the earthwork picture becomes clearer. Easement and Right-of-Way limits are finalized. Maintenance of Traffic plans solidify MOT approaches. The plan sets begin to take form within your organization's applicable styling and protocol. Title sheets are drafted. Layouts are formatted. The plans are beginning to look like real plans.

If your project meets the applicable federal and organizational thresholds, a Value Engineering (VE) study may need to be completed. This can range from a straightforward to involved examination of the plans by independent reviewers to identify risks and cost-saving opportunities. Such a review is appropriate at this time in order to ensure any selected recommendations can be successfully integrated into the plan before the right-of-way and easement lines are finalized. VE is further discussed in Chapter 8, *Managing Quality*.

During the early part of this phase, the PM should spend time evaluating project risks that may impact the project footprint. One major focus of this phase is to finalize the right-of-way and easement lines. Details like signs, signals, lighting, and landscaping may seem premature at this stage, but the PM needs to ensure their conceptual solutions do not adversely impact the proposed right-of-way and easement lines.

With the additional design information and project limit becoming defined, the necessity and extent of invasive investigations becomes clear. Did the historical survey determine follow-up digs are necessary? Does it appear you need test holes in any location to nail down potential utility conflicts? Where does the structural engineer need soil borings to advance bridge abutment foundation design? And so forth. This additional information will enable the respective discipline designers, engineers, and scientists to refine their designs with increased clarity. It is important to determine this information now since the resultant decisions and actions may influence the RW and Public Hearing (PH) plans.

The additional design solutions and field data should push utility conversations. Utility designations should be completed, along with any needed test holes. This is also the appropriate time to initiate conversations with impacted utility owners. Schedule a Preliminary Utility Field Inspection meeting to discuss possible conflicts and resolution strategies. Early coordination can go a long way to expediting preparation of requested utility easements.

Concurrent with all of these activities is continued environmental progress. Pending project specifics, there may be a whole host of permits and studies that need to be conducted, reviewed, and approved. A critical part of most every project is the Draft Environmental document that must be completed prior to Public Hearing. Do not try to navigate this regulative minefield by yourself. Rely on your organization's or team's environmental experts. Actively develop a productive partnership that is commonly focused on responsibly progressing the project within its budget, scope, and schedule constraints.

One primary focus of the Preliminary Design phase is public participation. Stakeholder input and involvement is crucial. All tasks in this phase are working toward the public involvement and securing Design Approval. The collaborative coordination of interdependent engineering discipline tasks remains a top PM priority. Proactive Risk Management and effective Communication are critical to success in this phase as the plans advance to the point of RW plan submission. This means all aspects of the plans should be sufficiently advanced such that all limits of disturbance, easements, and rights-of-way are firmly established. While specifics within these boundaries will continue to evolve, the project and acquisition footprint, limits of disturbance, should be settled. As such you should have all needed easements and rights-of-way defined, this should include any needed private utility easements (blue lines). Any needed design exceptions or waivers should be secured. Locations for all Storm Water Management (SWM) basins or facilities should be known. Env should have completed the Draft environmental document. Additionally, to that end, as you triage and prioritize discipline specific issues, it is especially important that you press to fully grasp the horizontal plan impacts. Is the redesigned stormwater basin changing its footprint? Does shifting the traffic signal pole to avoid underground utilities shift the sidewalk enough that the slopes are not able to tie back into the existing landscape within the proposed easements? Do the updated Maintenance of Traffic and Sequence of Construction plans fit within the proposed Limits of Disturbance? And so forth.

Public Participation may begin with an organization choosing to hold one or more Citizens Information Meetings (CIM) prior to the Willingness or Public Hearing. CIMs are voluntary, and as such may assume a variety of formats to satisfy differing objectives. A formal presentation format followed by an orderly question and answer session may work best for controversial projects with lots of public interest. An open house format may be preferred for smaller projects with less public dissent, while allowing staff to connect with the adjacent landowners and address their specific concerns. Meanwhile, a town hall format may be preferred to workshop the problem or gain feedback from local businesses. Whatever the format or objective, be intentional. These are often held in coordination with other outreach efforts conducted through a variety of virtual and in-person initiatives. Effective CIMs can be instrumental in facilitating a smooth Public Hearing. Public involvement is further detailed in Chapter 9, *Communications.*

While it may be different pending your project specifics and your applicable regulations and organizational expectations, many projects have a Willingness or Public Hearing. A Willingness is a commitment your organization publicly posts that you are willing to host a public hearing should it be requested. If any citizen does request a public hearing, the organization may meet with them to see if they can address their concerns. If they do not, then the organization must hold a public hearing. If no one responds to the Willingness within the specified time, then the organization can assume the project has public support and they are not required to hold a public hearing.

There are two types of Public Hearings: Location and Design. Location is predominantly used when there is roadway on a new location. Design is a chance for the public to comment on the design solution, be it on a new or existing alignment. Pending the project, the Location and Design Public Hearings may be combined. The format of a Public Hearing can vary pending your specific guidelines and expectations. There are some commonalities. The public should have a chance to review the plans and submit comments. The organization should then respond to all received comments. This can typically be done in person, or by a variety of virtual methods.

Following public involvement and the Public Hearing, the project should pursue Design Approval. Prior to Design Approval, the plans should be updated to incorporate any design changes prompted by VE or public involvement efforts. Design Approval is a formal approval by the authorized entity that confirms consensus on the design solution.

The Public Need is established when the appropriate governing body formally grants Design Approval. The Public Need is required to allow government transportation organizations access to the available legal options of Eminent Domain, which will be further discussed in Section 2.5, *Detailed Design Phase.* As such, many of the rules and requirements surrounding Willingness, Public Hearing, and Design Approval are detailed in state and local legal codes. Be sure you consult your SMEs to ensure all appropriate rules and regulations are meticulously followed.

Figure 2.10 provides general guidance on key activities, milestones, and deliverables during the Preliminary Design Phase, divided by engineering discipline. While each discipline has its own critical path, their work is remarkably interconnected. Exceptional PMs understand and leverage these interdependencies to increase efficiency and drive productivity.

2.5 Detailed Design Phase

> *"Engineering isn't about perfect solutions; it's about doing the best you can with limited resources."*
>
> – Randy Pausch

Detailed Design spans from Design Approval through the 60% design meeting. This phase requires close collaboration with all involved disciplines as detailed design is added to the Public Hearing plans, advancing them from approximately 50% plans to approximately

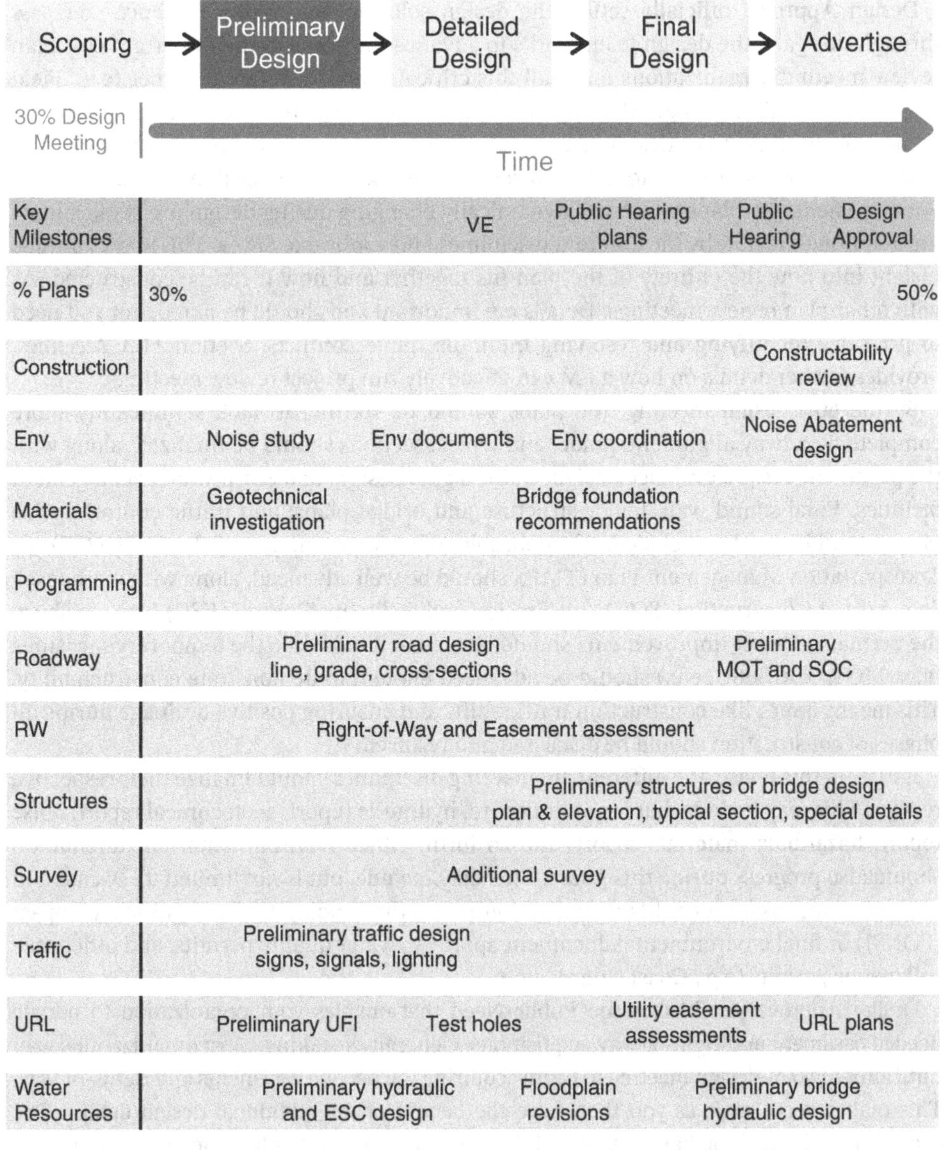

Figure 2.10 Preliminary design phase.

60% plans. The PM should work closely with the lead designer to ensure the different engineering disciplines' respective progress is consistent with, and does not conflict with, the overall strategic solution. As the plans dive deeper into the weeds and risks are worked to resolution, the PM is responsible to ensure the project remains within the defined triple constraint of budget, scope, and schedule. Any risks to the triple constraint should be actively tracked, and any resulting adjustments should follow your organization's applicable change management procedures.

Design Approval officially settles the design solution and approach. Once you pass through this gate, the design team works to advance the plans to the important 60% plan review meeting. Organizations may call this critical event by different names (e.g., Field Inspection, Detailed Design Meeting, 60% Staff Review Meeting, etc.). Regardless of the name, this meeting is a critical project development milestone. These review meetings provide the unique opportunity for all the disciplines' work to come together on a set of plans. For a moment, the plans which are dynamically changing during design are static, allowing a thorough review by the different disciplines and applicable SMEs. This allows unique insight into how the entirety of the plan fits together and how it can be constructed. As with other plan review meetings. Details are important and should be noted, but you need to prioritize identifying and resolving multi-discipline conflicts. Section 11.3, *Meetings*, provides further details on how a PM can effectively run project review meetings.

By the 60% design meeting, the plans should be starting to look significantly more complete. Roadway alignment, grade, and typical sections should be finalized, along with the pavement design, drainage design, traffic signal design, and Stormwater Management facilities. Final sound wall, major structure and bridge plans, and traffic control device plans should also be included, if applicable. The Maintenance of Traffic (MOT), or Transportation Management Plan (TMP), should be well advanced, along with a proposed Sequence of Construction (SOC) and Erosion and Sediment Control (ESC) plan. In short, the permanent asset improvements should be well defined while the temporary measures (e.g., MOT, SOC, ESC, etc.) should be advanced enough to demonstrate constructability. This means issues like construction traffic shifts and ensuring positive drainage during all phases of construction should be discussed and resolved.

Earlier in this phase, the different engineering disciplines should finalize their respective reports. This may include, but is not limited to, hydraulic report, geotechnical report, noise report, hazardous materials report, and so forth. Other Environmental documentation should also progress during this phase. This may include, but is not limited to, stream and wetland compensation credits or mitigation plans, the Finding Of No Significant Impact (FONSI) or final environmental document approval, water quality permits, and other miscellaneous permits (e.g., Coast Guard, etc.).

Design Approval establishes the Public Need that enables your organization to pursue needed easement and right-of-way acquisitions. Generally speaking, most organizations wait until after the 60% design meeting to begin acquiring the needed easements and rights-of-way. This makes good sense as you then have the benefit of the additional design details. The exception to this is if you have any Total Takes, which is when it has been determined that you require the entirety of the parcel. This can involve relocations of a business or resident. Due to the involved nature of the relocation process, it can be beneficial to begin the total take acquisitions during this phase, which is typically before one begins partial take acquisitions.

During this phase the PM will work with the designers, surveyors, and Right-of-Way SMEs to prepare to begin partial take acquisitions after the 60% design meeting. This typically involves preparation of the Right-of-Way plans (or individually prepared plats if your organization acquires by plats), updating property owner information on the right-of-way data sheet, and initiating or coordinating with utilities. The more complicated the utility situation, the earlier coordination should begin. If not before, during this stage, you or your RW SMEs should coordinate with all public and private utilities whose assets will be impacted by your project. The PM should schedule a Utility Field Inspection (UFI) meeting in which all utilities come together to discuss the planned improvements, potential conflicts, and proposed relocation plans. Pending the project, it can be helpful to walk the site together. Plans should be distributed to the utilities before the meeting so those attending the UFI meeting can be informed and any risks or conflicts can be discussed in a cooperative manner by all involved parties. Take this opportunity when all are in the same room to press issues to resolution.

Figure 2.11 provides general guidance on key activities, milestones, and deliverables during the Detailed Design Phase, divided by engineering discipline. While each discipline has

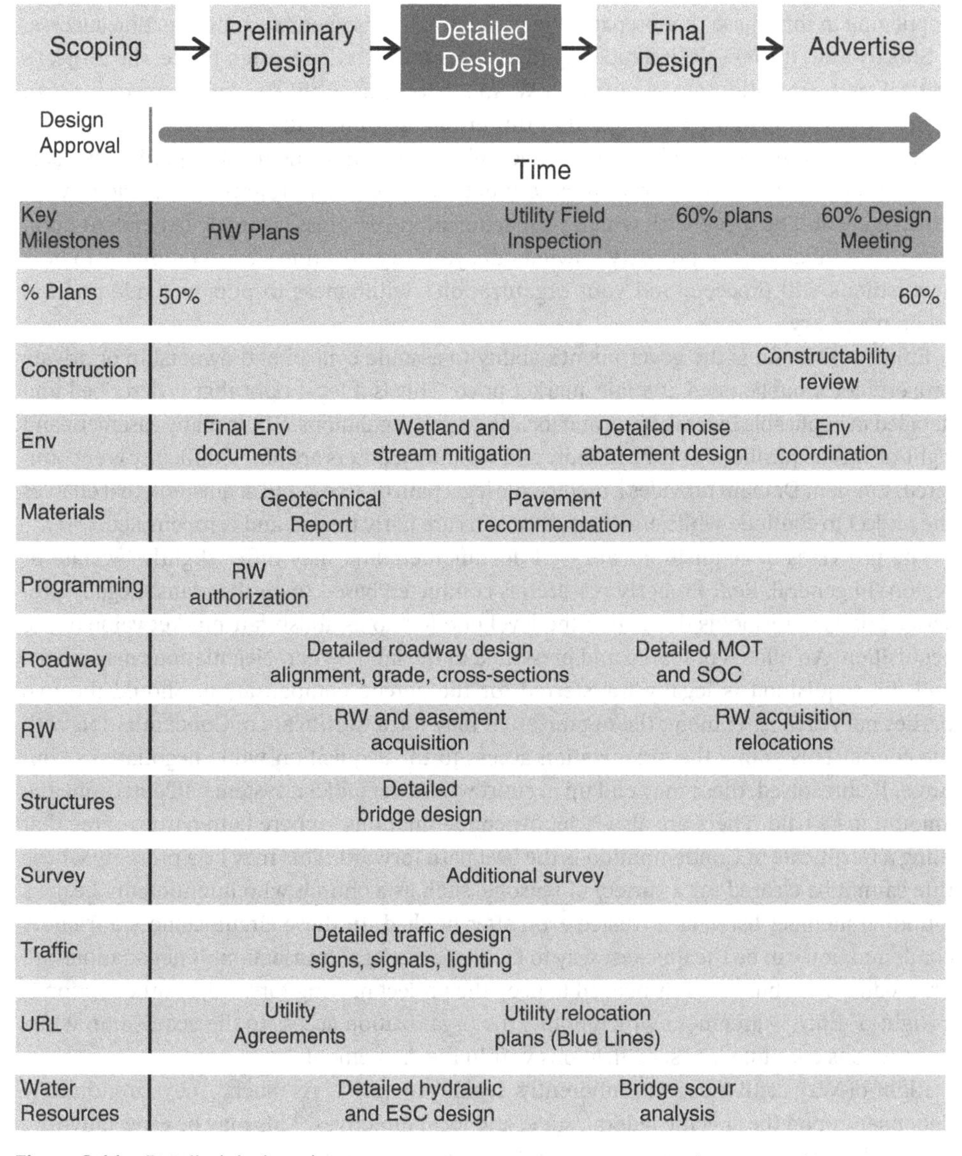

Figure 2.11 Detailed design phase.

its own critical path, their work is remarkably interconnected. Exceptional PMs understand and leverage these interdependencies to increase efficiency and drive productivity.

2.6 Final Design and Right-of-Way Acquisition Phase

"A goal without a plan is just a wish."

– Larry Elder

This phase advances the plans from 60% to 90% complete. However, there is so much more happening in this phase that prepares the plan for Advertisement and construction success.

Shortly after the 60% design meeting that closes the Detailed Design phase, the project's critical path transitions to acquiring RW and relocating utilities. An organization can benefit from formalizing the handoff of this critical path in order to set realistic expectations of budget, scope, schedule, and deliverables. This can be the most unpredictable part of the project development process as so much of the critical progress is outside of your direct control. The speed with which your acquisitions progress is largely dependent upon the parties involved, the property's ownership, your organization's expectations as to how acquisitions will proceed, and your organization's willingness to pursue condemnation when necessary.

Eminent Domain is the government's ability to assume control and ownership of private property for a public need at a fair, market price. This is a legal right that is described and detailed in applicable federal, state, and local laws and regulations. While many easement and right-of-way acquisitions may be friendly and voluntary, others are not. If difficulty is encountered, Eminent Domain provides a predictable legal path of access and acquisition that enables the project to continue while ensuring all citizens are fairly treated and compensated.

The processes of acquisitions are well documented, and may differ slightly by state or region. In general, Real Property research is conducted based upon the plans. Legally prescribed dollar thresholds determine the level of effort to establish fair market value of the acquisition. An offer is prepared and presented to the land owner. Negotiations may ensue, and the acquisition is legally transferred for the agreed compensation. Should the two parties not reach agreement, the organization may file a Certificate of Condemnation with the courts. This grants the organization access to the acquisition while negotiations continue. If unresolved, these may end up in court where the judicial system will determine the amount to be paid. There are also "friendly condemnations," where both parties agree that filing a Certificate of Condemnation is the best path forward. This may be a property whose title cannot be cleared for a variety of reasons, such as a church who intentionally keeps a founding member listed as a trustee even after he died. In these circumstances a friendly condemnation can be the quickest way to facilitate the legal transaction. If negotiations are proceeding well, but access is needed to keep the project moving forward, another option is a Right-of-Entry Agreement which grants the organization access to the acquisition while negotiations continue, per specifications within the Agreement.

Right-of-Way activities are inherently legal processes. As such, they are directly dependent upon the specific federal, state, and local directives. This may be especially true should there be residential or business relocations. Some states may also have other legal considerations in acquisitions, such as allowances for loss of business during construction.

Because of the need to ensure all the i's are dotted and t's are crossed in exactly the right order, Right-of-Way acquisitions are best left to those professionals who know and understand the governing regulations, specializing in this uniquely focused aspect of project development.

As PM, your role in acquisitions is likely to coordinate and facilitate, more than to actually do the related tasks. As such, it is imperative the designers and RW team members are well connected. The design within the designated easements and rights-of-way may change, but if the evolving design solution prompts any movement of permanent, temporary, or utility easements or rights-of-way, you need to immediately bring the RW SMEs into the discussion. Similarly, if the acquisition negotiations introduce new conditions or constraints you need to bring the appropriate engineering discipline SMEs into the discussion. Any design changes prompted by the negotiations and settlement terms should be reflected and incorporated into the plans.

Once all rights-of-way and easements are cleared, utility relocations may begin. This may require separate approvals by your organization, one for each separate utility relocation agreement. Consult your experts as to who pays what for utility design and relocation costs. This can vary greatly pending the legal instrument by which the utilities' assets are in your rights-of-way.

Utility relocations can frustrate project progress due to occasionally unpredictable response times. A utility's relocation prioritization, especially those at their own expense, may not match your own. There are other real considerations that may be out of your control. If your relocation crew is union and they go on strike, your job sits. If there is a natural disaster (e.g., a hurricane or other natural disaster) and your crews are pulled to another state to assist in emergency relief, your job sits. This can introduce real schedule and budget risk into your project. Wise PMs leverage their organization's URL experts and their existing professional relationships with utility contacts to effectively influence progress. Early and consistent communications of expectations and project schedule constraints can go a long way to forming successful partnerships.

This same mindset can also be beneficial when the project touches or crosses a railroad. The presence of a railroad immediately introduces significant schedule and budget risk into your project. In larger part, this is often due to unpredictable responses as it can seem they dance to the beat of their own drummer. A bridge engineer once described a typical roadway-railroad interaction with following example. Imagine you are playing ball in your backyard with friends and it accidentally goes over the privacy fence into your neighbor's yard. You go to their front door and a grumpy curmudgeon opens the door. You explain the situation, and his response is "so your ball is in my yard, forgive me if I don't share your urgency in retrieving it for you", and then shuts the door. Obviously, there are talented, cooperative people in rail who want to successfully partner with roadway projects. However, it can sometimes feel like this example. In large part, this is due to the nature of their rights-of-way. They were often here first, and may have special legal status. They also have their own workflows that may not fit into your plans or schedules. Suffice it to say, if you have rail on your project, you need to remain proactive and vigilant as any rail related task can quickly become a critical path task. Seek out the experience and wisdom of those with previous experience with the rail company on your project. These can often be the structure and bridge engineers. Rail introduces significant risks on your project that must be actively monitored and managed.

Many organizations have mechanisms to extend the window for acquisitions and utility relocations to Advertisement, Award, or even into Construction, if needed. However, these approaches introduce significant schedule and budget risks into the project. While appropriate in certain extenuating situations, they should be actively avoided.

While rights-of-way and easements are being acquired and utilities are being relocated, the design is concurrently being finalized. This includes, but is not limited to, roadway, bridge, major structures (e.g., retaining walls, etc.), traffic control devices, landscaping, Erosion and Sediment Control (ESC) plans, Maintenance of Traffic (MOT) or Traffic Management Plans (TMP), and Sequence of Construction (SOC). Another concurrent path is to prepare all that is necessary outside of plans to proceed to Advertisement. This includes, but is not limited to, securing all necessary permits and ensuring all wetland, stream, and water-quality credits are purchased.

Once all these concurrent paths are complete, the project is ready to proceed with the 90% Design meeting. Some organizations refer to this as the 100% Design meeting, which is often called the Pre-Advertisement Conference (PAC). This is the final design meeting and is often the handoff from the designers to the contracting agents in the Advertise Plans phase. It is a critical opportunity to openly discuss the constructability, work zone regulations, MOT/TMP, ESC, SOC, communicate environmental permit constraints, expected Construction and Engineering Inspections (CEI) efforts, and remaining project risks.

This is also the time when the designers prepare and submit the Plans, Specifications, and Estimate (PSE). The specifications should be all that is needed for the contracting agent to prepare the Advertisement package; this includes any special provisions. Pending your organization's workflow, this may be the window in which your Construction division reviews and approves the bid specs and special provisions. As PM, if you know your project has unusual or uncommon special provisions, it is a good idea to vet these early through your reviewing and approving entity so as not to hold up progress at this point. The Estimate should be the final submission estimate, which is the most accurate estimate and last estimate that will be prepared before the Engineer's Estimate.

The 90% design meeting is one of the critical milestones of design. As such, the PM should ensure that the project is still progressing within its approved budget, scope, and schedule. If there are any discrepancies (e.g., final submission estimate exceeds the budget), prompt action should be taken to resolve any remaining issues so the project can proceed to Advertisement.

Figure 2.12 provides general guidance on key activities, milestones, and deliverables during the Final Design Phase, divided by engineering discipline. While each discipline has its own critical path, their work is remarkably interconnected. Exceptional PMs understand and leverage these interdependencies to increase efficiency and drive productivity.

2.7 Advertise Plans

"There are no shortcuts to any place worth going."

– Beverly Sills

This final project development phase takes the final plans, advertises the project, and extends through project award. This can be a very busy time for the PM, but in a very different way than every other phase. The steps in this phase center around procurement. The Advertise Plans phase name may more accurately be called construction procurement. In

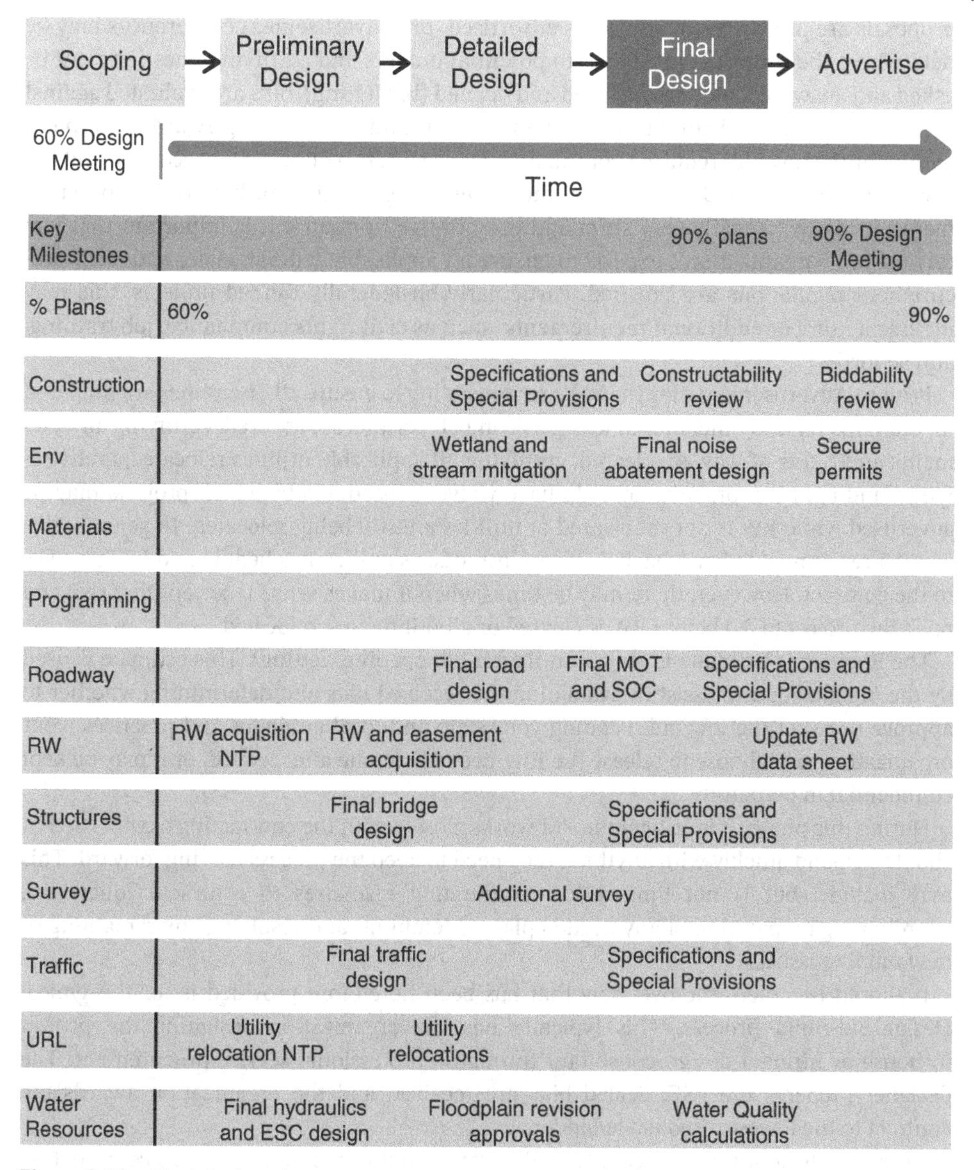

Figure 2.12 Final design phase.

essence, the plans are being packaged for Advertisement (Ad), and then shepherded through procurement until a contractor is secured and the bid has been Awarded by the governing authority.

It begins with the final design meeting that ends the previous phase. In it, the final sealed and signed Plans, Specifications, and Estimate (PSE) are prepared and submitted to your organization's contracting agent. In some organizations this may be your procurement office, in others it may be your construction division. Regardless, their function is much the same. The bid package is assembled, a biddability review is conducted, bid proposals are prepared, the project is advertised, pre-advertisement conferences may be held where the project is presented to potential bidders and clarifying questions can be asked and answered, bids are received and opened (bid letting), bids are evaluated against the sealed Engineer's Estimate, a business decision is made regarding award recommendation for the low bid, required contractual paperwork is secured and processed, and the Award is presented and approved by the applicable governing authority. This procurement process is typically very strict and prescriptive in nature. It is important that you rely on your organization's experts to ensure all applicable federal, state, and local procurement regulations are satisfied. Particularly on federally funded projects, this may include a number additional requirements, such as civil rights compliance, job training, and so forth.

Prior to Advertisement, it is the PM's responsibility to ensure all disciplines are prepared for Ad. This typically includes at least ensuring all permits are secured, certifying all easements and rights-of-way are cleared, certifying all applicable utilities relocated, and verifying funding. Most organizations do have waiver procedures by which projects may be advertised while RW is not yet cleared or utilities are still being relocated. In general, this is not a recommended approach as it can introduce significant schedule and budget risks to the contract. However, there may be times when it makes sense to accept that risk and push the project to Ad before RW is cleared or all utilities are relocated.

The Engineer's Estimate is sealed until after bid opening (letting). This estimate is used by the organization to assist in evaluating the received bids and determining whether to approve or reject the low bid. Pending your state and local guidance and practices, your organization may choose to release the Engineer's Estimate after Award, or it may be kept confidential in perpetuity.

During this phase, it is critical the PM works closely with the contracting agent. The PM should respond quickly with anything they need to keep the process moving forward. This may include, but is not limited to, coordinating responses to contractor questions, coordinating preparation of any needed plan addendums, and resolving any remaining or new funding issues.

It should be noted, the overview that has been heretofore provided is for the typical Design-Bid-Build process. This typically has the organization designing the project in-house or hiring a design consultant through a professional service procurement. The designer prepares the PSE. Sealed bids are received and the organization awards the contract to the lowest responsible bidder.

At its conclusion, the project is typically transferred to the organization's construction division for contract administration and oversight. Organizations should have clear descriptions of this handoff, including all applicable roles and responsibilities. This extends to who has the responsibility for the project's budget, scope, and schedule. For some organizations the PM for project development extends into construction and they shepherd the project from cradle to grave, from initiation through final closeout. Other organizations transition project responsibilities during this time.

Figure 2.13 provides general guidance on key activities, milestones, and deliverables during the Advertise Phase, divided by engineering discipline. While each discipline has its own critical path, their work is remarkably interconnected. Exceptional PMs understand and leverage these interdependencies to increase efficiency and drive productivity.

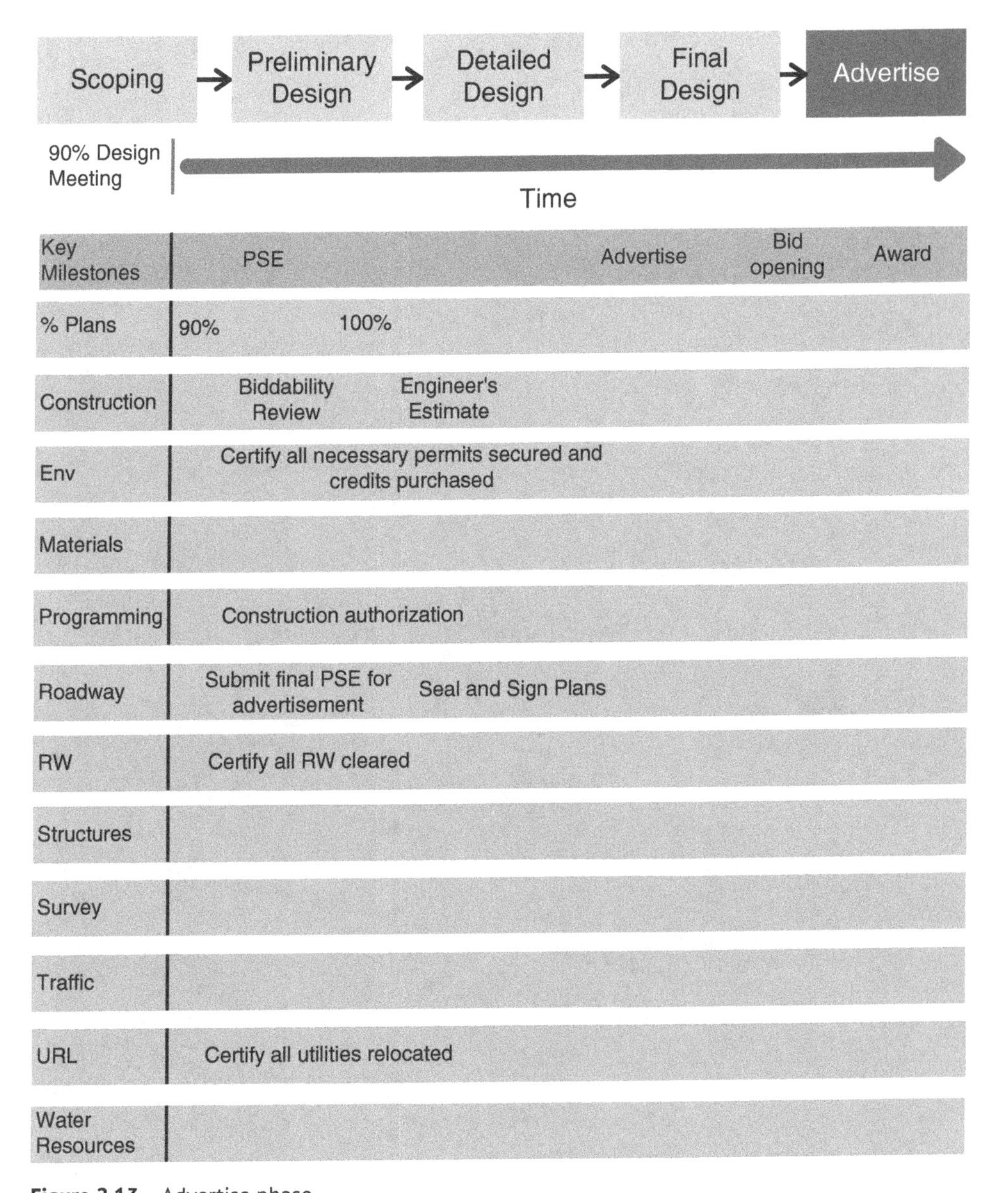

Figure 2.13 Advertise phase.

3

Budgets and Estimates

"Engineer (noun): A person who does precision guesswork based on unreliable data provided by those of questionable knowledge. See also: Wizard, Magician, Magic 8 Ball."

– Anonymous

3.1 Overview

"Follow the money."

– Anonymous

3.1.1 Purpose

The big picture in Transportation Program and Portfolio Management is that your organization should strive to ensure the right amounts of the right type of money are in the right place at the right time to keep the projects moving forward. In order to effectively do this at the program and portfolio levels, you need accurate project schedules and estimates. This requires consistent attention throughout the project development process.

Coexisting with this objective is the reality that transportation owner-operator organizations (e.g., DOTs, local governments, etc.) are not banks. Beyond planned reserves, surpluses and shortfalls pose complications. Shortfalls can be challenging when you don't have enough money to advance and construct projects as planned. Surpluses are to be avoided in part due to the moral obligation of public agencies to be responsible stewards of public funds. The public money is to be used for public good. These realities effectively make transportation owner-operators funding organizations. Money comes in, and money should go out in a similar fashion. In order to do this well, more mature organizations embrace the cyclical process shown in Figure 3.1 of selecting the projects, programming the money, executing the program, closing the project(s), and reallocating resources.

The triple constraint of budget, scope, and schedule are inherently dependent upon each other, directly constraining the others. You can't strain all three at the same time. Figure 3.2 illustrates the old adage that you can only have two of the three: good, fast, or cheap. It can be fast and cheap, but then quality will suffer as the scope is compromised. It can be fast and good, but the cost will be high. It can be cheap and satisfy the scope, but don't expect it to be fast.

Transportation Project Management, First Edition. Rob Tieman.
© 2023 John Wiley & Sons Ltd. Published 2023 by John Wiley & Sons Ltd.

Figure 3.1 Transportation programming cycle.

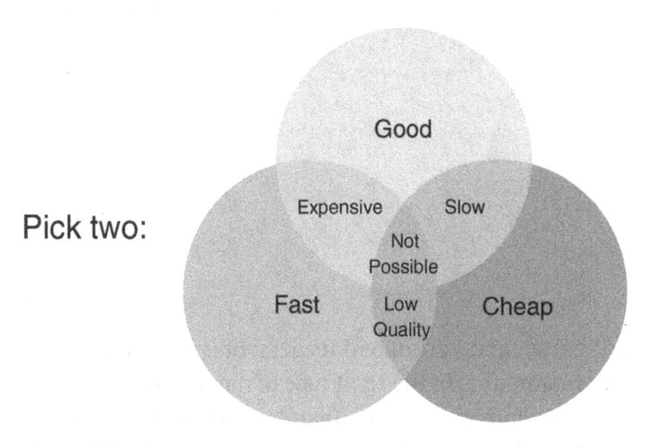

Figure 3.2 Pick two.

In transportation development, time is money. The schedule is only but so flexible without blowing the project budget. The project scope can only be reduced to a minimum level without sacrificing the intent and benefit of the project. And at the program and portfolio level there are many more needs than available dollars, so project budgets can be tight. These three constraints are operating in an environment of scarcity. Managing the triple constraint can be challenging as none of the three can be pulled too tight without impacting the other two beyond acceptable thresholds.

Within this environment, managing the Budget can prove the most challenging as project costs can vary widely over the life of a project given changing market conditions. That is why, although there are many different purposes for cost estimates, perhaps the most important is to enable the PM to actively manage the triple constraint throughout the project to ensure there is enough money at the right times to keep the project moving forward according to the schedule. As risks evolve and market conditions fluctuate, this can seem like you are chasing a moving target. But this is the job of a PM. And there is perhaps no worse situation than bringing a project to Advertisement, only not to be able to Award it due to lack of funds.

3.1.2 Budgets vs. Estimates

While Budgets and Estimates are certainly related, in transportation projects, they are two very different entities.

Budgets are not to exceed total cost of the project. This may be the obligated amount, the allocated amount, or the amount of real money available to complete the project. Federal Obligation is the legal commitment requiring the federal government to reimburse the state for the federal share of a project's eligible costs. Obligated funds can then be Allocated for spending purposes. It is possible for obligated funds to be allocated across multiple projects. It is simpler when the total obligated amount equals that which is allocated. Projects may be funded by a multitude of federal, state, regional, or local sources. Each different funding source, or color of money, may have different restrictions or requirements associated with the use of those funds. The total project budget is the combination of all these sources. Similar to schedules, the PM can best manage their project when there is one project budget, as opposed to different budgets for different audiences. Transparency builds trust and can focus the team to work within common constraints and toward a common goal.

Pending the funding source(s) and your organization's protocol, project budgets may be set and revised at various times. Some budgets are set at time of project application. Others at Scoping. Still others are finalized just before Advertisement. In most cases, a preliminary project budget is established early on in the project planning, and then refined as it progresses through project development. As PM, it is imperative that proper change management be practiced with each budget adjustment. Work closely with your programming resources to ensure there are adequate funds to support any budget adjustment. Work with your design team to ensure the budget adjustment is actually necessary. Your first job as PM is to operate within the existing triple constraint. If you can adjust the scope or schedule in order to maintain the existing budget, while not sacrificing quality or critical improvement benefits, you should do so. Adjusting the triple constraint is certainly allowed and appropriate at times, but only after proper change management has been practiced. This includes adding betterments to the project, such as a waterline extension to promote economic development, which can increase the total project budget even if paid for by a third party. This also means adjusting the budget downward if project savings are realized so that those surplus funds can be put to work by your organization to make a tangible impact in your community.

Estimates are the quantitative determination of the probable cost. Engineers and estimators use provided tools, historic bid tabs and cost data, market trends, and professional judgment to make their best educated guess on what the project may likely cost.

The remainder of this chapter primarily addresses the different types of estimates and their limitations. It is correctly said that estimating is as much art as science. I would readily accept the opinion of an experienced, gray-haired engineer over that produced from a shiny-new parametric estimating tool. Of course, ideally you would have that experienced gray-haired engineer using the new tool to produce an even more accurate estimate.

The big-picture purpose of the estimate is to help you, as the PM, ensure there is the right amount of the right kinds of money in the right places at the right time to keep the project moving forward in the most efficient and responsible path. Estimates should be updated at all major project milestones (e.g., major design meetings and public hearing) or when there are significant changes to the plan or project risks. This provides the PM an invaluable opportunity to check the project against the triple constraint. If the project estimate dramatically changes, you should promptly address this with the programming resources to verify and/or update the project budget.

3.1.3 Process – How to Set a Budget

Pending the requirements of the funding source or your organization, there may be a variety of ways a project sets its budget. Some funding sources rely heavily on preliminary or planning estimates to set the project budget. If this is required to secure the necessary funding, so be it. But this approach introduces inherent and substantial risks that will be addressed in more detail in Section 3.3, *Types of Estimates*.

In general, it is preferred to set the project budget at scoping. Each organization may have their own estimating and budget procedures, but Figure 3.3 outlines the recommended steps which are to properly set a project budget that is examined throughout project development, and adjusted, if needed, using approved change management.

1) Define the Scope
   The first step in setting a budget should be defining the project scope. Prior to scoping, at best you have preliminary estimates that are used for initial programming of funds or funding application submissions. The reality is that the only sure way to increase the accuracy of an estimate is to advance the plans further into design and burn down project risks. You can't make meaningful progress on the budget until you know what you are going to build. Additionally, one of the main focuses of scoping is to qualitatively identify all project risks, and then quantitatively convert them to time and money to responsibly adjust the scoping schedule and budget before being baselined. Chapter 4, *Scope*, details how the scoping process establishes the triple constraint of a project's budget, scope, and schedule, in which the PM is charged to operate.
2) Generate Cost Estimate
   The next step is to prepare a cost estimate, in today's dollars. This is a preliminary estimate, as the plans have not yet progressed to the point that construction quantities are available. This should include the Base cost, Allowances, and risk-based contingencies, all of which are described in Section 3.3, *Types of Estimates*.
3) Determine Schedule
   Time is money. You can't make meaningful progress on the budget until you know how long it will take to develop and deliver the project. Finalizing the schedule is an essential part of scoping, and should be done before setting the budget. Additionally, an organization may need to consider other time commitments regarding the project completion. Projects may be expedited to meet political or strategic time commitments, but accelerating a project development or delivery schedule comes at a cost that should be reflected in all estimates and budget.

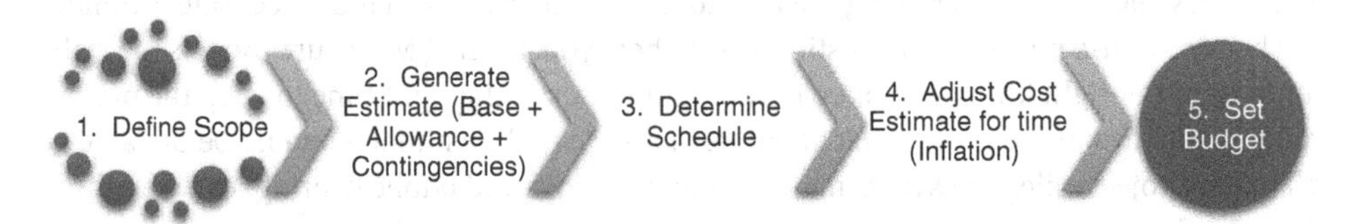

Figure 3.3 Process to set budget.

4) Adjust the Cost Estimate for time
   In order to set the budget, you need to adjust the current year dollar estimate to account for inflation. Inflation is a critical component of all cost estimates.
5) Set Budget
   Once you have defined the scope, established the schedule, and generated the cost estimate, you are ready to make an informed decision and set the project baseline budget.

It is important to note that many projects do not follow this logical progression. When the set budget process is completed out of sequence, it frequently launches the project on a challenging trajectory. For example, a budget that is set to available funds must adjust the scope and schedule accordingly. This can look very different from responsible value engineering and project change management. Similar challenges can arise when a budget is set too early. A project budget that is set based upon a preliminary funding application is inherently constrained in real ways that directly impact the development of the project. The triple constraint of budget, scope, and schedule is no longer equal as the budget becomes fixed, based upon a preliminary estimate that cannot reliably provide the necessary level of accuracy or confidence. As such, any risks or subsequent project change must be completely absorbed by changing the scope or schedule. Some projects can effectively do that, others cannot without sacrificing the core purpose and need of the improvement.

3.2 The Basics

"Estimating is what you do when you don't know."

– Sherman Kent

3.2.1 Organizational Consistency

From a project perspective, the purpose of estimates is twofold: (1) ensure there is the right amount of the right type of money in the right place to keep the project moving forward in a responsible and efficient manner, and (2) confirm the project-related construction procurement is fair and reasonably priced before proceeding.

From a program and portfolio perspective, estimates are critical data points to ensure: (1) there is the right amount of the right type of money in the right place to keep the project moving forward in a responsible and efficient manner, and (2) your organization's multiyear improvement plan can be balanced.

Notice none of these purposes focus on producing the correct estimate. The range of construction bids received for a project often vary by a far greater percentage than the organization's internal estimate success thresholds. Section 3.5, *Performance Metrics*, will address the accuracy of an estimate. In general, informed organizations are programmatically less concerned with the estimate accuracy, and more concerned with its precision.

Consistency matters. Are your organization's estimates reasonable, defendable, and repeatable? Does your organization have standards and guidance on estimating processes and procedures that are routinely and consistently followed? As a public institution that stewards public funds for public good, transparency is foundational to establishing trust and building credibility. More importantly, it is the right thing to do. You can't reasonably compare projects for prioritization or application selection unless there is a consistent estimating methodology. You need to be able to compare apples to apples. This is often more challenging than you might at first expect.

3.2.2 Critical Estimate Components

Outside of inflation, a project cost estimate consists of four basic components: Base Cost, Allowance, Contingency, and Management Reserves. An easy way to think about this is that each of these represents a different part of a project's knowledge quadrant regarding project costs, as shown in Figure 3.4. The knowledge quadrant divides knowledge into four distinct regions: things we know we know, things we know we don't know, things we don't know we know, and things we don't know we don't know.

- Base
  The Base is the portion of the estimate where a probable cost of development and construction can be reasonably determined. These are costs of known items whose requirements have been determined and quantified at the current level of project development. These are the known-knowns.
- Allowance
  An allowance is an amount included in the estimate to account for costs of known items whose requirements are as of yet undefined. This placeholder-like cost component covers the known-unknowns. An allowance is different from a contingency. As the design progresses, the allowances (known-unknowns) will be converted to the known-known costs that are accounted for in the Base. Examples of allowances may include, but are not limited to, an underground utility extension, quantity of unsuitable soils, E&S measures, MOT, Water Quality Credits (SWM), and Wetland/Stream mitigation costs. If your organization does not separate out allowances, these risks should be considered in the contingencies.
- Contingency
  A contingency is an amount added to an estimate to account for identified and unidentified risks whose likelihood of occurrence and significance of impact are uncertain. These are the unknown-knowns. Contingencies can be assigned to tasks, projects, phases, or engineering disciplines within phases. This is dependent upon which level risks are assigned by your organization at any stage of development. As a project develops, contingencies should decrease as risks should be more fully understood and properly addressed.

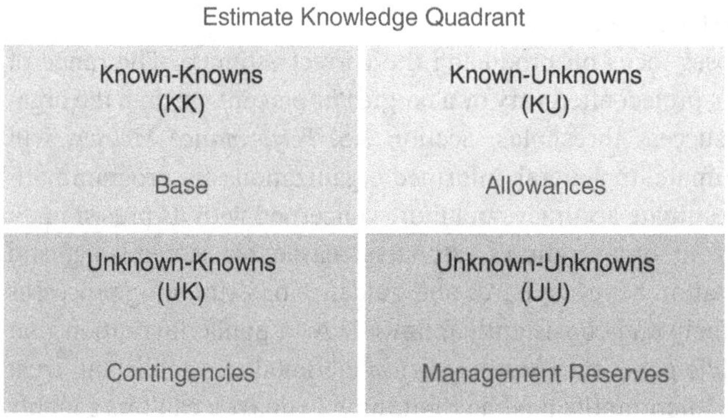

Estimate Knowledge Quadrant

Known-Knowns (KK) Base	Known-Unknowns (KU) Allowances
Unknown-Knowns (UK) Contingencies	Unknown-Unknowns (UU) Management Reserves

Figure 3.4 Estimate knowledge quadrant.

Contingencies can be calculated sums, but are often percentages of the base, allowance, or their sum. Your organization should establish and maintain guidelines on acceptable contingencies. All contingencies should be justified in the Basis of Estimate, which is a document that identifies the logic, data, methodology, assumptions, constraints, and calculations used to develop the project estimate. This document should also include the estimate's confidence level, which is the calculated or assigned probability that a result will be within the anticipated accuracy range. Some mature organizations have more comprehensively addressed this by adapting a risk-based estimating approach, which is discussed in Section 3.3, *Types of Estimates*.

- Management Reserve
  Management reserves are monies an organization may set aside to address unforeseeable circumstances that might increase the cost of a project. These are the unknown-unknowns, such as acts of God, historic storms, labor strikes, etc. Typically, senior management controls these funds and must approve releasing any management reserves to a specific project for a specific situation. Many organizations do not have dedicated Management Reserves. In those cases, contingencies can be adjusted accordingly to account for the unknown-unknowns, or the organization makes an intentional or unintentional decision not to account for unknown-unknowns in the estimate or budget.

3.2.3 Assembling the Estimate Components

Estimate components should be methodically and predictably assembled to create the project estimate at each milestone. It is critical that designers and estimators consistently adhere to your organization's guidance to ensure allowances, contingencies, and inflation are consistently considered in all projects. If not, it is easy for contingencies to be hidden inside other numbers and counted multiple times. This can quickly complicate competitive applications for specific funding sources, as you want to compare apples to apples in order to be fair. This can also be extremely problematic when managing programs and portfolios as reliable and consistent estimates work hand-in-hand with reliable and consistent schedules to ensure the right kind of money is in the right place at the right time to keep the project progressing in an efficient and responsible manner.

Your organization may have slightly different definitions or guidance; below is a general summary on how to properly assemble the estimate components. First, you calculate the estimate to complete the project if executed as planned. It should be unbiased and neutral, not overly optimistic or conservative. The base cost should include the base, or defined, items, and all allowances, but should not include identified risks as they will be addressed within contingencies. The base cost should be in current year dollars, and not include Inflation. The base cost = base + allowances, and shall specifically exclude contingencies and inflation. Contingencies should then be added to the base cost which produces the estimate of expected cost. Some organizations may then add management reserves to your estimate, as shown in Figure 3.5.

As the project develops, the accuracy of the estimate should improve as the plans advance and risks are reduced. This will be evident in your estimate components. As plans are refined with increasingly detailed design, the known-unknowns of allowances become known-knowns of the base cost. As risks decline, the unknown-unknowns of contingency should decrease, as depicted in Figure 3.6. Note the schematic below does not include

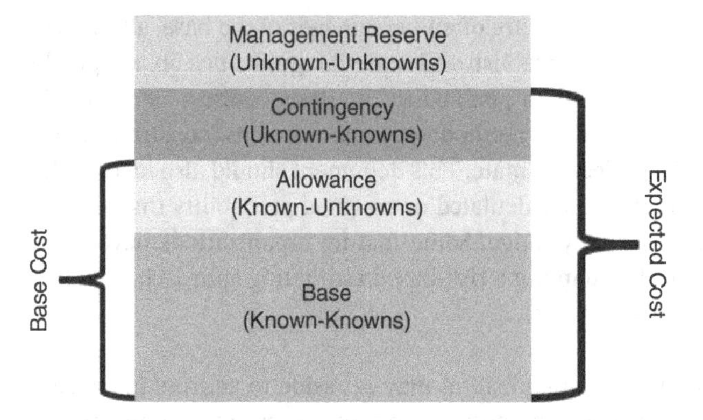

Figure 3.5 Cost estimate components.

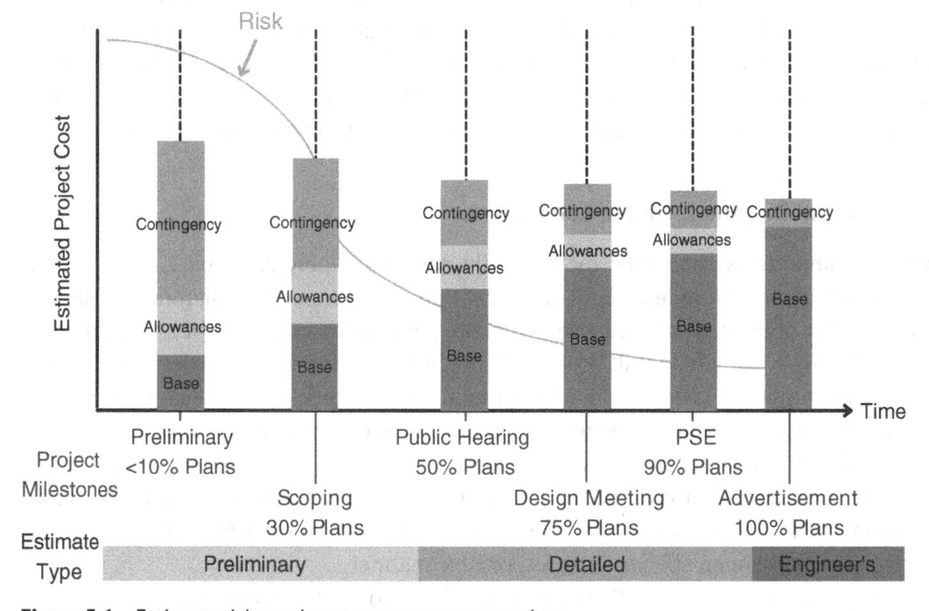

Figure 3.6 Estimate risks and cost components over time.

management reserves, as many organizations do not have this as a separate cost estimate component.

3.2.4 Inflation

When the estimate is complete, then inflation should be added. Inflation is the rate at which the purchasing power of money falls as the currency value decreases over time. More mature organizations may automate the addition of inflation into an estimate based upon the PE, RW, and CN phase dates in the project schedule. This automation enables an organization to control the utilized inflation rate and ensure it is fairly and consistently applied across all projects. If your organization does not have these capabilities, then your estimate should manually add inflation rates, as approved by your organization.

The selected inflation rates may have various sources, be it historical values, national, regional, or local indices and trends, or customized rates. While you certainly want to use the most accurate inflation numbers, you won't know if they are accurate until the time has passed. As such, the chosen inflation rate should be reasonable, defendable, and consistently applied across the organization. This consistency of use is imperative for efficient program and portfolio management.

Pending the complexity of your project, and your organization's processes, you may use different inflation rates for different aspects of the estimate. For instance, you may have one inflation rate for construction costs and another for rights-of-way. Or you may choose one inflation rate for bridge steel and another for all other construction materials. The organization should also intentionally choose the point at which the inflation is calculated (e.g., beginning of the phase, midpoint of the phase, etc.). This approach can provide a more tailored and precise estimate. If not dictated by your organization's practices and protocols, be sure to document all inflation decisions in the Basis of Estimate.

3.2.5 Cone of Uncertainty

The Cone of Uncertainty is a graphic representation of an organization's expected and accepted estimate accuracy throughout the project development process. This powerful tool should be a foundational element to every organization's estimate program. The specified ranges can be sourced from industry standards, historical data, peer organizational performance, or leadership expectations. Regardless of the source, an organization's Cone of Uncertainty should be intentionally set as it establishes the organization's estimating expectations throughout the life of a project. Figure 3.7 shows a typical Cone of Uncertainty.

As the design advances in the project development process, risks should decrease. Advancing the plan and reducing risks allow the estimate accuracy to improve, thus the cone of uncertainty narrows as the project moves to advertisement. While the plus/minus percentages shown in the figure may be general rules of thumb, your organization should determine its own acceptable variances at each stage of plan development. Some organizations may embrace risk-based estimation. This approach may necessitate different cones of uncertainty for varying levels of project complexity.

The Cone of Uncertainty reflects an organization's risk tolerance and the expected confidence level of all estimates. Having this documented in such a way that it is consistently used throughout the organization can be critical in managing leadership's estimate expectations when the inevitable challenging situations occur. It is said, hard cases make bad law. There is great value from a project, program, and portfolio perspective to relying on an established Cone of Uncertainty. How do you know if you have hit the target, if the target is not defined? How do you know if you have been successful? You cannot consistently frame the conversation with leadership, stakeholders, or team members as to define good or successful estimates without an established and approved cone of uncertainty.

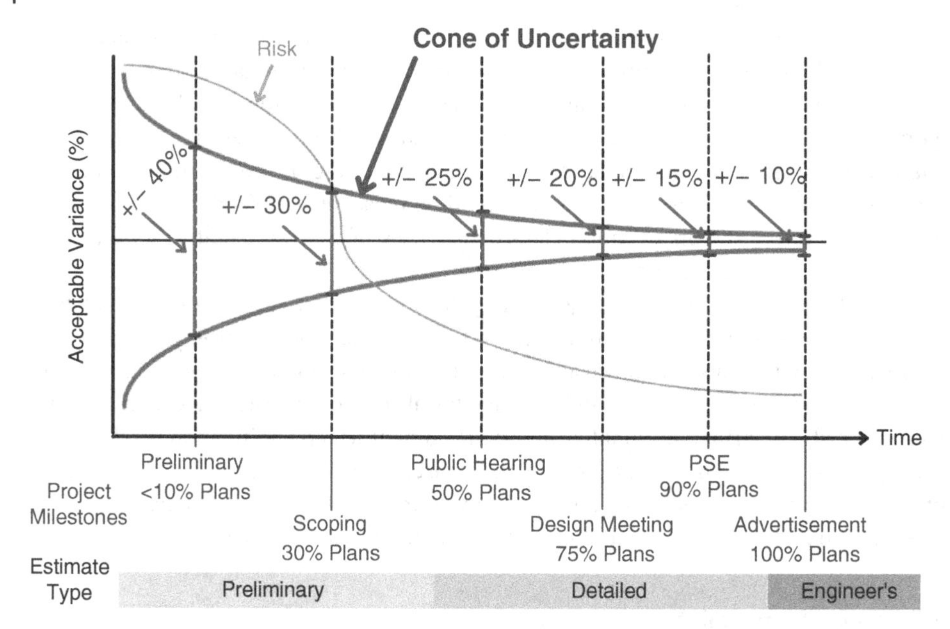

Figure 3.7 Cone of uncertainty.

3.3 Types of Estimates

"Estimating is as much art as science."

– Anonymous

Before a project is active, it is a candidate project. Different organizations may use different descriptors, but these are project ideas that may be researched and refined to wide-ranging levels of maturity. There are a variety of ways in which candidate, or potential, projects are chosen over others to become active projects. Application-based selections require potential projects to submit an application in order to be considered. These applications are often then scored according to preestablished criterium to prioritize selection and funding. More mature prioritization models often include some benefit–cost analysis. Some projects are driven by an organization's responsibility to maintain, repair, upgrade, or replace existing transportation assets. Many other projects are selected by direct government intervention, be it legislative or executive direction. This can be at the federal, state, regional, or local levels. Regardless of how a project become selected and funded, suffice it to say, there are many more needs and wants than available resources.

The field of transportation planning focuses on this important process of identifying potential transportation improvement ideas, researching them to ascertain their feasibility, and preparing planning level cost estimates. These planning level, or conceptual, estimates are often used as placeholders in long-range plans, or in evaluation as to whether a candidate project should be selected for funding. These order of magnitude estimates provide an early, imprecise idea of the time and money required to complete a project. Due to the vast amounts of unknowns and assumptions at this state, rule of thumb accuracy levels of these estimates may range from –25% to +75% of the actual budget.

While this writing acknowledges the significance and implications of preliminary estimates, and the critically important role they play in a transportation program or portfolio, we will focus our attention on estimates after a project is active. Once a project is selected, prioritized, put in the organizations, and perhaps state and federal, multi-year transportation plan, and then funded, it moves from the Planning arena to the Project Development arena, which is where we will concentrate our discussion.

For active projects, there are three unique kinds of estimates during project development: Preliminary, Detailed, and the Engineer's Estimate. Each is differentiated by the accuracy of the data inputs. To oversimplify, Preliminary estimates are those formed before quantities are known, Detailed estimates are those prepared using construction quantities, and the Engineer's estimate is that against which construction bids are considered when determining whether to recommend Award of contract.

As the project advances, Preliminary estimates are replaced by Detailed estimates. Just before Advertisement for construction, the Engineer's estimate is prepared. It is worth noting that the transition from Preliminary to Detailed estimates is rarely clean, meaning there may likely be a transition period where portions of the estimate are sourced from measured quantities, while other portions are still calculated using preliminary estimate tools. As the design advances, project risks should be decreasing and allowances should be clarified, thereby increasing the accuracy of the estimate. Traditionally, Preliminary estimates are the least accurate, while the Engineer's estimate is the most accurate, as shown in Figure 3.8.

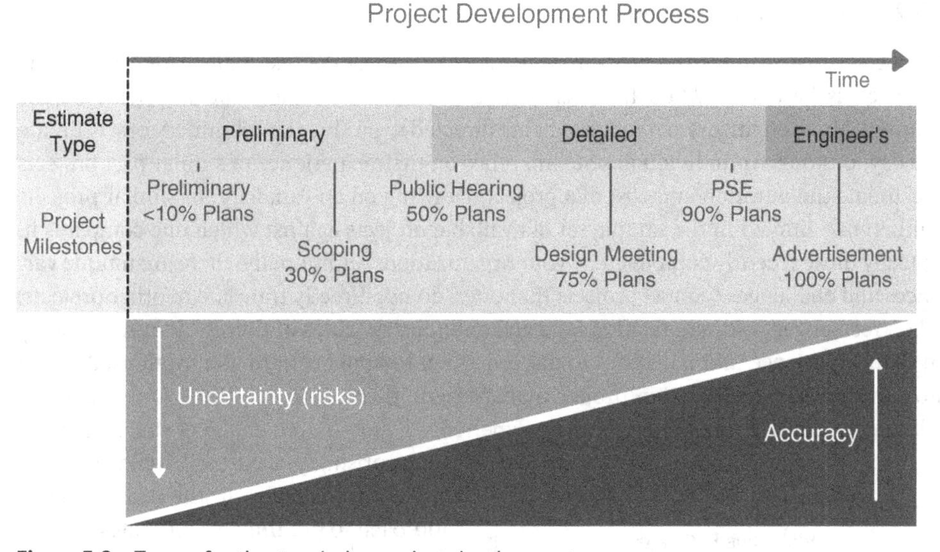

Figure 3.8 Types of estimates during project development process.

3.3.1 Preliminary Estimates

Preliminary estimates are those created before the design has progressed to the point where you have an itemized list of materials and projected quantities. These can be tricky business. They are often crafted by Planners or Engineers that may not have detailed design or construction experience. Complicating matters is that Preliminary estimates are often viewed or used in ways that cavalierly disregard a preliminary estimate's purpose and inherent limitations. A politician may latch onto the first number they hear and never forget it. The media keeps referring back to a preliminary cost estimate as they scrutinize a troubled project. Various funding sources even lock in the preliminary estimate listed on the application as the project budget. These examples all have others taking action based upon the underlying assumption that a preliminary cost estimate is both precise and accurate. In reality, while sophisticated organizations may achieve precision, the accuracy of a Preliminary Estimate is inherently limited.

Preliminary estimates are essential to plan and program funds for all required phases of a project: Preliminary Engineering (PE), Right-of-Way Acquisition (RW), and Construction (CN). It should be noted that some organizations may consider four phases, adding Utility Relocations (URL) between RW and CN. For the purpose of this writing, we will consider the three approved federal phases (PE, RW, and CN). This assumes URL is part of the RW phase. Most transportation programs are multi-year plans, and require placeholders until a project is initiated. Many funding source applications also require a preliminary cost estimate that is included in the selection criteria. Preliminary cost estimates are also invaluable in forecasting future workloads and resource allocations. These are examples of reasonable purposes for preliminary estimates.

Figure 3.9 shows the four main approaches to creating a preliminary cost estimate. Each has its advantages and limitations.

3.3.2 Analogous Estimating

Analogous estimating is the proverbial "back of the envelope" estimate. This "top-down" approach is best used when there is limited project information and a quick result is required. Many engineers can relate to being directed to produce a preliminary cost estimate in a day, or hour. An individual, or team, relies on their experience of similar past projects to estimate the duration and cost of a project. Drawing on an "analogy" of similar projects is inherently limited to the sample set of available projects against which one compares it, typically those recently completed by your organization. Additionally, there are unique variances and challenges to most projects that often do not directly translate to other projects. When recognized, estimators often fall back on industry "rules of thumb" that rely heavily on their experience and judgment to discern when to apply certain assumptions. As with most aspects of transportation project management, there is no substitute for experience. While a gut-check cost estimate from an experienced, gray-haired engineer can sometimes be surprisingly accurate, Analogous estimating is considered the least reliable approach to preliminary estimates.

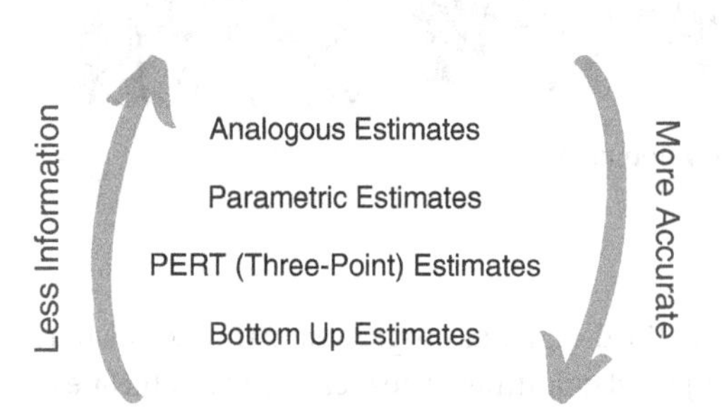

Figure 3.9 Types of preliminary cost estimates.

3.3.3 Parametric Estimating

Parametric estimating leverages historical data of recent projects, combining it with statistical, scalable calculations. As such, if done correctly, it is inherently more accurate than Analogous Estimating. Many organizations create parametric tools that vary in complexity from Excel spreadsheets to complicated databases. The idea is an estimator can enter known attributes (e.g., length of roadway, number of lanes, number of traffic signal poles, etc.), and the tool will convert these to defined unit costs and generate a preliminary estimate. Many parametric tools also consider other project factors as integrated multipliers (e.g., urban-vs.-rural, road classification, pavement section, terrain, likelihood of encountering unsuitable materials, known presence of rock, etc.).

When creating a parametric tool, one must keep the costs current on the internal tables. The units by which a cost is calculated must also be scalable in order to generate consistent, accurate, and defendable estimates. For instance, a project requiring a bridge with an 80-foot span, it is likely not twice the cost of a bridge with a 40-foot span. Business rules regarding scaling of units should be supported by historic data, and integrated into the tool. It should be noted that creating and maintaining an effective parametric estimating tool can represent a significant effort. However, many organizations have determined this is an investment worth making.

Those using parametric estimates would be wise to remember that these tools are just focused algorithms. As such, garbage in = garbage out. The user must understand the underlying business rules and be familiar enough with the tool to enter the input variables in the intended manner. Two estimators using the same tool can produce two very different estimates if they interpret or define the input parameters differently. Parametric estimating tools typically do not handle risks well, often requiring the user to manually input risks with an assigned value. Whether documented or not, every parametric tool has a long list of assumptions that generally apply for a typical project. In reality, there is rarely a project where all these assumptions are true. Additionally, some parametric tools are not updated or equipped to handle more innovative solutions well (e.g., roundabouts, diverging diamond interchanges, evolving types of environmental basins or stream mitigations, etc.). Understanding the limitations of the parametric estimating approach, and your agency's specific parametric tool and its business rules and assumptions, will better enable you to leverage its powerful benefits and interpret its useful results.

3.3.4 PERT (Program Evaluation and Review Technique) Estimating

PERT, or Three-Point Estimating, builds upon the previous two approaches by statistically considering risk. The intention is to reduce various estimating biases and lower uncertainties of estimate assumptions. In this approach, one must first generate three separate estimates for the same project, often utilizing the previous two methods. These are the Most Likely Cost (Cm), the Optimistic Cost (Co), and the Pessimistic Cost (Cp). The Most Likely Cost represents the typical workflow, where all goes as usual for your organization. Think of this as your true best guess. The Optimistic Cost is the estimate should all the stars align and all the known risks vanish into thin air right before your eyes. The Pessimistic Cost is the estimate for the worst-case scenario where every risk hits the fan. Once you have these three estimates, you plug them into the following formula: $Ce = (Co + 4 \times Cm + Cp) / 6$.

This simple equation generates the Expected Cost (Ce), or the PERT estimate. While more time intensive in that one needs to generate three separate estimates, this approach is more reliable than the previous two methods in that it normalizes the biases to create a more accurate estimate.

3.3.5 Bottom-Up Estimating

Bottom-Up Estimating is the most time consuming, and accurate, of the preliminary estimating techniques. In this approach, the Estimator leverages the Work Breakdown Structure to identify the lowest level of work packages. They then approach the Subject Matter Experts (SME) for each segmented unit. These are often the individuals who will actually be doing the design. The SME then evaluates and estimates their piece of the puzzle. The individual estimates from the lowest level then bubble up and are combined to form the preliminary estimate. The Estimator, or someone on the team, needs to carefully coordinate this effort, identify redundancies, and proactively address cross-discipline conflicts.

3.3.6 Detailed Estimates

Detailed estimates are those created by using anticipated construction quantities and unit cost prices to calculate a total estimate. Most detailed estimates leverage or use other resources. These may include historic bid tabs, 2D, 3D, or AASHTO design software and estimating tools.

It should be noted that detailed estimates evolve over time. While the arithmetic of multiplying the quantities by the unit price to calculate the line-item costs is simple, its implementation can be a bit more nuanced. As the design progresses, projected quantities will change. There will also be some line items that are bid as lump sum (e.g., mobilization, MOT, etc.) that requires some finesse. During earlier stages of design, there may also be an abundance of risks that may be considered with Allowances or Contingencies. This can be especially challenging in earlier Detailed estimates where one may feel they are doing a hybrid of quantities and preliminary estimate approaches. As the project advances closer to Advertisement and risks decrease, the simple math mechanics of conducting a Detailed estimate becomes much easier.

It should be noted many projects seem to conform to the 80–20 rule, where 80% of the anticipated costs are in 20% of the line items. In these instances, focus your time accordingly on those more critical line-item quantities and costs. Also be aware of the context of the bid tab information you are using. Are the line items defined by the same standards, or are different work items included? Are the referenced historic bid tabs low-bid only, or do they capture all submitted bids? How are special provisions reflected in the historic bid tabs? Even though the math is simple (line-item cost = quantity × unit cost), there is still plenty of room for sound engineering judgment.

At every stage, it is important to try to consider all project costs. It is not uncommon for more inexperienced engineers to unintentionally omit line-item costs that impact the total cost. Examples may include environmental credit purchases or mitigation costs, rented police presence during construction, construction CEI, state costs to oversee administration of a locally-administered project, and so forth. Many organizations have checklists that can be extremely helpful. These checklists can also be invaluable in bringing consistency to an organization's estimates. This consistency should bring defendable precision to the estimates, while hopefully also increasing estimate accuracy.

3.3.7 Engineer's Estimate

The purpose of the Engineer's estimate is to evaluate the low bid and determine whether or not to recommend Award to the governing body. This is very different from the purpose of previous project estimates, primarily being: to ensure there is the right amount of the right color of money at the right place at the right time to keep the project moving forward. While the Engineer's estimate may not be the only consideration as to whether or not to Award (e.g., not enough allocations, contractor determined to be incompetent, etc.), it is typically the major tool used to evaluate the reasonableness of the low bid amount.

The Engineer's estimate should be the most accurate of all estimates. This makes sense. The Engineer's estimate is typically prepared concurrent with, or immediately after, Advertisement. This estimate is prepared using the same Plans and Specifications that contractors are using to prepare their bid. As such, it should be clear who will bear all remaining project risks. Other critical construction details should also now be settled. Examples may include construction duration, hours of operation, defined staging areas, miscellaneous or unusual restrictions, and so forth.

It should also be noted that an Engineer's estimate attempts to evaluate costs as a contractor would. This means it should consider factors that extend beyond quantities and unit costs. These may include men, equipment, materials, production rates, proximity to asphalt plants or suitable material, accessibility of defined staging areas, and so forth. As such, Engineer's estimates are typically done by an organization's most experienced estimators.

Since the purpose of an Engineer's estimate is to be that by which submitted bids are compared, it is common for these estimates to remain sealed. Pending the organization and applicable state guidance, some may unseal this estimate after bid opening, after Award, or keep them sealed indefinitely.

3.3.8 Risk-Based Estimating

One may assert, at its core, all project management fundamentals are rooted in risk management. Likewise, one may assert at its core, all estimating fundamentals are rooted in risk management. This probabilistic and determinist mindset is solidified under the umbrella of risk-based estimating.

A Risk is an identified uncertain event or condition that if it is realized, could have a positive or negative effect on the project's objectives. All significant scope assumptions should be included in the risk register. Negative risks that could adversely impact a project are threats, while positive risks that could benefit the project are opportunities. Risks should be accounted for within the estimate, typically in allowances or contingencies.

Risk-Based Estimating is a methodology that adjusts the Base Estimate to account for project-specific risks in order to produce a more informed probable cost estimate. Guidelines for applicable and acceptable contingencies and allowances are specified for various project risk complexities, which are based on historical data. AASHTO has defined a Risk-Based Estimate process wherein Risks are analyzed and Contingencies assigned based upon the following three project types:

- Type I – Risk-Based Percentage Contingency
- Type II – Risk-Based Deterministic Contingency
- Type III – Risk-Based Probabilistic Contingency

Type I, a risk-based percentage contingency approach is the simplest form of risk analysis, where contingencies are assigned based upon provided percentage guidelines. The percentage guidelines often have low-, medium-, and high-risk values at various stages throughout the project development process. The default contingencies used for estimates should be the medium percentage guideline. Any variance should be justified within the Basis of Estimate. Any contingency percentage that is outside of the low to high percentage guideline range may require organizational approval and should be justified in the Basis of Estimate.

Type II, a risk-based deterministic contingency approach determines contingencies through the use of a risk assessment that identifies all applicable projects risks. A specific contingency is then assigned to each Risk based upon the probability of occurrence and significance of impact. Applicable details should be included in the Basis of Estimate.

Type III, a risk-based probabilistic contingency approach is reserved for the most complex of projects. Risks on any project warranting a Probabilistic Risk analysis should be individually evaluated, often utilizing a Monte Carlo simulation. Such risk-evaluations are rare, highly complex, and should be conducted by experienced subject matter experts.

The essence of risk-based estimating is to assign project and/or task contingencies based upon project-specific risks. This often takes the form of a pre-set table of acceptable contingencies based upon a selected risk category (e.g., low, medium, or high). The overarching strategy is that estimates will be more accurate if they are tailored to consider the unique project- or task-specific risks.

It should be noted that most organizations that embrace risk-based estimates require a risk matrix and risk register in their basis of estimate documentation. A risk matrix is a chart used to qualitatively analyze individual risks by rating the risk's probability against its impact. A risk register is a dynamic tool used by PMs and project teams to actively track and monitor project Risks throughout the life of the project. Risks are addressed in more detail in Chapter 6, *Managing Risk*.

3.3.9 Estimate Range vs. a Single Number

In general, humans are terrible at estimating just about anything. At a large training event I once attended, the instructor gave a test to the group where we were to write down our answers in a range with 90% confidence level of accuracy. The general questions included the year Mozart was born, length of the Nile River, diameter of the moon, and so forth. It was astonishing how poorly the group of accomplished engineers scored. We obviously were far more confident in our estimating ability than we perhaps should be.

Beyond the results, one reason I found this illustration fascinating was that we were to provide answers in ranges. Engineers are typically preconditioned to solve for x. While there may be ranges, most of us were trained that there is typically one right answer, or one most probable answer. It should be noted that some research suggest humans are not inherently inclined to estimate most anything with a singular number, rather opting for a probabilistic range.

Bringing this concept back to transportation project estimates, should the estimate be a number or a range? Which makes more sense? Which does your organization encourage or require?

There are good reasons for a single number. It is difficult to program a range. Most organization's multi-year funding plans require a number. Furthermore, most organization's multi-year funding plans cannot accommodate a range. Likewise, most funding, planning, and monitoring systems require a number and cannot accommodate a range. There are also practical considerations. Most politicians or news outlets don't say, "this project will likely cost somewhere between $2.42 million and $4.07 million," instead opting for, "this $3.2 million project...." Single numbers are easy to understand, simple to convey, and pragmatic for associated systems and funding plans.

There are also good reasons for ranges. Estimate ranges allow for consideration of multiple streams of risk, which can more effectively shape stakeholder expectations and lead to improved decisions. These can be especially useful on larger megaprojects or more complex projects where the optimistic and pessimistic estimates can vary widely. As the project advances, this range should shrink until the Engineer's Estimate is typically a single number. While most experienced engineers and program managers agree an estimate range is perhaps more appropriate, system constraints and stakeholder expectations often drive an organization to revert back to a single number in their estimates.

3.4 Managing the Budget

> *"Give me six hours to chop down a tree and I will spend the first four sharpening the axe."*
>
> – Abraham Lincoln

3.4.1 Cost Management Plan

The Cost Management Plan is the part of the Project Management Plan that details how project costs will be estimated, budgeted, managed, monitored, and controlled throughout the life of the project. Some mega-projects, or more innovative delivery methods, may require a project-specific cost management plan. For most transportation projects, a tailored cost management plan is not required. This is predominantly because these workflows are typically already well established, be it by precedence or applicable guidance and regulations.

At its core, public transportation organizations are funding entities. Money comes in, and money goes out to build and maintain transportation and related assets. As such, public transportation organizations should strive to be responsible stewards of public funds. In most circumstances, much of the money flow is regulated or directed to promote or demonstrate transparency, responsibility, and a balance of authority. Consequently, much of what would normally be included in a cost management plan already exists in different formats in different locations, often spread across an organization's divisional lines. This is particularly true for how allocations, obligations, and budgets are established, modified, and approved, which can even be dictated by the funding stream that is being leveraged.

The issue of estimate guidance is often a different story. While it may seem more straightforward, that is not always the case. Regardless, your organization should have documented expectations for the who, what, when, and how of project estimates: who does them, what they are to look like, when they are required, and how they are to be submitted and/or inputted into your organization's system(s).

Remember the big picture of transportation is to have the right money in the right place at the right time to keep projects moving forward. Estimates, allocations, obligations, and budgets play an incredibly important role in achieving this objective.

Any project-specific cost management plan should include a section focusing on change management. Not only is it the right thing to do, it is incredibly important to maintain the financial integrity of the project.

As with most bureaucratic workflows, an organization should be careful not to venture into the space where you pass beyond the point of diminishing returns. The best of intentions can quickly lose favor when the juice isn't worth the squeeze. An organization's estimate guidance and requirements should be scaled to be reasonable and appropriate for the different kinds of estimates throughout the life of the project.

3.4.2 Estimate Validation

There is a growing trend in the transportation industry toward application-based selection processes to prioritize and select projects for funding. Many organizations see real advantages to this approach, including increased public transparency, making project selections more objective and less subjective in nature, and being able to direct and emphasize the shape of future projects by defining and tweaking the selection criterium. Many of these processes include some form of a benefit–cost analysis. In these instances, it may be deemed responsible for the administering organization to validate the submitted project estimates.

Organizations should not underestimate the resources and effort required to validate all submitted estimates. This can be especially concentrated when there is a rush of submitted applications right before the submission deadline. The validation effort can even be a multilayered validation, pending the size and complexity of the project. For instance, the first layer of validation may be thought of as Quality Control (ensuring the validity of the data). The second layer of validation is often analogous to Quality Assurance (ensuring the integrity of the process). It is important to realize that generating a cost estimate, and validating a cost estimate are two very different tasks. It is like designing the roadway plans or reviewing someone else's work to ensure compliance. The best reviewers were often first designers who had actually done the work at an earlier time. Similarly, the best estimate validators are often those that have direct experience generating cost estimates. It is important to note that estimate validation relies upon the foundation of the cost estimate itself: what is and is not included, project and estimate assumptions, the estimate's incorporation of risks, its dependency upon a well-crafted Scope, and an accurate Schedule. This can put estimate validators in a precarious position if they doubt any, or all, of these foundational assumptions.

One important aspect of estimate validation is to predetermine the actions to be taken when the validation is complete. What if the validation determines the submitted solution design is not tenable or constructable? What if the validation suggests the submitted estimate is woefully low? What then? Does your organization have the ability and authority to revise the estimate or reject the application? These situations will occur, and it is imperative you understand the dynamics and know the answer to these questions before they arise.

3.4.3 Managing the Triple Constraint

The primary job of the PM is to advance the project within the confines of the triple constraint: Budget, Scope, and Schedule. How do you do that? A successful PM is vigilant in confirming the scope, monitoring the schedule, updating the estimate, and verifying the budget again and again throughout the life of the project, as shown in Figure 3.10. When this circular path is consistent, the project can progress in a healthy manner. Inharmonious inconsistencies represent project issues that must be addressed, the sooner the better. Entropy is alive and well in that if a discrepancy is ignored, the project will inherently digress to increasing levels of disorder.

Figure 3.10 Managing the triple constraint.

3.4.4 Change Management

Tony Robbins said, "*Change is inevitable. Progress is optional.*" This may be especially true as it applies to project estimates. The reality of the triple constraint dictates you cannot change one of them (budget, scope, and schedule) without impacting at least one of the other two. Of the three, one could assert Estimate, or budget, changes are the hardest to capture. Why?

Scope changes are usually directly reflected in the design plans. If you extend the sidewalk another 200′, it is right there for all to see. Most scope changes create ripple effects that must be coordinated and managed by the PM across discipline lines. If the sidewalk is extended, perhaps there are environmental documents or permits that need to be modified, perhaps traffic needs to now consider another crosswalk at the intersection, perhaps ADA requirements on the extension now require additional utilities or storm drains to be relocated, perhaps you need to initiate enhanced stakeholder outreach to update the impacted neighborhood and their HOA, and so forth.

Schedule changes are also typically transparent within the project team and to leadership. If you are supposed to reach a milestone by a certain date and don't, it is a binary result. You didn't make the date, so you adjust the schedule. Most public transportation projects have various internal and external touchpoints throughout the project development process. It becomes readily apparent when a schedule slips and these touchpoints are delayed or need to be rescheduled.

Budget changes can be relatively straightforward in a top-down situation when a budget change is driven by funding or allocation changes. For instance, if funds are added to a project due to a program surplus, then the project has an increased budget. Likewise, if money is pulled from a project for what could be a variety of reasons, the budget is reduced and the PM is forced to react and respond accordingly.

Budget changes can be a bit more complicated in bottom-up scenarios, when they are driven by changes in the estimate. There may be many reasons for this. Quality estimates are time consuming to generate. They are also snapshots of the developing plans that are continuing to evolve and increase in detail. As such, most projects generate meaningful estimates only at key project milestones, which typically may include: planning (before project selection), scoping (or 30% plans), Public Hearing, 60% plan, 90% plans, and 100% plans. Pending your organization's workflow, there may be interim milestones as well. Other organizations may require current estimates, meaning they at least have to be updated annually for instance. It is evident that estimates are captured at less frequent intervals than the scope or schedule is evaluated. There may also be significant organizational or project pressure to keep the design within budget. Additionally there may be behavioral pressures that can be exacerbated by an organization's culture or performance metrics. This can manifest itself in many ways. Pending your organization's estimating guidance and requirements, and given the approximate nature of estimates, particularly if applying risk-based estimating, allowances, and contingencies, there is some room for interpretation of the numbers. For instance, two estimators will calculate very different estimates if one is inclined to be conservative with their numbers, and the other more aggressive. This is sometimes exacerbated in that most truly don't really understand the money flow anyway.

This can lend itself to the extremely undesirable result of letting a project and not having the budget to award it. The project perspective may then encourage overestimating the costs to ensure there is enough money to award. However, this can cause complications for the program and portfolio, including that these surplus monies could be used to fund other projects. As such, PMs and discipline leaders should strive to generate realistic estimates. If these estimates exceed the budget, it is far better to realize you may be short on money early on so necessary adjustments or tough decisions can be made.

Ideally, project change management workflows and procedures should be well defined at the organizational level. This consistency of process can bring clarity and predictability to submissions and approvals of budget changes, and do so in such a way as to make the appropriate adjustments to scope and schedule. Ideally, your organization may have a change management form that documents significant changes (as defined by established thresholds) and their resulting impacts to the triple constraint. Mature organizations often have this form go through a defined approval process so that leadership, key stakeholders, and team members are aware of the changes and ramifications thereof.

3.4.5 Formatting Estimates

Many organizations find substantial benefit in standardizing the format of their estimates. This can be a deceptively challenging task when one considers the range of estimates throughout the life of a project, and the various internal and external entities that are preparing estimates for any organization. Much of the discussion within an organization when determining a standardized format will likely concentrate on the desired level of granularity. On one end of the spectrum, you need four numbers for federal authorization: preliminary engineering (PE) phase estimate, right-of-way (RW) phase estimate, construction (CN) phase estimate, and total project cost estimate. At the other end of the spectrum is the engineer's estimate with cost item details. This task can be complicated by different stakeholders' priorities. Early estimators may be most concerned with risk categories, and ways to consistently account for base costs, allowances, contingencies, and inflation. Meanwhile many construction estimators have trouble relating to any estimate that does not contain full quantities.

If an organization chooses to pursue standardized estimate formats, they will hopefully settle on something that works well for them. One way to balance the level of detail conundrum is to subdivide each phase into discipline-specific costs, which can often facilitate a more workable cross-walk of costs throughout the lifecycle of project estimates. This is especially true should your organization or state's cost items be divided in a similar manner. Granted, the crosswalk may not be 100% accurate (e.g., curbing for traffic control may end up in MOT or safety as opposed to roadway costs, etc.), but it can be a great place to start. It is worth acknowledging that there are likely exceptions to every rule (e.g., special provisions may need to be individually addressed and subdivided into the discipline summaries).

While this effort can be a challenge to create and implement, there are real benefits. The biggest is arguably the consistency it can bring to estimates, particularly as it relates to base, allowances, contingencies, and inflation. Without a standard format, it can be difficult to ascertain exactly where the allowances, contingencies, and inflation are being included, what exactly they are to be covering, and whether multiple layers of contingencies or inflation are included. A standardized format can be especially valuable when conducting estimate validation as it relates to funding applications. In addition to bringing estimate transparency and order, it can also serve as a valuable checklist that can become an important component of the basis of estimate (estimate supporting documentation). Ideally, this consistent estimate data can be captured and leveraged for progressive data analytics that would empower more precise and accurate estimates while enabling more efficient worklfows.

3.4.6 Who Owns the Contingency

Who owns the contingency? This simple question is difficult for most organizations to answer. More accurately, it is a question most organizations would struggle to answer consistently. This may be especially true at different levels and in different areas. The executive leader may think they do. The programming folks may think they do. The accountants likewise. Districts or other subsections may consider the extra time and money at their discretion. The PM may think they own it. Even task owners or consultants may think they own any contingencies on their efforts.

The reality of who owns the contingency has far-reaching implications. Problems are inevitable if this is not well defined. Many more mature organizations take a project-centric approach. Under this paradigm, the project contingencies are owned by the organization. Project surpluses are returned to the organization for reprioritization and reallocation. If contingencies are owned, or perceived to be owned, by PMs or District, estimates can be artificially inflated to effectively act as an additional program contingency across other projects within their purview and/or control. Lack of clarity encourages competing objectives that can manifest themselves in actions at different level that are inherently counterproductive. Fundamentally, each project should stand by itself. While program and portfolios can, and should, be balanced across projects, this should be done at the program or portfolio management level, not the project management or District level. A consistent answer to this basic question is foundational to organizational data quality and the integrity of the project, which enables strong programs and portfolios.

3.4.7 Estimates and the Public

As PM, you are responsible for the accuracy and timeliness of project estimates. This has direct impacts on the programming of funds and the ability for your project to efficiently and responsibly progress. Estimates can also play a powerful role in stakeholder reaction to the project. The media will inevitably cite the project cost when providing an update or examination of a project. This anticipated project cost may be based on the budget, but is often based upon the most recent estimate. When discussing project estimates with the public, it is important to remember a few basic principles.

First, most of those gathering, reporting, or consuming information on your project likely do not readily understand the different cost components of base, allowance, contingency, management reserves and inflation. Furthermore, they are likely not familiar with your organization's estimating procedures or acceptable accuracy ranges, as expressed in your organization's cone of uncertainty. Attempting to explain this may quickly overestimate their curiosity. Most just want a single number.

Second, anchoring is real. Anchoring is a cognitive bias where one unintentionally assigns undeserved weight to preliminary information when making future decisions or evaluating future actions. An example is when one becomes attached to a specific estimate number without considering the associated Accuracy Range or Confidence Level of the estimate. This often surfaces later in development when the estimate is rightfully adjusted, and suddenly the project is viewed as unexpectedly expensive or a bargain. Imagine a reporter, or Board Member, questioning you as to why your project has risen in cost by 30% halfway through design. Anchoring is rooted in the assumption that the earlier estimate was more accurate than the accepted Accuracy Range at that point of project development.

Thirdly, optimism bias is real. This is the cognitive predisposition to assume your chances of realizing negative risks are lower than reality, coupled with the assumption that your chances of realizing positive opportunities are greater than reality. This natural and institutional bias can unintentionally set unrealistic expectations, both at the project and program levels. This often occurs within organizations when discussing high-profile projects, or recovery plan for projects that have slipped. The stars rarely all align.

3.5 Performance Metrics

"*The budget evolved from a management tool into an obstacle to management.*"
– Charles Edwards

3.5.1 Overview

The purpose of performance metrics should be to monitor progress, enhance efficiency, and enable continual refinement of an Agency's practices and procedures in order to more closely align with an organization's mission statement, and effectively achieve its mission critical objectives. With regard to budgets and estimates, the purpose should be to more precisely and accurately estimate project costs so as to enable an organization to more efficiently program available funds and execute the program in a way that optimizes value-added improvements to the community. Like many aspects of transportation project management, this is much easier said than done.

As with all performance metrics, it is imperative to carefully and meticulously define exactly what you are measuring. The "what" you are measuring should be consistent with the "why" you are measuring. Once that is decided, you should then deliberately choose the "how," "who," and "when," all of which should be consistent with the "why." It is easy for organizations to become so involved in the details of the metric that they unknowingly drift away from "why" they were wanting to measure it the first place.

3.5.2 What Is an Accurate Estimate?

What is an accurate estimate? This simple question can be a difficult to answer.

At its core, performance metrics compare datapoints to a baseline. Each organization should intentionally decide what data point(s) and baseline(s) they will use in this comparison. These decisions will inherently empower or constrain the resulting insights the performance metrics provide. They will also shape the organization's behavior as it relates to estimates in expected and sometimes unanticipated ways.

Each performance metric should start with a clear purpose for the metric. Just because something can be measured, doesn't mean it should be measured. Does the performance metric add value? What is the purpose of your budget performance metric? Another simple question that can be difficult to answer.

Ideally the data point(s) and baseline(s) against which they are compared are consistent with the metric's purpose. Many times, they are not. Frustrations and skepticism take hold when metrics are used to draw conclusions that reach beyond the inherent constraints of the metric inputs and logic. This can be especially challenging with estimates.

So how does an organization determine if their estimates during project development are accurate? It is practical to start with the end in mind. What target are you trying to hit? The final submission detailed estimate the designer prepares that accompanies the Plans, Specifications, and Estimate (PSE)? The sealed Engineer's, or Evaluative, Estimate? The low bid? The mean (average) of the submitted bids? The median (middle) of the submitted bids? The final total construction costs after the project is complete? What you choose will constrain the use and conclusions of the metric.

In many ways, the final total construction cost makes the most sense. It is the most holistic measure, truly comparing apples to apples – your estimated total cost versus the actual total cost. While this may also provide the best opportunity to produce effective lessons learned that could ultimately improve your estimating accuracy, there are significant timing issues. By the time the project is constructed, most people frankly don't care anymore. Impatience and apathy may readily ensue. The project development phase has long been done and they are on to new challenges. The organization is likely much more focused on project close-out at that point than determining how accurate an estimate was that may have been prepared years before. It can seem at best an afterthought to examine and rehash the estimate history on completed projects, often years after the estimate was prepared.

The other end of the spectrum is to create metrics that remain relevant during project development. This means you have to use baselines that cannot extend beyond the bids. This can be extremely effective if the purpose is to focus careful estimating during project development. However, is it really measuring estimate accuracy? What if the plans are lacking and there are an abundance of construction change orders? At the time of bids, you don't know the total project cost. Designers and estimators may shift their actions based upon the prevailing metric. In this instance, the preliminary engineering estimates may look through the bid lens to determine accuracy. This may cause other funding issues if these same estimates are used to program total project costs.

Some organizations prioritize ensuring there be enough money to award the project. In these cases, estimates should target the low bid. AASHTO defines accuracy as the measurement of how close estimates are to the Award low bid. Typically, this is an organization's most accurate estimate, that being the Engineer's Estimate vs. the low bid. This is a reasonable metric as this comparison is already used to help make the business decision whether to Award or not. If the intent is to measure estimate compared against the industry, then the organization should perhaps compare their estimate(s) against all submitted bids. Some organizations want to see how they are compared to what they previously promised. In these cases, future estimates are compared against previous estimate baselines, such as scoping. In these instances, the baselines are often referred to as the project budget. It should be noted that the term "budget" can often mean very different things to very different entities within the same organization. There is tremendous benefit in gaining concurrence across the organization on some of these commonly used terms. Some organizations want to emphasize balancing programmatic funds. In these cases, scoping estimates should be compared against the final total project costs.

In most cases, the estimates should tie to real money, that being bids or total costs. Comparing estimates only to other estimates can provide more immediate feedback, but is inherently constrained to each estimate's underlying assumptions. Since these are not tied to real money, they may provide limited value to the organization in their programming of funds, and can more easily be manipulated.

Once your organization determines the data point(s) and baseline(s) they will use in the metric, the logic must be settled as to what is considered accurate. What delta percentage is good enough to be considered a good or accurate estimate throughout the different stages of project development? Some organizations may choose to utilize an accuracy range that expresses both the high and low acceptable estimate accuracy boundaries. These change over time, becoming more stringent as the plan develops and risks are addressed. An organization's Cone of Uncertainty should graphically depict the Accuracy Range throughout the project development process. Other organizations may also choose a metric with a sliding scale of accuracy. This allows each subsequent estimate to be compared to the previous baseline estimate. This can work quite well provided the metric ties back to reality of bids or total cost.

3.5.3 Last Thoughts

Every now and again, an estimate is completely blown. When this occurs, it is critical for organizations to understand why. Equally important is for the organizations not to overact and pursue reactionary system-wide changes. It is imperative to discern if this is a project or program issue.

Project busts should be handled at the project level. When examining a project, there are often too many cooks in the kitchen, all looking for someone else to blame. In these instances, it is important to discern the truth of the matter. Were the numbers arbitrarily set for political reasons? Was there professional incompetence, oversight, errors, or omissions? Were the right people included in the discussion, and at the right times in the process? Were there unusual risks that were not properly identified or quantified? Were there unforeseen market conditions or other circumstances that legitimately were outside of your control?

Whatever the reasons, you owe it to your Agency, your employees, and your citizens to determine why there is an unforeseen disruption to the public project utilizing public funds. As the reasons become apparent, I encourage leadership to prioritize advancing the project, if reasonable, and then focusing on incorporating lessons learned into your organization's best practices. Remember, the estimates may be sound, and perhaps the bids are the outliers. This certainly doesn't mean that you shouldn't revisit the estimates to improve your processes, but unless you are experiencing systemic busts, try to avoid organizational overactions. If the bids are not in line with the estimates, then the organization must make a business decision on how best to proceed.

Program busts should be handled at the program level. If project estimates are consistently off by variances beyond industry standards, or your organization's expectations as defined in your Cone of Uncertainty, then the current estimating systems, environment, tools, and procedures should be examined and refined, as needed. In these circumstances, it can be beneficial to take a deep breath and approach the situation in a proactive, and not reactionary frame of mind. Organizations should strive to ensure corrective measures are consistent with the overall purpose of estimates. Any new or revised performance metrics to track or drive these changes should also be consistent with the organization's why, what, who, when, and how of estimating to add value to the organization.

4

Scope

"No matter how good the team or how efficient the methodology, if we're not solving the right problem, the project fails."

– Woody Williams

Scoping is a task, and a process. All project development schedules should include a scoping task, which may be the most important task in the schedule. It represents the culmination of the scoping process, or phase, of the project. The deliverable for this task is often an approved Scoping Form or Report. This should be so much more than an obligatory form or rubber-stamped approval. Successful scoping charts a course for success. Conversely, inadequate scoping ensures complications.

At its core, Scoping establishes the project's budget, schedule, and design content. You cannot close scoping until all three of these are resolved. The formal approval of scoping then grants the project approval to proceed. The established budget, scope (design content), and schedule are reflected in the project's Triple Constraint.

One of the foundational elements of successful project management is effective risk management. This is as much a mindset as it is a prescribed list of activities. Great PMs are exceptional at viewing their world through risk management-colored glasses, which grants them an anchored budget-scope-schedule perspective to innovation and problem solving. This is perhaps never more evident than in Scoping.

Generally speaking, Scoping should define the problem, solution, and project approach. But Scoping extends beyond confirming and clarifying the project's purpose and need. You should assemble the project team and name discipline leads. Initiate the environmental review process and determine the required permits and environmental documentation. Identify and engage key stakeholders. Specify applicable design criteria. Perform survey and other essential field investigations. Evaluate performance-based engineering alternative solutions. Proactively identify design, environmental, right-of-way, utility, and construction constraints. Determine project delivery method (e.g., Design Bid Build, Design Build, PPTA, etc.). Determine level of public involvement (e.g., Post a Willingness, Location Public Hearing, Design Public Hearing, Combined Public Hearing, etc.). Work with your team to identify, evaluate, and prioritize qualitative project risks. Quantitatively convert these qualitative risks to project impacts in terms of time (Schedule) and money (Budget). Adjust the project Schedule and Budget accordingly. Gain confirmation on required resource

Transportation Project Management, First Edition. Rob Tieman.
© 2023 John Wiley & Sons Ltd. Published 2023 by John Wiley & Sons Ltd.

availability and concurrence from leadership on the resultant project package of Scope, Schedule, and Budget. Some organizations require pre-scoping worksheets be completed by each involved engineering discipline. This provides an effective way to engage key team members, document discipline specific risks and associated schedule and budget adjustments. This should all then be documented in the Scoping Form or Report that is approved by the appropriate persons, positions, or entities. This final scoping document becomes the basis for all future change management discussions.

On larger or more complex projects, other documents that may be generated during this phase or incorporated into Scoping discussions include Project Charter, Project Management Plan, Issues Log, Assumption Log, Risk Register, Risk Report, Stakeholder Management Plan, Stakeholder Engagement Plan, Stakeholder Register, Scope Management Plan, Requirements Management Plan, Requirements Traceability Matrix, Schedule Management Plan, Cost Management Plan, Basis of Estimates, Quality Management Plan, Quality Control Measurements, Risk Management Plan, Resource Management Plan, Project Team Assignments, Resource Calendars, Project Calendars, Work Breakdown Structure, Communications Management Plan, Performance Metrics, and Schedule and Budget Scoping baselines. There is a point of diminishing returns on the value added in the formal documentation of many of these efforts for smaller projects; however, these areas should all be considered during Scoping for all projects. Most of these items should live in the Project Management Plan.

4.1 Scoping Process

"Slow down. Calm down. Don't worry. Don't hurry. Trust the process."

– Alexandra Stoddard

4.1.1 Introduction

Scoping is a process. This process starts early, typically contains two key milestone meetings, the Project Scoping Kickoff Meeting and the Preliminary Field Inspection (30% Design Review) Meeting, and concludes with formal scoping approval with well-defined deliverables.

Unofficially Scoping starts at project conception. A project idea is first imagined to solve a problem or create new opportunities for improvement. These first thoughts begin to shape the Purpose and Need. Planning resources may proceed to investigate and justify the candidate project with studies and projections. Often candidate projects must comply with previous Planning studies and identified corridor improvements. Preliminary schedules and cost estimates begin to take shape.

4.1.2 Project Initiation

Pending the organization and funding, there may be different paths for a project idea to become reality. Be it political or operational in origination, most selected projects have a strong project sponsor advocating on its behalf. Eventually, some approval must be granted by the governing body or funding entity that endorses the project and commits funding for its execution.

A growing trend in the transportation industry is criteria-based project selection processes. Applications are prepared and submitted that detail project specifics. The applications are then objectively scored based on predetermined factors. These may include safety, congestion mitigation, economic development, accessibility, land use, environmental quality, and other local or regionally specific priorities. A Benefit/Cost (B/C) ratio is often calculated, which can be the basis for scoring the projects for selection. The B/C can be weighted by criteria factors given preestablished regional priorities. This approach strives to replace a black-box political project selection with a transparent, data-driven strategy that is public facing.

One advantage of criteria-based project selection is the emphasis on reducing costs in order to increase the application's B/C which increases odds of selection. Performance-Based Practical Design Solutions, as detailed in Chapter 10, *Controlling the Project*, Section 10.4, *Balancing Innovation – Performance-Based Design*, are pursued. From the onset, the focus is to find innovative and cost-effective solutions that solve the problem at hand. In a financial environment where there are more transportation needs than available funds, public agencies should be especially vigilant in their obligation to be responsible stewards of public funds and use them in a way that realizes a significant return on their investment, in terms of safety and system improvement.

One disadvantage of criteria-based project selection is the need for accurate and advance information on the application. In order to fairly evaluate the benefits and cost of each proposed project, some meaningful preliminary work must be performed. While plans and invasive field investigations are typically not completed for the application, other typical scoping activities are advanced. The problem and proposed solution need to be well defined. Project risks should be identified. A schedule and estimate are prepared. This information, as well as the project approach, is often then summarized within the application.

Owners can spend significant time and money in order to prepare a strong application. If the project is selected, much of this effort is not wasted as the project will begin already deep into the scoping process. However, if the project is not selected, this becomes a learning experience with a deliverable of perhaps a head start on the next cycle's applications. Once the applications are submitted, there is much work to do. Applications need to be reviewed for completeness. Estimates need to be validated. The applications need to be scored. Administering the entire process and program represents a significant effort in terms of time and money. If this is not properly planned and allocated, the drain on resources will adversely impact other ongoing work. Pending the size of the program, some have instituted a two-phase approach in an attempt to more responsibly allocate limited resources for both the applicants and the reviewing entities. In this approach, a more limited application is first screened. Then those candidate projects advancing to round two can then be further evaluated to complete the full application.

4.1.3 Consultant Procurement

If utilized, the Scoping process advances during consultant procurement. While some consultant contracts are structured to be paid in terms of time and materials, most are not. Generally, the owner begins by crafting a Request for Proposal, or Task Order, that describes the problem to be solved. For project-specific selections, design consultants research the project to prepare proposals that detail their ability to solve this situation by demonstrating their capabilities and success in similar past efforts. The owner reviews the submitted proposals and short-lists a few firms for interviews. The selected firms continue to research the project and brainstorm how best to sell to the owner that they are best qualified to develop and deliver a successful solution. Even if the owner does not allow project-specific solutions to be presented in the interviews, the firms will likely still discuss their approach to solution development and highlight major challenges of this project that they are uniquely prepared to overcome. The owner then chooses the top ranked firm, and they begin to negotiate the consultant contract. For task orders, the owner-operator leverages an existing on-call contract. For both project-specific selections and task orders, this agreement details the consultant's project scope, design standards, project assumptions, success criteria, deliverables, anticipated public involvement, the development schedule, and the consultants' cost. In order for the consultant to accurately prepare a cost estimate for their services, they thoroughly identify and evaluate project risks. Involved disciplines detail their anticipated work. A workflow is refined. Potential solutions are vetted. This entire process rests on the fundamentals of the Scoping process: clarify the Purpose and Need, identify risks, refine the solution, and delineate time and budget constraints. For many owners, the consultant contact is the basis for the Scoping Form or Report.

4.1.4 Project Scoping Kickoff Meeting

A Project Scoping Kickoff Meeting is typically the first interdisciplinary team meeting. This is the first meeting that integrates the consultant's and owner's teams after the consultant contract is executed. If a design consultant is hired, their first official meeting once under contract should be with the PM. Goals for this first meeting should include clarifying working relationship expectations. This allows both parties to discuss and decide the day-to-day operations and logistics of communications, invoice processing, deliverable delivery, and so forth. It should be run by the Owner PM, and should be scheduled almost immediately after contract execution. The first meeting where the consultants would meet the rest of the Owner team will likely be the Project Scoping Kickoff Meeting.

The PM has much to do prior to this meeting. Reviewing all project documentation. Bringing the consultants, if present, up to speed. Identifying functional leaders, key stakeholders, and critical team members. Initiating scoping discussions with discipline leads. Initiating field data collection (e.g., surveying, traffic data, geotechnical evaluations, subsurface utility investigations, wetland flagging, etc.). Visiting the site. Confirming project limits. Ensuring environmental has begun preliminary reviews. Evaluating delivery methods with leadership. And confirming project funding and schedule expectations. The PM should then schedule the Project Scoping Kickoff Meeting, which all essential team members should attend.

The PM should run the Project Scoping Kickoff Meeting. This can be challenging since it can include a large number of attendees, not all of whom know each other. Make an agenda, and stick to the scheduled topic timeframes. Provide an overview that focuses on the *WHY* of the project. Detail the desired outcomes of this meeting. Have everyone introduce themselves and describe their role. Discuss risks, and review the framework by which the scoping process will be executed. Be clear with next steps, and clearly delineate who needs to provide what to whom by when.

4.1.5 After the Kickoff Meeting

Following the Project Scoping Kickoff Meeting, work begins in earnest. Each discipline should forge ahead to evaluate the identified risks and develop 30% design plans. This is an exciting time in the project development. Data is coming in and being evaluated. Strategic design decisions are being made, even as different disciplines are chasing down answers to their own critical questions. The PM needs to ensure everyone is providing what others need when they need it, in order for the project to progress. The PM should work closely with the lead designer to ensure the right people know the right information at the right times so wise decisions can be made. You can think of each discipline as being its own train on its own track. The PM needs to ensure the trains are all making concurrent progress such that they all arrive at the 30% Design station at the same time. This is complicated by the fact that some of these trains rely on input from other trains along the way in order to chart their course of action. For instance, the hydraulic report can be critical to the roadway approaches and bridge design. Or environmental permit constraints may impact limits of construction which may directly impact RW and construction costs. Or chasing an adequate outfall may require additional field survey, RW, and environmental impacts.

Successful PMs understand the significance of information during this early and critical phase of project development. Like a pebble thrown into a still lake, the PM should grasp the ripple effects of information and decisions. When they read the preliminary geotechnical report that details extensive unsuitable materials, they need to ensure the road designer is aware and the appropriate mitigation strategies are included in the Scoping cost and schedule estimates. When a homeowner mentions their septic drain field is actually much closer to the roadway than the records indicate, they need to ensure it is accurately located and alert the utility engineers to determine if it will be adversely impacted and need to be replaced. When the Subsurface Utility Survey shows a major natural gas line traversing the project, they need to ensure the road designer can provide adequate cover and can work with the hydraulics engineer so the enclosed storm sewer system can work around the existing utility. In these instances, there is no substitute for experience. The more projects you develop and construct, the more you will be able to quickly discern the intended and unintended impacts of information at this early stage of development. Experience brings a practiced perspective to triage urgencies, assign priorities, and bring the right people in the loop at the right time. Exceptional PMs will not know all the answers, but they do know what questions to ask.

4.1.6 Citizens Information Meeting

There are times when it is advantageous to schedule a Citizens Information Meeting (CIM) between the Project Scoping Kickoff Meeting and the Preliminary Field Inspection (30% Plan Design) Meeting. Reasons may include the complexity of, or the community's familiarity and acceptance of, the proposed solution. Or the number or influence of impacted stakeholders. Or the density of the project corridor. Or the political sensitivities of the project. Hosting a CIM at this stage is a calculated risk. Given the lack of information that can be presented, there may very well be more questions than answers. It adds tasks that will delay the project schedule and may mobilize stakeholders in a direction that is contrary to the project's purpose and need. However, if all goes well, it can build a foundation of open, honest, and credible communication that can accelerate stakeholder cooperation. The decision as to whether to host a CIM at this point should be carefully evaluated. While stakeholder outreach should be underway at project inception, in most cases public meetings are best held after completion of Scoping.

4.1.7 30% Design Review Meeting

The 30% Design Review Meeting may also be called the Preliminary Field Inspection (PFI) Meeting or 30% Staff Review meeting. This is one of the foundational project milestone meetings. From the PM's perspective, the project is being concurrently developed, each discipline running its own train down its own track. While there are certainly touchpoints of coordination, it is at the milestone meetings that all the concurrent tracks converge to produce milestone design plans. The first one is at 30% Design. It is imperative the PM frame the proper team expectations regarding this plan submission. They are 30% plans. This means 70% of the plans are not yet known. The intention of the 30% plan is to confirm a unified direction moving forward. The altitude is high, and with each plan submission we descend further into the weeds of details.

30% plans should answer the fundamental questions that enables a more complete picture of qualitative risks. A rough line and grade (horizontal and vertical alignment of the roadway) should be established. An adequate outfall should be demonstrated, along with sufficient calculations to demonstrate a workable stormwater management solution. Existing wells and septic drainfields should be located such that adverse impacts can be evaluated and necessary waterline or sanitary sewer extensions are understood. Planned utility improvements should be identified. Adjacent projects and planned developments that will impact site, traffic flow, or the project's purpose and need should be evaluated. The level of public involvement should be decided. Preliminary rights-of-way and limits of construction should be delineated. The Maintenance of Transportation (MOT) plan should be developed such that it answers the big picture questions, such as will we need to close the road, is a detour necessary, what temporary pavement may be needed to secure temporary travel lanes, etc. The Sequence of Construction (SOC) plans should be developed such that it answers the big picture questions, such as can this solution be built, can the necessary construction equipment access the site, etc. Environmental preliminary reviews should have confirmed the needed permits and environmental document type.

The PFI (30% Plan Review Meeting) should be run by the PM. Prior to the meeting, the 30% plans and supporting documents should be submitted and circulated to reviewing entities with enough time for a meaningful review. Ideally comments would be assembled and circulated before this meeting. This allows for more productive discussions at the meeting. The PM should take special care to keep the meeting discussion focused, and at the right altitude. This is PFI (30% Design Plan Review Meeting). Comments should be at this level. Stormwater and other utility details will not be known. For example, while there should be enough data to demonstrate an outfall will daylight and the stormwater strategy is viable, the design will not yet be done. Critical elements of the PMP should also be updated, along with other project documents such as the Assumption Log and Risk Register.

Prior to the meeting, assign a scribe to take notes. It can be a large invite list, and the event may spawn many sidebar conversations. The PM cannot effectively run the meeting and take notes at the same time. Following the meeting, promptly distribute minutes that includes contact information for all attendees. As with all meetings and most communications, be sure you acknowledge others' contributions, and say *Thank You*!

4.1.8 Completing Scoping

Scoping is not complete until the time, money, and content (problem and solution) are finalized. This takes the form of Budget, Scope, and Schedule. This can only be completed once risks have been quantitatively identified and then qualitatively converted to time and money.

Closing Scoping involves four key steps, as shown in Figure 4.1. All of these are important, and until all four are done, Scoping is not complete.

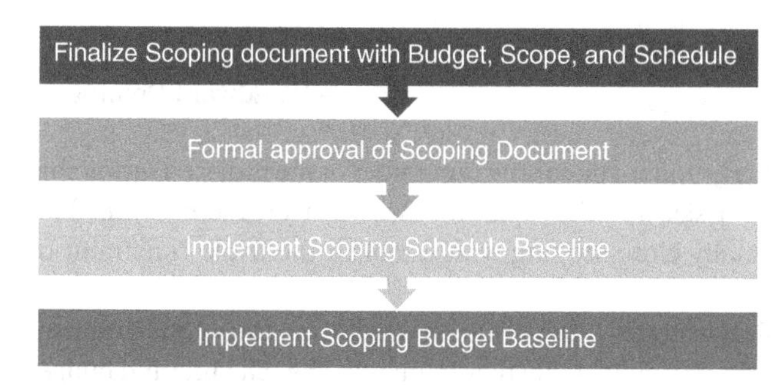

Figure 4.1 Closing scoping.

All of the details determined and the decisions made must be documented in the final Scoping documentation. Many transportation system owners have standardized Scoping Forms that simplify this effort. If no existing and accepted forms are available, a Scoping Report should be prepared in its place. This document should contain all critical project information, such as project numbers, length, termini, and the like. A detailed listing of project risks should be included, typically by discipline. Delivery method should be specified, along with the public involvement requirements. A Scoping Schedule should be finalized. A Scoping Estimate should also be finalized, along with appropriate funding details. Applicable supporting documentation, such as all discipline-specific Scoping summaries, should be included or attached to the final Scoping document. This Scoping document is an essential project document. The PM should bring this to every subsequent milestone meeting, and use it as a guide to ensure the project is progressing as intended. If the project is deviating from this document, the PM should initiate Change Management procedures.

The Scoping Form or Report is not official until it is approved by the appropriate authority. This approval may be required by different levels and different organizations. For instance, a locality may approve its Scoping Report, which is then submitted to the state DOT for various approvals. Often the cost or complexity of the project dictates at what level approval is required. This should inherently convey the importance of the Scoping process in setting up the project for development success. The final Scoping document is approved when the final signature is secured.

The third and fourth steps of closing scoping are to implement the Scoping Schedule and Scoping Estimate. This will often involve adjusting schedules in whatever scheduling software or tools are being used. Similar adjustments will need to be made with the Estimates. More mature organizations may leverage these Scoping Schedules and Estimates in performance metrics to gauge future progress. This Schedule and Estimate are also key management principles to help the PM effectively manage the project and future expectations.

Closing scoping will transition the project from the Scoping Phase into the Preliminary Design Phase of project development. Scoping clarifies the project path of the journey forward. At this point the PM has the Budget, Scope, and Schedule triple constraint clearly defined. Time to get to work!

4.2 Schedule and Budget Baselines

"In God we trust, all others must bring data."

– W. Edward Deming

Scoping positions your feet securely under you for the remainder of the project. During this process you qualitatively identify risks, and then quantitatively convert them to time and money impacts. When you formally close scoping, you establish the triple constraint of budget, scope, and schedule.

The scope portion of this triple constraint is the confident understanding of the problem, solution, applicable design criteria, and deliverable requirements. The schedule and budget are reflected in the Scoping Schedule and Budget baselines. These are often the baselines by which your actual task completion dates and future estimates are measured against to evaluate performance through the remainder of project development. As such, focused care should be given to their formation.

Organizational guidance should be used in determining how to weigh risks and their impacts. It is not reasonable or realistic to assume 100% of your schedule or budget risks will be realized. Similarly, it is not reasonable or realistic to assume 0% of your schedule or budget risks will be realized. There are established mathematical models to assist in determining reasonable ways to normalize risk which are described in Chapter 6, *Managing Risk*. Regardless of the method employed, the organization's culture and risk tolerance play an important role in determining how the identified risks are incorporated in the Scoping Schedule and Budget baselines. Pending the project specifics, there may also be external factors (e.g., political pressures, funding requirements, special events, etc.) that drive the scoping schedule or budget baseline. A PM's greatest challenge can be juggling these non-technical constraints that conflict with the technical realities. In these circumstances, managing expectations is key to success. This topic is discussed in detail in Chapter 10, *Controlling the Project*, Section 10.1, *Managing Expectations*.

Whether by an organizational schedule template, past experience, or a project-specific effort, the Work Breakdown Structure (WBS) should be established during Scoping. Many transportation organizations have established schedule guidance. This can range from previous project knowledge to formal schedule templates by project type. During scoping, the PM should carefully examine the schedule and WBS, which may result in adding or deleting tasks. More likely, you will adjust the established task durations to account for risks, opportunities, and resources. This is not an opportunity to arbitrarily extend the schedule to create a generous buffer. That is neither right, nor especially clever. Rather the scoping schedule baseline is an opportunity to right-size the schedule to the specifics of the project. Remember the big picture of project schedules is to balance driving project development with current realities, while ensuring the right amount of the right type of money is in the right place at the right time. This requires schedules to be honest, credible, and achievable. Arbitrarily extending the schedule out of ignorance, incompetence, or undefined uncertainties, as opposed to evaluated risks, often cause funding issues later on that will unnecessarily delay the project.

Condensing or extending project schedules from the initial template should be justified with an associated risk. At scoping you should revisit your project's assumption log. All assumptions have an associated risk. When a risk is identified, it is typically qualitative in nature. Those with experience and expertise will more accurately convert these risks to quantitative schedule adjustments. All assumptions and decisions should be appropriately documented in project documents. The PM should then adjust the schedule accordingly before it is set as the Scoping Baseline Schedule.

Although both are established at formal scoping acceptance, the Schedule Baseline should be finalized before the Budget Baseline. This is because the base cost estimate (expressed in today's dollars) will need to be adjusted for contingencies and inflation, both of which are dependent upon when the various project development and delivery activities occur. Contingencies, management reserves, and inflation should be applied to the project per organizational guidance. More details can be found in Chapter 3, *Budgets and Estimates*.

4.3 Roles and Responsibilities

"*Sometimes a player's greatest challenge is coming to grips with his role on the team.*"
– Scottie Pippen

It is said the most efficient form of government is a benevolent dictator. Conversely, the most oppressive form of government can be a tyrannical dictator. The difference is determined by the respect for, or abuse of, absolute power. To temper this risk, democratic republics, democracies, and most other western forms of government have established a clear distinction between the separation of powers. These checks and balances strive to limit abuse of power and ensure public good is pursued.

This same principle holds true in the planning, management, and execution of public transportation projects. There are separate and defined tasks and roles. Often these require vastly different skill sets and are performed by different people, in different divisions, in different organizations, all working together for the public good.

In the example above, efficiency can increase and delivery can be expedited if roles are consolidated under a benevolent dictator. Conversely, efficiencies can quickly go off the rails and abuse can run rampant if roles are consolidated under a tyrannical dictator. An experienced and talented PM can sometimes act as a benevolent dictator, streamlining operations and forwarding progress in a way that benefits all. A PM may also knowingly or unknowingly assume the tyrannical dictator role if they are inexperienced, incompetent, or have more nefarious motivations. Given public projects are using public funds that are to be expended for public good, it is typical a system of checks and balances are established to encourage responsible behavior and limit the opportunities for abuse.

There are defined roles within a project. Pending the size and complexity of the project, and your organization's requirements and practices, not all of these roles may be recognized or assigned. Likewise, some of these roles may be filled by the same person, especially on smaller projects or in smaller organizations. It is also possible there are redundant positions at different levels (e.g., federal, state, local, and consultant levels). Regardless, the function of each role may need to be done. And it is important each role perform that function, but stay in their lane to do so. Typical transportation project-related roles include:

- Project Sponsor – the individual whose support and approval are required for a project to start and continue, and is ultimately responsible for enabling project success
- Project Manager (PM) – the individual assigned by the organization to lead the team in developing and delivering the project within the established budget, scope, and schedule
- Project Champion – an informal role of one who makes the project success a personal responsibility by pushing the team, liaising with stakeholders on behalf of the project, and supporting the PM
- Functional Manager – one of authority over organizations, divisions, or disciplines with which the PM will work to facilitate successful project development and delivery
- Project Team Leader – one who supports the PM in leading an aspect of the project team in accomplishing a task or set of tasks
- Team Member – one who contributes by working to complete a project-related task, in full or in part
- Stakeholder – any individual, group, or organization who may affect, be affected by, or perceive to be affected by the development or delivery of the project
- Planning and Investment Manager – the individual who ensures appropriate funding is properly obligated, allocated, and positioned so the project can advance according to the established schedule
- Project Development Engineer – upper management individual responsible to execute development and delivery of the project in accordance with approved program
- Agency Liaison/Project Ambassador – one who represents the project or organization to other organizations, external agencies, the media, and stakeholders
- Senior Leadership – influential division or department directors
- Executive Leadership – individuals at the highest level of leadership within an organization, typically the top person in every major silo or hierarchical pyramid within the organization
- Project Management Office (PMO) – central body that establishes, administers, and maintains organization's project methodologies, procedures, and tools
- Subject Matter Expert (SME)/Technical Leader – individual whose expertise, typically discipline dependent, is critical to project advancement
- Governing Board(s) – authoritative state or local board that oversees projects, approves funds, awards contracts, and bears the ultimate financial responsibility for the project and program

There may be other individuals and organizations that impact your project. This can especially be true on the funding side where FHWA, Metropolitan Planning Organizations (MPOs), regional taxing authorities, and others may play a role. Many organizations strive to intentionally separate time and money roles on a project, ensuring different individuals fill the authoritative financial (Project Investment Manager) and project development (Project Development Engineer) roles. This approach ensures no one person has complete control of the project's financial resources. For instance, if the project schedule or budget changes, another individual must be involved in the allocation/obligation of additional funds and financial adjustments within the organization's multi-year improvement plan.

It is common that the PM may wear many of these hats at one time or another in the life of a project. Particularly on smaller projects or in smaller organizations, the most common role combination may be that of PM and lead roadway designer. In this circumstance, you need to remember these are two very different roles. When wearing one hat, remain true to that role. Same with the other. These dual roles can be challenging, and require a disciplined approach to maintain balance both in time management and project objectives.

Inherently the PM is responsible to facilitate and coordinate most project-related relationships and communications. As such, the PM is responsible to document the Project Team, Management Team, and key stakeholders in the Project Management Plan. This should be updated if or when project roles and responsibilities change throughout the life of the project.

4.4 Scope Management

> *"Management is, above all, a practice where art, science, and craft meet."*
>
> – Henry Mintzberg

Project Management Institute's (PMI's) *Project Management Book of Knowledge* (*PMBOK*) carefully lays out their six scope management processes that walk you from initial planning to scope determination to management and control during project development. While these processes are all directly applicable, the transportation project workflow necessitates some adjustment from PMI's traditional definitions. Figure 4.2 shows the six project management processes, a description of its purpose, and the desired outcomes, or deliverables, of each.

Proper Scope Management begins with establishing a Scope Management Plan. The purpose of this is to capture how the scope will be defined, approved, managed, and controlled throughout the life of the project. In many transportation organizations this approach is already documented and dictated in formal policies, tools, and practices that are outlined in established procedures. Additionally, many transportation organizations have established, detailed criteria of acceptable deliverables throughout project development and delivery. This established guidance should be leveraged whenever possible. There is no need to reinvent the wheel, or risk missing or violating key elements of organizational protocols. If a Scope Management Plan is being prepared, this section will often refer to established organizational standards and expectations. The notable exceptions are unique or especially complex projects or procurements that dictate new or unusual design consultant relationships or acceptance of risk. Large design-build projects are an example where the project will likely have a project-specific contract that dictates the terms of agreement in a way that may supersede other organizational practices and preferences.

In order to effectively manage anything, you must understand the objectives. Collecting requirements can be a daunting effort. Similar to other industries, the problem you are trying to solve needs to be defined. A solution approach should be determined and vetted to determine if it is reasonable before proceeding with a detailed design. Success criteria should be carefully determined. Many transportation projects then move beyond normal

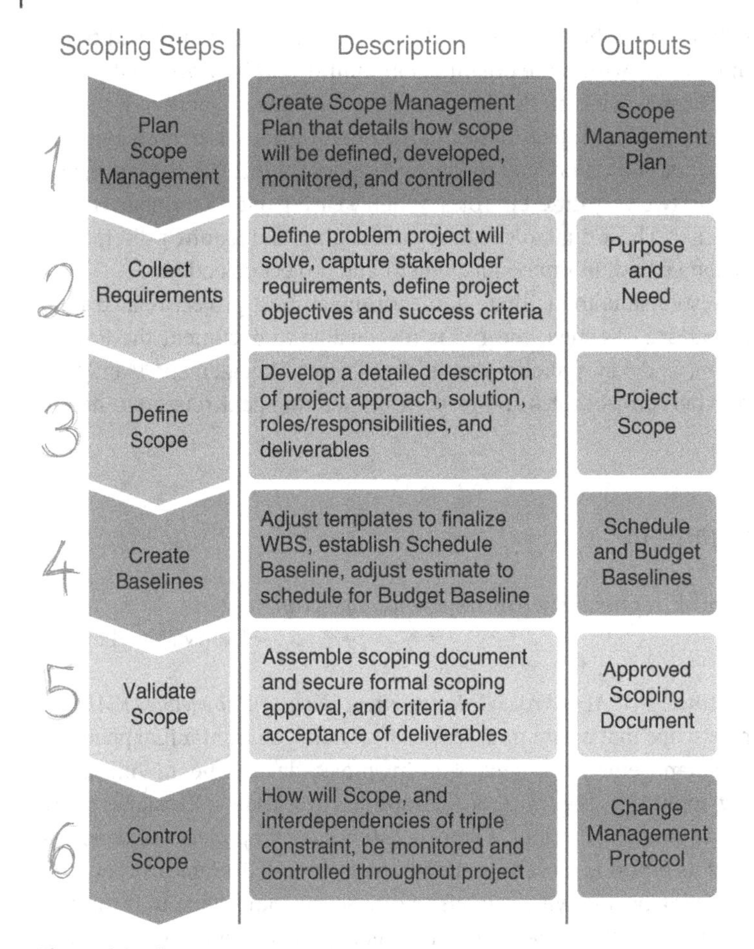

Figure 4.2 Scope management.

product specifications and begin stakeholder engagement. This effort can be critical to refining success criteria and managing expectations through the project. This process generates the Purpose and Need, which is critical to advance the project. This Purpose and Need can bear legal significance should the project require a public hearing in order to exercise eminent domain for acquisitions. A public need is required. The basis of this is the project's Purpose and Need.

After requirements are collected, the scope is defined. The scope details the project's description, project approach, solution, roles and responsibilities, and deliverables. In short, it captures all that must be accomplished in order for the project to be deemed a success. This is a technical document. While it will not include all design features or specifications, it will detail the minimum deliverables. For instance, the scope may not include the detailed pavement section or storm drainage design, but it may specify a Class II roadway with curb and gutter and an enclosed storm drainage system. The scope is the baseline of what is to be delivered. As PM, you need to understand it. You should frequently refer to it, and highlight it at major milestone team meetings. The scope is the basis by which the project is defined. Know it, and fiercely defend it.

Once the Scope is defined, you need to create the schedule and budget baselines. This effort is incredibly important. Remember one of the fundamental goals of transportation projects is to accurately schedule and execute the work so the right amount of the right type of money is at the right place at the right time for the project to efficiently advance.

Transportation professionals often use the same word with different meanings. Scope can refer to the process of collecting and examining project details to determine the selected solution. Scope can also refer to the document that describes the project details and success criteria. Scope may also be a specific project task in the WBS. If your project schedule includes a scoping task, it is not complete unless and until you have the full triple constraint complement of budget, scope, and schedule that may become the established baseline for the remainder of the project development. The complete scoping document, that includes the budget, scope, and schedule, should be reviewed and formally approved, per your organizational protocols. This formal approval is your organization's promise to the public regarding this project. It establishes the baselined triple constraint that is the basis by which all future change management considerations will be measured and evaluated.

Controlling the Scope is the only scope management process that is not in the project planning stage. Controlling the scope occurs continually throughout the life of the project. Success requires an observant and vigilant PM who is fiercely protective of the project's triple constraint while faithfully following your organization's change management procedures. If your organization does not have change management procedures in place, you should create them for your projects and document them in the Project Management Plan. While change may be inevitable, successful change management requires consistency and discipline. Change management is not only the foundational principle of controlling the budget, scope, and schedule, it is the quiet secret ingredient to almost every successful project. This is discussed in detail in Chapter 10, *Controlling the Project*, Section 10.3, *Change Management*.

4.4.1 Scope Threats

There are countless examples of scope threats that may endanger or hinder project advancement. As PM, you need to be particularly aware and wary of scope threats. Your job as PM is to develop and deliver the project within the established triple constraint of budget, scope, and schedule. If the scope needs to be changed, great. But do it correctly. Follow established change management procedures to evaluate the impact of the change and obtain formal approval. This isn't being absurdly bureaucratic; it is being realistic and responsible. Your job is to defend the triple constraint.

Scope threats are project risks. As such, they should be logged and addressed according to the project's risk management plans, tools, and procedures. Many scope threats fall into the following categories of Triple Constraint, Scope Creep, and Gold Plating.

4.4.2 Triple Constraint

With lengthy development durations, swaying electoral priorities, budgetary challenges, stakeholder involvement, adjacent developments, and a myriad of seemingly unrelated and unlimited factors, the one constant experienced PMs know very well in complex transportation projects is most often change. Chapter 1, *Project Management 101*, Section 10.2, *Triple Constraint*, details the unique relationship in project management between budget, scope, and schedule. These three constraints are inherently and interdependently connected. One cannot change without impacting the other two. It is rare that any transportation project will not face some stress on one of these constraints that will necessitate adjustment of the other two.

Be it via a well-intentioned, carefully-orchestrated deliberate action, or incompetence, the results may look much the same. If your budget is cut, you may need to accelerate your schedule or trim scope. If your schedule lags, you may need to increase your budget or reduce scope. If your scope expands, you may need to increase your budget and/or schedule.

As PM, you need to be fiercely protective of your triple constraint. Only a fool adjusts scope without considering and adjusting their budget and schedule. This line can be remarkably difficult to maintain in the midst of political and community pressures to expand scope. These requests often begin with noble intensions to practice responsive customer service that seeks to optimize stakeholder satisfaction. But actions have consequences.

It can be difficult to stand up in a public meeting and explain to a group of well-meaning citizens that their requested sidewalk extension that would connect a now existing gap cannot be accommodated because it is outside of the project scope, budget, and schedule. That task can become more daunting when the elected leader conveys you will reexamine that issue. The night may get even more difficult if an elected leader commits to installing it, when you know it is outside of the project boundaries and will require you to collect additional survey, modify all of your environmental permits, adjust drainage, acquire more rights-of way, move additional expensive underground utilities, and so forth.

Charles R. Swindoll said, "*Life is 10% what happens to you and 90% how you react.*" Regardless of the motivation of the changed condition or constraint, how you react is crucial. As PM, you must remain firm in practicing applicable change management procedures. Maybe adding that sidewalk is a great idea? And if the budget and schedule can accommodate that scope change, terrific. If they can't, agreeing to the scope expansion in hopes it will all somehow work out later is an irresponsible response that most often leads to disaster.

4.4.3 Scope Creep

Scope creep is often one of the most common challenges a transportation PM faces. As projects advance, additional stakeholders emerge, conditions change, and elections bring new leaders and priorities. Scope creep is especially dangerous when it occurs in gradual, minor, incremental changes that often are directly or implicitly accepted without a formal change management evaluation or approval. Although it may seem like you or your organization are being nimble in positively responding to stakeholder needs and desires, scope creep is always a real risk to your project success. Scope creep represents unapproved expansions in scope while not correspondingly adjusting the project's budget and schedule. Even if the resulting budget and schedule impacts are not immediately evident, they will inevitably appear, usually with a vengeance. Unlike fine wines, triple constraint issues do not improve with time.

Good, bad, and crazy project ideas can and will come from most any and all sources. Some of these are brilliant and innovate; others test the definitions of reasonableness and common sense. As PM, you should be the primary gatekeeper of the scope. Ideas should come through you. Your charge is not to express your preference, but rather to determine if the idea warrants further attention. If it does, then you should proceed with applicable change management procedures, as established and accepted by your organization. This includes evaluating the resulting impact to budget, scope, and schedule, along with other critical nontechnical considerations or success criteria. Unless you unilaterally control the budget, scope, and schedule, then you are not formally approving or rejecting the change. Your job is to ensure the potential change is properly evaluated, and if approved, appropriate adjustments are made to the budget, scope, and schedule to position the project for success.

4.4.4 Gold-Plating

Whereas Scope Creep involves new ideas to add to the project scope, Gold-Plating is modifying or improving upon existing scope components. This may seem less dangerous, but Gold-Plating can represent an enormous risk to your project if left unchecked. Imagine a project that is to construct a new four lane bridge over a state highway. The project scope clearly defines the structure's required dimensions, design criteria, and applicable standards. After public hearing, as the project advances to final design, an active community group decides the bridge should accommodate progressive bike and pedestrian accommodations. Local businesses submit a lengthy petition requesting the bridge be more aesthetically pleasing with some distinctive architectural features that retain and emphasize the unique character of the corridor. Soon the elected leader summons you into their office to explain their vision that this bridge will be the signature gateway to their community. You calmly explain that the project budget and schedule are based upon the scoped bridge. They reply, that hasn't changed, we still want that bridge, just with a couple additional features and improvements that will significantly boost stakeholder satisfaction.

The scope establishes the minimum and maximum of what should be delivered. Adding features or improvements to existing scope components that exceed standards or success criteria, as defined in the approved scope, represent real risks to the project budget and schedule. A community that just before advertisement desires more expensive street lights or decorative street signs and posts is requesting Gold-Plating. If managed correctly, your project estimate should not be able to absorb the costs of gold-plating existing scope components. And even if the budget can absorb the cost increase, as public transportation professionals, we are called to be responsible stewards of public funds. As such, changes such as these should be processed and formally approved using applicable change management procedures.

5

Schedule

"A schedule defends from chaos and whim. It is a net for catching days. It is a scaffolding on which a worker can stand and labor with both hands at sections of time."

– Annie Dillard

When a friend or family member has a baby, the two most common questions are: is it a boy or a girl, and what is its name. These are natural, expected, and eminently reasonable. Similarly, when discussing a transportation project, the two most basic questions are: how much will it cost and when will it be completed.

A well-crafted schedule creates and maintains order amidst chaos. The schedule is an essential component of the triple constraint. Not only does it shape expectations for development and delivery, but it is the fundamental framework by which resources are allocated, funding is programmed, cash flow is planned, and inflation impacts are incorporated into cost estimates.

5.1 Critical Path Method (CPM)

"How does a project get to be a year late? One day at a time."

– Frederick Brooks

While there are a variety of scheduling methodologies, the majority of transportation projects and programs are best run using the Critical Path Method (CPM). The CPM is a proven mathematically-based scheduling algorithm developed in the 1950s. It is a powerful approach that enables PMs to proactively manage their projects and promotes on-time and on-budget delivery. CPM scheduling analysis can quickly assess your current progress, where tasks can be run in parallel, the shortest time when a project can be complete, when and where resources will be stretched thin, and the most important tasks upon which you should focus to keep your project moving forward. The CPM is built upon three core foundational elements: tasks, task durations, and task interdependencies. Once the foundation is established, the model can then be enhanced by tailoring it to project specifics or organizational preferences.

Transportation Project Management, First Edition. Rob Tieman.
© 2023 John Wiley & Sons Ltd. Published 2023 by John Wiley & Sons Ltd.

5.1.1 Tasks

The first step is to define the list of all tasks required to complete the project. A Task is a discrete and defined process that can be assigned resource(s) and has associated costs. Each task, or activity, should be an action that pushes the project forward. These should not be Cover Your Ass (CYA) events or checks. Tasks should be a verb plus a noun (e.g., execute contract, deliver plans, secure Rights-of-Way, etc.). Tasks will then be ordered and connected in the schedule network with constraints and dependencies.

If starting from nothing, these would be logically arranged within a Work Breakdown Structure (WBS), which is a comprehensive hierarchical model of the deliverables that comprise the project scope. The WBS is the holistic project picture that is composed of smaller components. Effort within the WBS is defined by Work Packages, which are the lowest-level of deliverables where cost and duration can be estimated and managed. Most transportation organizations have the project activities already well defined in their Project Development Plan. More mature organizations may have project templates where these tasks are already selected and tailored for you based upon project specifics.

5.1.2 Task Durations

The Task Duration is the time it takes to complete a task. This is the delta, or difference, between the start and finish dates.

Most transportation organizations will have established standards, or accepted rules of thumb, for task durations, or span lengths for series of tasks. Predominantly these are based on historic precedents. This means past performances for similar tasks in similar projects are used as the basis for time estimates. Some organizations may have parametric scheduling tools where project durations are based upon project specific criteria. Others may do a three-point estimation, which applies a weighted average to the optimistic, pessimistic, and most likely time projections. Still others may do a full bottoms-up time estimate for each task. However, you determine your preliminary schedule, it is imperative that you tailor it to your own project specifics.

The typical duration unit for transportation projects is days. This may be Calendar Days, or the preferred Working Days. Using Working Days will require your organization to create a custom calendar that only includes days your organization is working, removing all weekends and holidays. While more effort up front, this approach is more accurate as the number of actual workdays in each month can vary widely. Each organization should intentionally define this duration unit, and strive to be consistent across all systems.

The duration units may be hours if an organization is using the scheduling tool to fully resource load each task. This means the organization is using the scheduling tool to also track time, resources, effort, and money attached to each task. Some larger design-build projects, or mega-projects, may do this. They can have a team of scheduling experts to create and manage an extremely detailed project-specific scheduling model that tracks time, cost, and effort. Practically, most transportation organizations separate their scheduling, financial, and time tracking functionalities into different systems. Additionally, most transportation projects do not benefit from this extreme level of detail. However, all transportation projects benefit from a CPM schedule. Pending the size and complexity of the project, the schedule could have ten or hundreds of tasks. Determining the expected schedule level of granularity is an enterprise-wide decision each Agency must make. Eventually you will practically reach the point of diminishing returns, where adding additional tasks and resources actually reduces productivity. Experience will help you find this sweet spot where your tasks capture the essence of your project development, and enable you to drive productivity.

More mature organizations may include other events on their schedules that are not tasks. A Milestone is a specific schedule event. These may serve as triggers to involve others or decision anchors within the project development process. Milestones typically have zero durations, no resources, and no work. While they do not impact the project schedule, they represent significant project progress points worth tracking or celebrating. Some milestones may represent a project Gate, which is a process through which a project must successfully pass to move to the next authorized phase of work. A Summary task, or Parent task, is a created task that visually represents the combined information of a set of related subtasks. Summary tasks are for reference use only, and are typically displayed in BOLD print in the project outline. In general, avoid assigning resources to Summary tasks, and do not enter dates or change constraints on summary tasks. Let the scheduling engine automatically roll up this information, including the duration, from the subtasks.

5.1.3 Task Interdependencies

Once all tasks have been identified and durations determined, they then need to be sequentially positioned to represent the intended workflow. All tasks need to be somehow connected to at least one other task.

Task interdependencies are captured in the following four types of relationships between tasks. These dependencies link two tasks and determine when the Successor task may start. A Predecessor task comes immediately before another task and determines the start or finish date of the following activity based upon the schedule logic. A Successor task comes immediately after another task, and depends upon the task immediately preceding it.

Figure 5.1 shows the Finish to Start (FS) relationships, where the Successor task cannot start until a Predecessor task has finished. FS is the most common, and preferred, relationship.

Figure 5.2 shows the Start to Start (SS) relationship, where the Successor task cannot start until the Predecessor activity

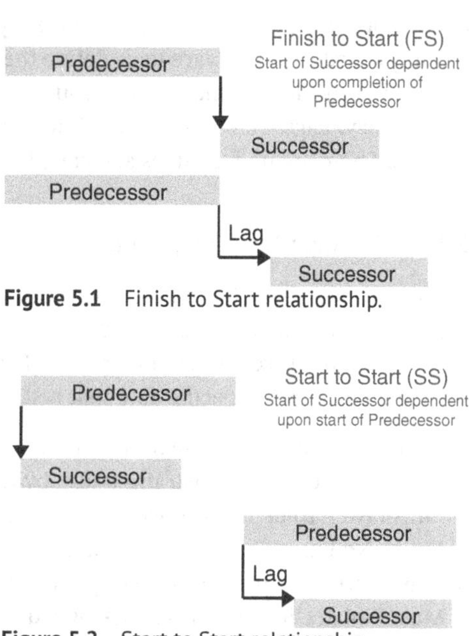

Figure 5.1 Finish to Start relationship.

Figure 5.2 Start to Start relationship.

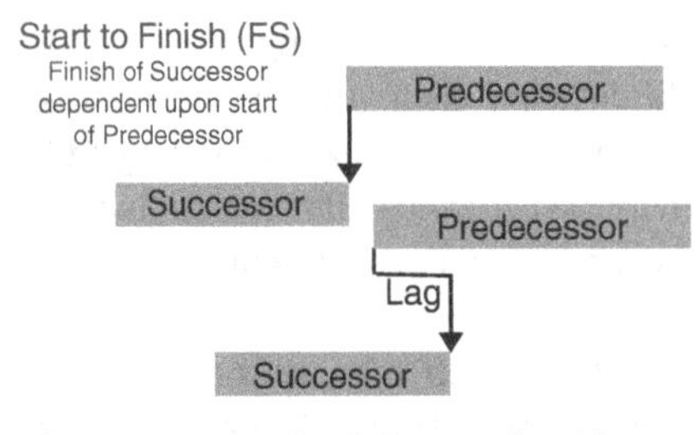

Figure 5.3 Start to Finish relationship.

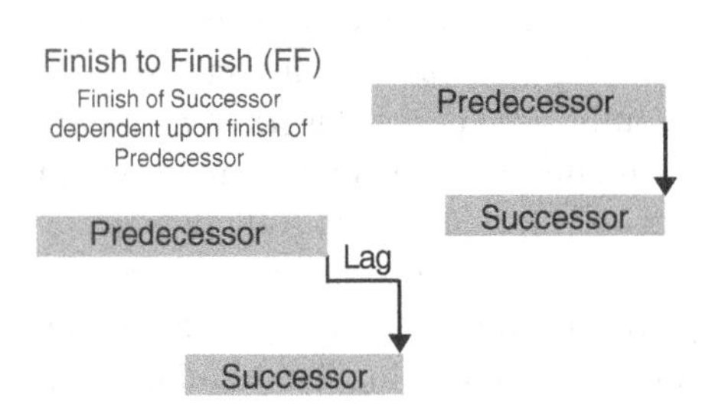

Figure 5.4 Finish to Finish relationship.

has started. SS relationships often have associated leads or lags.

Figure 5.3 shows the Start to Finish (SF) relationship, where the Successor task cannot finish until the Predecessor task has started. This is the least commonly used task relationship.

Figure 5.4 shows the Finish to Finish (FF) relationship, where the Successor task cannot finish until the Predecessor task has finished. FF relationships often have associated leads or lags.

Incorrect relationships can cause inconsistencies in the CPM schedule. Two of the most common are a Loop and Open End Task. A Loop is a logic error where a Successor task attempts to start prior to a Predecessor task. An Open End Task is any project task without a predecessor or a successor task. When present, both of these must be fixed in order for the scheduling engine to properly run the algorithm.

5.1.4 Leads and Lags

Leads and lags are ways to offset the schedule logic in order to further customize the schedule beyond relationship dependencies. Lags are delays in the successor tasks and can be applied to all relationship types. Leads are accelerations of the successor tasks and can be used only on a Finish-to-Start dependency relationship.

Many scheduling software tools define these both as lags. Lags have a positive time unit and extend the dependency. Leads would then have a negative time unit and advance the dependency. Lags are often inserted into schedules to accommodate internal process constraints. There are no resources associated with leads and lags.

5.1.5 Common Task Constraints

Each task may be further tailored with constraints. A Constraint is a limitation placed on the project that must be accommodated during development and delivery. There are eight types of common schedule constraints. The following six types are "hard constraints," meaning they have a specific date that restricts the task:

- Start No Earlier Than constrains a task by requiring it to start after a specified date.
- Start No Later Than constrains a task by requiring it to start before a specified date.
- Finish No Earlier Than constrains a task by requiring it to finish after a specified date.
- Finish No Later Than constrains a task by requiring it to finish before a specified date.
- Must Start On constrains a task by requiring it to start on a specified date.
- Must Finish On constrains a task by requiring it to finish on a specified date.

The following two types are "soft constraints," meaning they do not have a specific date restriction associated with the task:

- As Late As Possible dictates the successor task starts as late as possible. This constraint intentionally consumes all available float. This is typically used in backward scheduling where you begin with the final deadline and work backward to the project start.
- As Soon As Possible dictates the successor task starts as soon as possible based on relationship dependencies. This is typically used on forward scheduling where you plan the tasks based upon an early milestone. Most public transportation projects should be scheduled using this constraint as it introduces flexibility by intentionally preserving float while it minimizes the risk of schedule slippage.

5.1.6 Float

Float, or Slack, represents schedule flexibility. This is the amount of time a task can be delayed without causing project delays. Technically slack is associated with inactivity, meaning an activity can start later than originally planned; and float is associated with activity, meaning an activity can take longer than originally planned. Practically, most use these terms interchangeably.

There are different types of float, and it is important to understand which type is being discussed. Free Float is the time a task can be delayed without impacting the early start of successive tasks. This is the time a task can be delayed with no downstream schedule consequences. Total Float is the time a task can be delayed without impacting the project completion date. The Total Float is shared across the series of tasks. Project Float is the time a project can be delayed without impacting an externally imposed project deadline set by executive leadership. Independent Float is the time a task can be delayed if all predecessor tasks are as late as possible and all successor tasks are as early as possible. Interfering Float is the time a task can be delayed without delaying the project's planned completion date. Tasks on the critical path, by definition, have no float.

Negative float, or slack, indicates there is not enough time scheduled for the completion. It represents the amount of time beyond the project completion date a task requires. The most common cause for negative slack is when task constraint dates are introduced that cannot be met given task durations or schedule logic. When negative float, or slack, is present, the schedule must be adjusted to eliminate it. Critical path tasks with negative slack are called Hypercritical.

5.1.7 Critical Path Basics

The Critical Path is the sequence of project tasks that most quickly finishes the project. This means the chain of tasks that has zero float. As such, any delay to a critical path task will delay the project completion.

Any task with zero float is a Critical Task. The Critical Path is the sequence of critical tasks, which is the shortest time possible to complete the project. Any delay of any critical task will impact the critical path and delay the project.

Near-critical path is a series of activities with very small amounts of total float. A near-critical path may become a critical path if its float is exhausted. Near-critical tasks are those activities with very small float. These should be watched closely as near-critical tasks can quickly become critical tasks, which may have far-reaching project implications.

5.1.8 Network Path

The project schedule should be consistent with, and not conflict with, your organization's Project Development Process (PDP). The PDP are the rules by which you play the transportation game. It is imperative the schedule accurately reflects this established workflow. If it does not, then you are playing a losing hand, trying to fit a round peg into a square hole.

The Network Diagram is a visual representation of how the project tasks are ordered and related. This workflow is a powerful project management tool that can capture available flexibility and enable streamlining the sequences of critical or parallel tasks. In order for the project to be successful, these tasks must be in the same order as the PDP so that the dominoes can fall in the needed sequence.

A Network Path in a series of sequential tasks within the network diagram. These can be useful when considering a specific engineering discipline or other logically related tasks. Path Convergence is when a network path has multiple predecessors. Path Divergence is when a network path has multiple successors.

5.1.9 Three Sets of Dates

To begin to grasp the power within the CPM, it is important that you understand there are three sets of start and finish dates for each task: Planned, Current, and Actual. Understanding the purpose of each is essential to successful schedule management.

- Planned dates are the dates a task is planned to occur. These are the baselines, by which performance should be measured against in applicable performance metrics.
- Current dates are the current schedule of when tasks should occur. These are the most up-to-date estimates of when a task will start and finish. They use the last actual dates and then dynamically adjust the remaining schedule based upon established schedule logic.
- Actual dates are the dates the task actually happened.

5.1.10 Forward and Backward Pass

The Critical Path Method uses an algorithm to determine the Critical Path. Most all CPM scheduling is now done by specialized software. But there is value in quickly reviewing how one would go about calculating these values by hand.

To begin, let us assume you have determined your project has seven tasks that can be arranged in the following workflow due to their task dependencies, as depicted in Figure 5.5. Let us also assume you have already determined each task duration in days.

For this exercise, each task is represented by a six-sectioned box that is partitioned with the following assignments, as shown in Figure 5.6.

The Forward Pass starts at the beginning of the project and calculates each task's early start and early finish dates. Early Start (ES) is the earliest date a task can start given the schedule's logic and constraints. Early Finish (EF) is the earliest date a task can logically finish given the schedule's logic and constraints.

Numbers will be placed in each box. For transportation projects, the units for all six boxes are typically in days. Refer to Figure 5.7 as you read the description below.

One potentially confusing element is that you count each workday. This means ES and LS are the beginning of the workday, while EF and LF are the end of the workday. For example, the ES for Task A is Day 1 and the task duration (D) is 2 days. To calculate the EF,

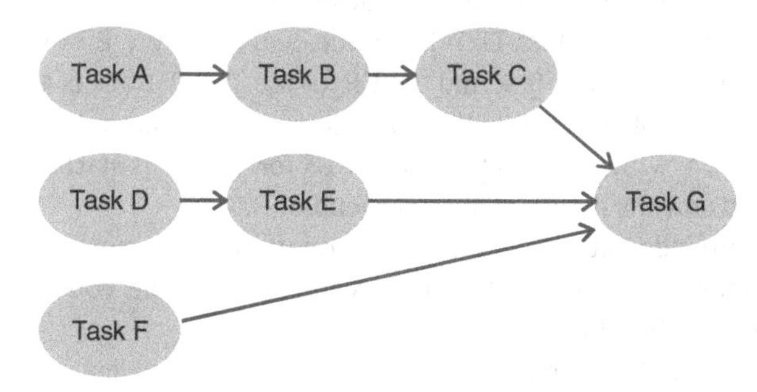

Figure 5.5 CPM example workflow.

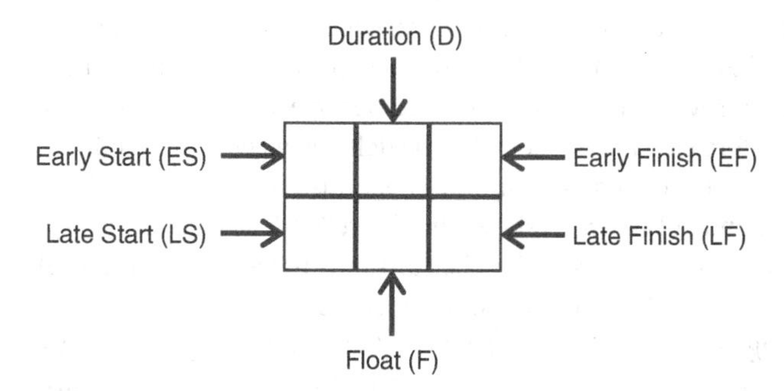

Figure 5.6 CPM task detail box.

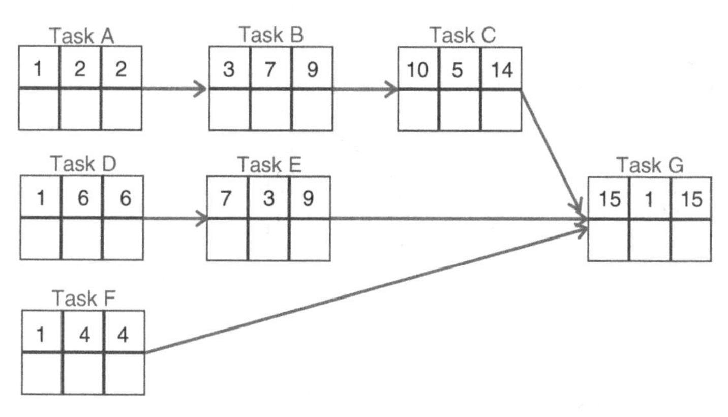

Figure 5.7 CPM example forward pass.

you add the D to the ES (EF = ES + D). So, the EF for Task A is Day 2. This is because the task begins (ES) at the start of Day 1, then add a duration of two full workdays, which is all of Day 1 and all of Day 2. The EF is the end of Day 2. This thinking cascades to subsequent tasks, so the ES of Task B (the immediate downstream task) would be start of Day 3. Similarly, the EF for Task B is end of Day 9, so the ES for Task C is the start of Day 10.

This workflow has three separate parallel paths that converge at Task G. Since Task A, D, and F have no start constraints, the ES of each is Day 1. They could all start at the beginning of Day 1. You calculate the EF of each task in the manner described above until you complete all concurrent paths that converge to Task G. When done, you know the EF for the Task A track is 14, the EF for the Task D track is 9, and the EF for the Task F track is 4. The workflow logic dictates Task G cannot start until all the previous tasks are complete. As such, the latest EF is end of Day 14, which means the earliest ES for Task G is start of Day 15. With a duration of 1 day, this means the EF of Task G is end of Day 15.

These forward pass calculations provide you with valuable information that is summarized in the Figure 5.7. Perhaps most important, is that it is required to enable you to complete the backward pass that determines float.

The Backward Pass, as shown in Figure 5.8, starts at the end of the project and calculates each task's late start, late finish, and float. Late Start (LS) is the latest date a task can start without delaying the rest of the project. Late Finish (LF) is the latest date a task can finish without delaying the rest of the project. The LF often equals the EF; however, there are times when the LF may be an external constraint. Examples may be, needing to complete the road improvement before the start of the college football season to support the increased traffic flow, needing to complete the improvement to enable a scheduled special event, or needing to complete the improvement before election day. Float is the delta between LS and ES.

In our example, the EF of Task G is the end of Day 15. This is also the latest Task G can finish, so the LF of Task G is the end of Day 15, as shown in Figure 5.8. The LS is the LF – D, which is End of Day 15 minus one workday, so LS is the start of Day 15. Float is then calculated by subtracting ES from LS. For Task G, F = 0 days since ES = LS.

For all immediately preceding tasks, the LF is the day prior to the following LS. So, if the latest Task G can start is beginning of Day 15, then the latest Tasks, C, E, and F can finish

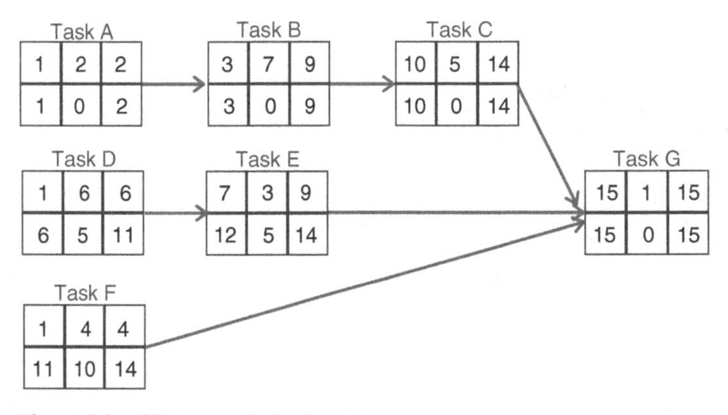

Figure 5.8 CPM example backward pass.

is end of Day 14. Subtract the duration of each of these tasks to determine the respective LS. Remember working days, so Task C, for instance, has a LF of end of Day 14, with a LS of start of Day 10 to accommodate the Duration of five working days. Repeat these backward pass steps for each task in each parallel path.

For Task E, you can see the LS is start of Day 12 while the ES is start of Day 7. Subtracting these two values gives you a float of 5 days. Similarly, Task F has a Float of 10 days.

Once the forward and backward pass have been completed for all tasks, you can begin to harness the power of CPM. All tasks with F = 0 are critical path tasks, as shown in Figure 5.9. None of these can be delayed without directly delaying the project finish. As such, these deserve your prime attention. The critical path of these critical tasks must keep moving forward in order to achieve project success.

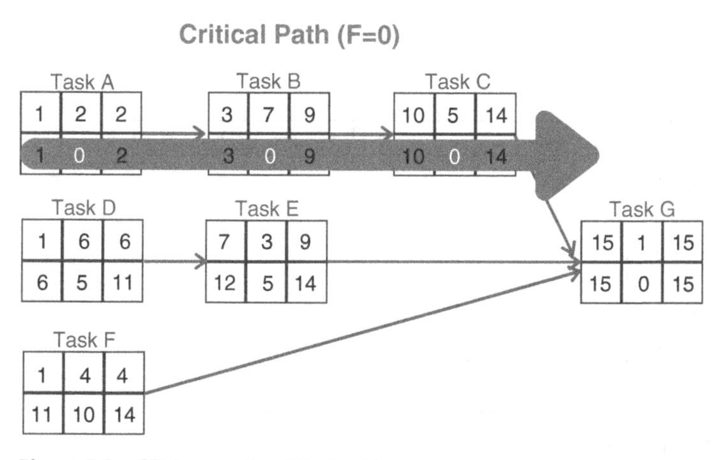

Figure 5.9 CPM example critical path.

Understanding the significance of the LS and F may provide insight into optimizing resource allocations. In this example, Task F may start anytime between Day 1 and Day 10 and still not be critical, but on day 11 task F becomes a critical path task. The resource for Task F may have other work for other projects that are more urgent. Grasping the collective intelligence and insights of tasks across project lines provides Program and Portfolio Managers with powerful insights that can dramatically increase efficiencies and optimize productivity.

As PM, you should regularly update and evaluate the schedule. Actual dates should be promptly entered. This allows the schedule engine to churn and update the downstream tasks based upon the established project logic. This is essential in order to determine if any near-critical tasks have become critical.

5.1.11 Gantt Chart

Most scheduling software applications are able to graphically display the schedule in a Gantt chart. These can be extremely useful in visualizing the workflow, highlighting the critical path, and managing the schedule. A Gantt chart is a bar chart where tasks are listed on the vertical axis and project time is the horizontal axis. Task durations are displayed as horizontal bars that are tied to start and finish dates on the horizontal axis, and connected with applicable relationships.

Pending the size of the project, the Gantt chart can become quite large. You may need a plotter to print it. But it is worth it to print it and display it. This encourages the schedule to become a concrete reality that all can see, subconsciously accelerating team buy-in. The schedule is no longer the PM's schedule, but the project schedule. Everyone should have ownership and be vested in its success. An updated Gantt chart can be a powerful visual anchor for the team that clearly shows where you are and what needs to happen when, to move forward. This simple act can build team trust and encourage collaboration as transparent expectations inherently encourage individual and collective accountability.

Figure 5.10 shows the Gantt chart for the provided example. Some CPM software may be able to be customized to show additional information such as task owner, percent complete, or other project data.

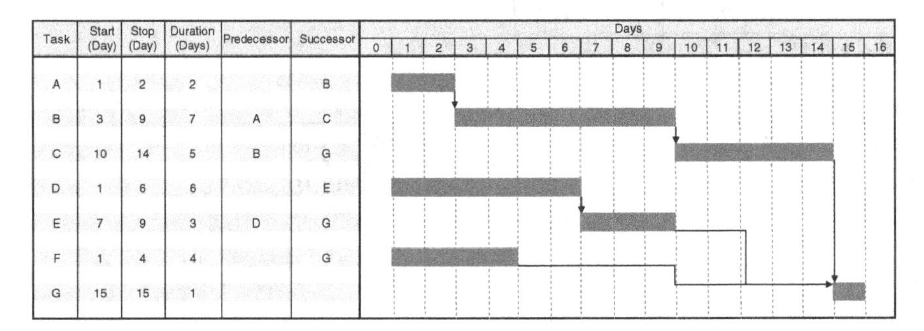

Task	Start (Day)	Stop (Day)	Duration (Days)	Predecessor	Successor
A	1	2	2		B
B	3	9	7	A	C
C	10	14	5	B	G
D	1	6	6		E
E	7	9	3	D	G
F	1	4	4		G
G	15	15	1		

Figure 5.10 CPM example Gantt chart.

5.2 Schedule Management Plan

"The bad news is time flies. The good news is you're the pilot."

- Michael Altshuler

A Schedule Management Plan describes how the project schedule will be managed during the project. The plan provides guidance on the organizational policies and procedures for planning, developing, managing, executing, monitoring, and controlling the project schedule.

Some typical sections may include: roles and responsibilities, update frequency, baselines, schedule change management, calendar definitions, performance metrics, control thresholds, tools, and reports. Like all other aspects of the Project Management Plan, this should be tailored to the size and complexity of the project, while adhering to your organizational expectations.

While larger design-build and mega-projects may have very detailed Schedule Management Plans, most transportation organization projects may not. Scheduling processes, procedures, and expectations are best set at the organizational program or portfolio level. Generally speaking, each project schedule should be treated similarly to every other project schedule within the same organization. This consistency across projects builds credibility while increasing efficiencies and productivity.

Your transportation organization may already have extensive guidance on how to develop, manage, monitor, and control your schedule. This can be especially true if your organization utilizes an enterprise-wide scheduling CPM software tool for all projects. However, just because the practices, definitions, and expectations may be so established that it is taken for granted, it doesn't mean they are not important. These are your rules of engagement. As PM, it is your job to know them and work within them. Even if documented ad nauseam in your organizational guidance and culture, there are four key schedule aspects that every PM should know and emphasize: Templates, Baselines, Monitoring, and Change Management.

5.2.1 Schedule Templates

As PM, you need to understand how your project schedule was formed. What tasks were included? What tasks were left out? How were task durations determined? How were task interdependencies and relationships defined? What are the assumptions, risks, and constraints?

At the program or portfolio level, many transportation organizations have already determined this for you. Many have already done the work of creating the WBS, identifying key tasks, establishing reasonable durations, and mapping the task relationships to create the project workflow. Whether it be detailed in guidance documents, expressed in other tools, or take the form of preselected project templates, many PMs never have to worry about this aspect as it is already decided for them. However, even if this is the case, the PM should still understand the schedule tasks, durations, and relationships. They are the details of the schedule ship in which you will set sail. They, along with the CPM scheduling software rules and restrictions, will provide insights on how you can best control the schedule throughout the project.

5.2.2 Schedule Baseline

Chapter 10, *Controlling the Project*, Section 10.7, *Performance Metrics*, details how the best performance metrics balance reflecting reality and driving productivity and innovation. Be it an individual or an organization, everyone likes to look good. Everyone wants to win. Pending your organization, you may have various degrees of performance metrics. In almost all of these cases, there is a baseline against which progress is measured to determine the level of success.

One of the most basic of project performance metrics is schedule performance. Was the project developed and delivered by when you said it would be? This comparison inherently relies on setting a schedule baseline. This baseline is a snapshot in time that is locked and used as the measuring stick moving forward. Some organizations have multiple baselines throughout the life of the project. There is likely a preliminary baseline which is established when the project is created. Most mature organizations allow the baseline to be adjusted at time of scoping. This is intentional. At Scoping, you should have the project problem and solution defined. You should have risks qualitatively identified, and then quantitatively converted to potential time and money impacts. As PM, it is imperative that you understand your organization's performance metrics, including when and how they capture your schedule baseline(s). Be deliberate and strategic in positioning your project for success.

5.2.3 Schedule Monitoring

As PM, you are responsible for the schedule. Your perceived success or failure in this responsibility may be directly tied to your organization's schedule expectations. Meaning, you need to know how often the schedule is to be updated. You should know who needs schedule status updates and when.

It can also be helpful to understand how the schedule data is used by others. For instance, pushing project Award back one month due to right-of-way acquisition issues may not matter to anyone, but it may become a huge deal if that month moves the Construction phase from one fiscal year to the next. That same month may impact anticipated workloads or billable hour goals of those administering and supporting the construction. Or that month may cause tremendous turbulence if a prominent politician promised the project would be awarded before the election, which is in two weeks. Situational awareness can be a PM's secret weapon.

5.2.4 Change Management

Schedules change. However, as Scott Adams keenly observed in his Dilbert comic strip, perhaps the eighth revision of the original schedule would be more appropriately called a calendar. While schedules should be periodically evaluated and perhaps adjusted, the fastest way to lose schedule credibility is to change your schedule baselines as a matter of convenience or in an attempt to hide incompetence. This is a seemingly expedient solution that fools no one.

Chapter 10, *Controlling the Project*, Section 10.3 *Change Management* discusses in detail the importance of practicing established and consistent change management protocol. This is essential to controlling the project's triple constraint. The PM is directly responsible to maintain the schedule, which includes practicing proper change management when needed. These procedures and expectations should be documented or referenced in the Schedule Management Plan.

5.3 Scheduling Issues

> *"Either you run the day, or the day runs you."*
>
> – Jim Rohn

5.3.1 One Schedule

It is important that each project have one schedule. Some PMs or organizations mistakenly think multiple schedules are advantageous. Perhaps they keep best- and worst-case scenarios, schedules they show the boss or leadership, others they share with the public, or artificially accelerated schedules to instill urgency to team members. Resist this urge.

Regardless of the reasons, keeping multiple schedules for different audiences creates confusion, reduces accountability, and invites criticism of individual or organizational credibility. This practice is an insecure time sink that creates more problems than it solves. Nothing good comes from multiple schedules.

The PM controls the schedule. The team members all should have access to it, as should leadership and key internal and external stakeholders. The specifics of the schedule that are public facing is a decision that should be made by the organization. However, public transportation projects are completed for public good using public funds. Most all project documents, including the schedule, are available for public consumption and scrutiny, even if only by a Freedom of Information Act (FOIA) request. Closely guarding the schedule to protect the PM or project, or for most all other reasons, is hugely counterproductive. The goal is to develop and deliver the transportation project for the public good.

The schedule is the organization's commitment to the public and team members. As such, every project should have just one schedule. The schedule should be transparent and available to all. This increases accountability and ensures consistency of message and purpose, while reducing the opportunities for conflict and misunderstandings. Having one schedule builds individual and organizational credibility and trust, and is foundational to the PM effectively managing expectations. From a program perspective, the one project schedule should be used to ensure proper funds are available at the correct times, manage resources and workloads, and improve project and program development efficiencies. And most importantly, it is the best way to develop and deliver the project on-time and on-budget.

5.3.2 Who Owns the Schedule?

Who owns the schedule? The PM manages the schedule. The PM is responsible for the schedule. Consultants and team members work to the schedule. But who owns the schedule? This is a tremendously important question, the answer to which has far reaching consequences.

Who owns the float? If a critical task finishes one week early, who owns that week? Does the task resource wait a week to say they are finished so they can work on other tasks? Does the task Functional Manager leverage that week so they can redistribute their section workload to better meet their team goals? Does the next task owner own that float and delay their start by a week? Does the PM keep that float for an additional buffer in case subsequent tasks fall behind? Does the Program Manager shift the float to later in the schedule to coincide with the PM's upcoming vacation?

What happens if a critical path task is a week late? Do all subsequent task owners maintain their planned task durations? If there a schedule contingency, or buffer, built into the schedule or certain tasks? Who owns the extra time? Who can decide when and how to use it?

Who owns the float? If the answer is not clearly defined and communicated from the organization's leadership, then by default the answer is everyone and no one. Everyone thinks they own the float, but no one actually does. This environment inherently encourages "game-playing." Task owners will artificially inflate their needed durations. PMs will hide time in tasks so they have their own buffers. And so on, perhaps at every level. The results are predictably inaccurate data and fictitious schedules. This is not upholding the public trust to be good stewards of public funds. Additionally, it is horribly inefficient.

In a project-centric environment, the organization owns the schedule. The organization owns the float. Similarly, the organization owns the budget and all contingencies. While Public servants, and those acting on behalf of the transportation organizations to advance the project, may be responsible for the schedules, they do not own it. As such, if a task owner finishes one week early, they are obligated to give that week back to the project. The overall schedule should move ahead one week, maintaining all downstream scheduled task durations and relationships. This is accomplished if all schedules have task relationships with "As Soon As Possible" constraints.

5.3.3 Control Thresholds

Control thresholds are established metrics or deviations that are monitored to trigger management action. These limits are important tools for a PM. They should be carefully selected and closely watched. Generally, there are two types of thresholds: predictive and corrective. Predictive thresholds are those that indicate the project is trending to bust. Corrective thresholds are those whose violations prompt immediate response.

Schedule thresholds are largely dependent upon your organizational culture and expectations. Some examples of common corrective thresholds may be any delay to the critical path, or a set percentage bust in critical path tasks. Meaningful predictive thresholds are typically harder to define. Some examples of common predictive thresholds may be project scoping, near-critical path tasks, or tasks that leverage critically constrained resources.

5.3.4 Fast-Tracking and Crashing

If you are behind in your schedule, there are generally two ways to compress a schedule without reducing scope. Fast-Tracking compresses the schedule by advancing successive tasks and running them in parallel. The concurrent execution of sequential tasks can be effective, but it will introduce new project risks as the Work Breakdown Structure logic planned from them to be sequenced. By definition, you need to compress critical path tasks. To control the associate risk with fast-tracking, the rule of thumb is to advance sequential tasks up to 33%, meaning you would begin the downstream tasks when the prior is 66% complete. Pushing beyond this 33% trigger may significantly increase project risk. Crashing compresses the schedule by adding resources to critical tasks. Accelerating the schedule will increase costs. When engaging in Fast-Tracking or Crashing, it is imperative to proceed very cautiously. You are tinkering with critical tasks. This can introduce significant schedule, budget, scope, resource, and quality risks to the project if things don't go well. With both, you will need to revisit the schedule as the downstream critical path and task floats will change. Pursuing these two strategies assumes the critical path tasks cannot be further pruned without sacrificing scope.

When considering adjusting your schedule, consider the kinds of task relationships. Generally speaking, there are two kinds of logic linking tasks. Hard logic are relationships that cannot be adjusted. Examples may be contractual obligations or physical constraints.

These are also known as mandatory logic or hard dependencies. Soft logic are relationships that are linked by preference, reason, or industry standards. These are also known as discretionary logic or preferred dependencies. When adjusting or resequencing tasks in your schedule, focus on those with soft logic.

5.3.5 CPM Scheduling Software

Most PMs will never calculate CPM by hand. More mature transportation organizations have some type of CPM scheduling software that they use to plan and manage their projects. There are many choices in the marketplace. Each may have their unique strengths and costs. Some are geared more to the novice user, while others require higher level of expertise to be proficient. Some are stand-alone applications, where others allow customization and integration with other organizational systems and data sources.

The choice of software should be an organizational decision. The choice should accommodate the full spectrum of their operations. Larger, more complex projects are nearly impossible to manage without specialized software. There are also tremendous program and portfolio management benefits as the data can be rolled up from individual projects for analysis and optimization. Additionally, a standard software with accompanying guidance and protocol introduces consistency and conformance with the organization's project management expectations. This can be essential for transportation organizations that experience staff turnover, lack adequate training, or have newer, less experienced, or less competent PMs.

As PM, learn the foundation and nuances of your organization's scheduling software. This is your job. Discover how to squeeze every bit of information from the CPM that it can provide. Take advantage of the instant analysis of float and critical path. Leverage the Gantt chart and other tools. Proactively manage your schedule.

5.3.6 Planning Fallacy and Managing Expectations

Transportation projects have a long history of being delivered late and over-budget. This begs the question, compared to what? Who sets the initial expectations against which the project is judged? And why hasn't our industry learned our lesson through the years?

Chapter 12, *Real-World Challenges*, Section 12.7, *Planning Fallacy*, details how overly optimistic preliminary project schedule and budget estimates unfairly saddle a project with unrealistic expectations. Once you understand the Planning Fallacy (the *what*) and the motivations behind such assertions (the *why*), you will be better positioned to effectively manage your project.

Some transportation organizations involve their PMs early in the project planning process. In these cases, this knowledge can be used to help positively influence and shape initial expectations. In other organizations, the PM won't be assigned the project until after it is funded and active. In these instances, managing the CPM schedule and following proper change management procedures are critical to successfully developing and delivering the project within challenging constraints. Failure to actively manage the schedule and project changes will quickly exacerbate the already challenging situation created by overly optimistic preliminary schedule and budget estimates.

5.3.7 Some Additional Thoughts

The simple example earlier in this chapter demonstrates the basics of CPM. Most projects have more than seven tasks. While some stripped-down schedules may have ten or less tasks, others will have hundreds. Larger Design-Build projects may have thousands.

Exceptional PMs leverage the many strengths and insights of CPM schedules to proactively and strategically manage their project schedule. CPM can be equally effective on both small and large projects, allowing a systematic approach to most all aspects of project management. The workflow is mapped. Critical tasks become evident. Task dependencies become visible. Float is calculated. This essential information enables the PM to control the project by optimizing time, cost, and resource efficiencies. It should guide the PM directly to those critical tasks that deserve your special attention.

The CPM allows you to easily compare planned vs. actual progress. This provides opportunities to predict, plan for, and respond to, tasks that run long. The impact of missed task deadlines or project milestones can range from commonplace or catastrophic. How can you tell the difference? CPM allows the PM to identify and leverage parallel paths. It also allows the PM to analyze the downstream impacts of these schedule busts, and determine if the late tasks change the critical path or critical tasks. This knowledge is essential to adjusting the schedule and resources to be able to successfully recover.

The PM is responsible for the project schedule. You can't be nimble, innovative, and proactive if you are flying blind. Knowledge plus action equals power. CPM is a powerful tool that can empower the PM to strategically manage the schedule. Embrace it. As PM, you need to actively and intentionally manage your schedule, or it will manage you.

6

Managing Risk

"If you don't invest in risk management, it doesn't matter what business you're in, it's a risky business."

– Gary Cohn

6.1 Why Risk Management Matters

"You cannot manage risk, or for that matter, build a risk management capability without first understanding the 'business value' of risk management."

– Pearl Zhu

Risk Management is more than a functional component of transportation project management; it is foundational to all we do. Risk Management is at the heart of Project Management.

There are many reasons why transportation projects are delivered late or over budget, but perhaps the biggest is the way in which many transportation organizations and PMs identify, monitor, and respond to risks. Upon reflection, this may be surprising as engineers should excel at managing risk.

At our core, engineers solve problems. We delineate the problem, define success, identify constraints, and then use logic, math, science, mechanics, and reason to solve it. Engineers think. One fundamental tenet of an engineer's approach is to identify and mitigate uncertainties. Most everyone knows there is a 50% chance a flipped coin will land on heads, but give an engineer enough details of the coin, the force, and other situational attributes, and they can calculate with much greater confidence if the coin will land on heads or tails. Be it significant figures, safety factors, manufacturing tolerances, or estimating assumptions; they all represent efforts to identify, quantify, and manage risk.

Any experienced engineer that prepares a schedule or cost estimate includes a list of assumptions. This is standard practice, and a great first step. But what you do with these assumptions matters even more. Each assumption is inherently partnered with an associated risk. And as the project progresses, risks evolve and develop.

Transportation Project Management, First Edition. Rob Tieman.
© 2023 John Wiley & Sons Ltd. Published 2023 by John Wiley & Sons Ltd.

Risk management is a foundational responsibility of every PM. In order for a PM to satisfy their primary objective, to execute the project within the established scope, schedule, and budget, they must understand, appreciate, and practice effective risk management.

The reality is, you and your organization already have an established risk tolerance and are practicing a risk management strategy. By deliberate intention or blissful ignorance, you and your organization fall somewhere on the spectrum shown in Figure 6.1:

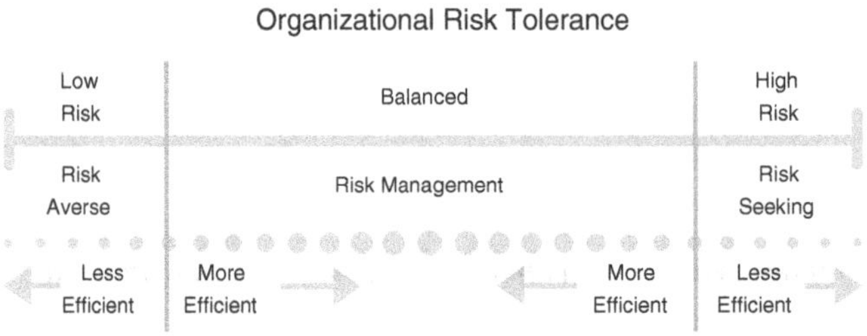

Organizational Risk Tolerance

Low Risk | Balanced | High Risk
Risk Averse | Risk Management | Risk Seeking
Less Efficient | More Efficient | More Efficient | Less Efficient

Figure 6.1 Organizational risk tolerance.

Some are so afraid of making a decision, or making the wrong decision, that they drift toward paralysis of analysis and become overly risk averse. Others are naïve, in denial, or just provide risk management lip service and allow risks to evolve on their own, hoping it will all eventually work out. Ironically, this can put them with those who actively seek risk. On transportation projects, being overly Risk Averse or Risk Seeking will put you in a defensive, reactive position with risk that ultimately jeopardizes your budget, scope, and schedule.

Ideally, you and your organization should fall in the balanced portion of this spectrum. Here you are actively managing risks throughout the project in a manner that allows you to be aggressive and proactive in risk responses, coupled with established change management procedures. This combination is foundational to your project's success.

Risk Management is more than a set of documents and processes. It is a mindset. For some, this conscious decision requires a massive paradigm shift. Risks are real and present, if you acknowledge them or not. Your identification, analysis, and response to risks directly impact your projects' and programs' level of success. Establishing and practicing an integrated and meaningful risk management strategy may be the single most important thing you can do to increase your on-time and on-budget performance.

An effective organizational Risk Management strategy should be scalable. While every transportation project should have a formal risk management plan, it may be tailored to the size, cost, priority, or complexity of the project. Mature organizations understand this and often establish thresholds for varying levels of risk documentation. You should ensure the juice is worth the squeeze. While many risk management tools may not be deemed appropriate on smaller projects, the mindset should remain. If you are not actively thinking about identifying and responding to risks with your team, you will soon be actively scrambling to respond to changes that jeopardize your project's budget, scope, and schedule. Unmanaged risks can directly lead to increased costs, unanticipated schedule delays, poor performance, lower quality, and damage to your and your organization's reputation. You may not be able to control all the risks, but you do control how you prepare and react.

Every PM should go through the Risk Management Project Workflow illustrated in Figure 6.2 on every project, through the life of the project.

Figure 6.2 Risk management workflow.

6.2 Risk Management Plan

"If you fail to plan, you are planning to fail."

– Benjamin Franklin

Transportation projects can extend in duration for years, often cycling through multiple PMs and critical team members. In these circumstances, the organization's risk attitude is crucial. The Risk Management Plan is essential to maintaining continuity and uniformity in risk management activities, and positioning the PM and project team to successfully develop and deliver the project on-time and on-budget.

The Risk Management Plan describes how the project team will identify, evaluate, respond, and monitor risk throughout the project. This formal document is often based upon an organizational template. It details the framework, processes, tools, and expectations as to how risk will be managed. Information contained therein should reflect organizational risk attitudes, and comply with applicable processes and procedures.

Risk Management Plans typically include:

- Definitions of Risk Probability and Impact
- Assumptions
- Risk Strategy or Approach
- Methodologies
- Roles and Responsibilities
- Risk Identification
- Risk Assessment and Analysis
- Risk Response
- Risk Monitoring and Controlling

Risk Management tools and details that are included or referenced within the Risk Management Plan typically are:

- Risk Breakdown Structure
- Probability Impact Matrix

- Accuracy Estimates
- Risk Register
- Assumption Log
- Schedule detailing frequency and timing of risk management activities
- Risk reporting format

The Risk Management Plan should be developed during early planning phases of the project and finalized before scoping is complete. The result should be incorporated into the Project Management Plan. It should be noted that like other aspects of the PMP, an organization may have all of this already documented within their existing policies and procedures.

The PM ultimately bears all responsibilities for the Risk Management Plan. They shall guide its planning, and formally approve it. The PM shall regularly review and update this plan. The PM must document all risk and risk responses, per the plan. The PM must communicate applicable risk information to their superiors and to the project team, secure decisions and concurrence on risk responses, and ensure each risk response is appropriately and effectively implemented. The PM should ensure there is appropriate budget and time for risk identification, analysis, response, and monitoring activities. The PM also plays a vital role of promoting a balanced Risk Management mindset within the project team that promotes optimization of resources, time, and money. One way to accomplish this is to make Risk Management a standing agenda item at all regular team meetings.

6.3 Risk Identification

> *"PMs are the most creative pros in the world; we have to figure out everything that could go wrong, before it does."*
>
> – Fredrik Haren

A Risk is any uncertain event, activity, or condition that can impact the project in a negative or positive way. Negative risks are threats. Positive risks are opportunities. Residual risk is risk that remains after a risk response strategy has been implemented. A Secondary risk is a risk created by your response to another risk. These risks represent the intended and unintended consequences of your efforts to actively manage other risks. Often secondary risks are less serious than primary risks, but not always. Both residual and secondary risks can be negative or positive. Risk identification is the ongoing process of identifying and examining risks and their effects on project objectives.

Effective Risk Identification begins with establishing and ensuring mutual concurrence on project fundamentals. The purpose and need. The project context. The project's Scope, Budget, and Schedule. The project's success criteria. If these are not defined, understood, and accepted, you will not be able to effectively identify and manage risks.

This Progressive Elaboration is when projects incrementally develop in small steps. This is particularly common in the planning phase of a project, as expectations and specifics continue to be gathered and clarified. In these instances, risk identification should be continually revisited and refined.

Many organizations have predetermined checklists of risk categories to consider. These are often associated with Scoping forms, estimate preparation, schedule assumptions, or risk management specific procedures. Some standard industry strategies to employ during risk identification include the following acronyms:

- PESTLE (Political, Economic, Social, Technological, Legal, and Environmental)
- TECOP (Technical, Environmental, Commercial, Operational, and Political)
- VUCA (Volatility, Uncertainty, Complexity, and Ambiguity)

Risks can also be multidimensional and cumulative in nature. They can interact with each other and amplify in consequence. Sometimes, you need to bundle them to reasonably assess their impact. On public sector projects, risks can frequently interact in unpredictable ways that create turbulence in the critical path, which can dramatically slow the project.

Risk Identification should begin early in the project planning process. The PM should meet with the entire project team and key stakeholders to brainstorm potential project risks. Each team member brings a unique perspective and bears the responsibility to identify risks within their own areas of expertise. During brainstorming, all risks should be logged without judgment. The identified risks will then be assessed and analyzed. Organizational and available industry lessons-learned logs can also be leveraged. The PM is ultimately responsible to initiate and maintain the project's risk register.

This effort should include reviewing project assumptions and constraints. An Assumption is a presumed or necessary condition that will enable successful or predicted completion of the task, objective, or project. Every assumption has an associated risk. Every risk has an associated assumption. Assumption = Risk. Risk = Assumption. Many PMs fail to acknowledge this fundamental truth. The PM should regularly cross-check project assumptions with the risk register, and vice versa.

Larger and more complex projects may warrant more formal risk identification measures. A Risk Workshop is a formal, multiday event that facilitates collaboration among selected team members, Subject Matter Experts (SMEs), and stakeholders to discuss and identify project risks. Think of a formal Value Engineering (VE) proceeding, but focused on risk. An invite list is carefully selected, a formal agenda is followed, meetings are held, and it culminates with submission of a final report. On very large projects, the total duration from start to finish may be many months. All projects will benefit from an organized risk identification process. Organizational guidance should determine the appropriate granularity of this effort based upon established project thresholds.

To be effective at focusing on risks and identifying potential problems that aren't yet there requires a creative imagination and a healthy dose of skepticism. If you are, by choice or nature, an optimist; this can be a tedious and disheartening task. Others thrive on discussing what may go wrong and what impending calamities are waiting just around the next corner. Pessimists, in addition to the steady stream of problems that regularly assaults PMs, can make it easy to develop a defeatist or apathetic attitude. Exceptional PMs do not tolerate or accept complaining. They are achievers who thrive on overcoming obstacles and solving problems. Risk management is not choosing to focus on reactive negativity, but rather proactively advancing and positioning the project for success.

Perhaps the most critical milestone in transportation project development is Scoping. When scoping is complete, the problem should be defined and the solution decided. The technical requirements, budget, and schedule should be set for the remainder of the project. Should any changes occur to the triple constraint thereafter, applicable change management procedures should be followed. The foundation of Scoping is to advance and refine the design so you can identify, analyze, and quantify project risks. By close of scoping, you should have completed a prioritized Risk Register, determined risk responses, and adjusted your project's time (schedule) and money (budget) to accommodate identified risks. This positions your project for success. After scoping, the PM transitions into Risk Monitoring and then repeats the Risk Management Project Workflow presented in Figure 6.2.

6.3.1 Knowledge Quadrant

The Knowledge Quadrant defines information and risks by separating them into four distinct categories that balance knowledge and awareness, as shown in Figure 6.3. PMs can use this approach to more efficiently brainstorm ideas and risks, process information as it emerges and evolves, and quickly determine how best to move forward.

Known-Knowns (KK) represent our knowledge. These are things we are aware of and understand. Some examples include the consultant Scope of Services, project scope, organizational policies/procedures/processes, design standards, past experience, lessons learned, and skilled expertise. If not handled correctly, some KKs may become risks. KKs should be addressed in Project Planning.

Unknown-Knowns (UK) are answers to questions we don't realize we already know. These are things we understand, but of which we are not aware. UKs should be explored and leveraged by networking intentional knowledge management and sharing. In larger organizations, UKs should also be addressed with focused data mining efforts. The PM is responsible to facilitate team knowledge sharing across discipline lines. For example, there may be no need for designers to struggle with constructability SMEs to reaccess equipment access to minimize disturbance for a stream restoration project if there can easily be surplus wetland disturbance under the permit due to design changes elsewhere on the plan.

Known-Unknowns (KU) are identified risks. These are things we are aware of, but don't understand. Some examples may include assumptions, unstable requirements, knowledge gaps, uncertainties, risk confidence levels, changing requirements, and inexperienced staff. KUs should be identified and actively managed and monitored via Risk Management. Adam Grant said, "*If knowledge is power, knowing what we don't know is wisdom.*"

Unknown-Unknowns (UU) are emergent risks, meaning they can only be recognized after they have occurred. These are things we are not aware of, and don't understand. These are things that jump up and bite you. Independent of

Knowledge Quadrant

Known-Knowns (KK) Things we know we know	Known-Unknowns (KU) Things we know we don't know
Unknown-Knowns (UK) Things we don't know we know	Unknown-Unknowns (UU) Things we don't know we don't know

Figure 6.3 Knowledge quadrant.

the consequences, UUs can produce two undesirable results. Their unexpected nature can cause stress and confusion while simultaneously lowering others' confidence in your ability to lead. Whether reasonable or not, it is often assumed the PM should have anticipated the risk. Should an issue emerge from the UU quadrant, it should be promptly addressed with developed project team initiative, creativity, and resilience. UUs can be accommodated for in Design Margins, Schedule Float, Contingencies, and Management Reserves.

6.3.2 Uncertainty and Risk

Many PMs mistakenly equate uncertainty and risk, which can cause unnecessary confusion. Uncertainty is not assuredly knowing the result of, or recognizing the existence of, multiple possible outcomes. Risks can positively (opportunities) or negatively (threats) impact your project. Some uncertain outcomes may be risks, some may not. You can have uncertainty without risk; but you cannot have risk without uncertainty. Uncertainty is not an unknown risk. Just because you don't know what will happen, does not automatically mean it is a risk.

Risks can be identified, monitored, and managed. Uncertainty is not controllable, and cannot be managed. Risks can be measured and quantified, while uncertainties cannot. Uncertainties should be considered and monitored to identify if and when they warrant being elevated to a risk.

6.3.3 Risk Register

Throughout the project, as risks are identified, they should be logged on the Risk Register. This should be a living document that is regularly reviewed and updated. The PM is responsible for the creation and maintenance of this document that should be consistently shared and reviewed with the project team. Pending the size and complexity of the project, this is typically monthly or quarterly, and at each project phase gate.

A Risk Register, or Risk Log, is a PM tool to log and track project risks, the potential impact and likelihood of occurrence, and the risk response strategy. This document will enable the PM to better determine and manage risks and project contingencies, assist with effectively communicating and managing team and stakeholder expectations, and better position your project for success. On larger or more complex projects, a risk database may be employed.

Risk Registers are typically formatted as a spreadsheet, with an identified risk on each row filling columns with important risk specifics. Figure 6.4 provides some guidance on specifics that could be captured for each identified risk.

A Risk Breakdown Structure can be a helpful visual tool to capture and organize risks. It can take many different formats given the project specifics. Most commonly it may look like an organizational chart or a spreadsheet with nested columns or rows. Regardless of this format, the intention is the same, to arrange the risk in hierarchical categories. This can bring clarity and focus when many risks are identified, and better enable proactive responses by increasing organization.

Suggested Columns	Description
Risk ID	Unique ID (number) for each risk – this is for internal tracking purposes
Name	Name to which risk will be referred
Date	Date risk was identified
Type	Is this a negative (threat) or positive (opportunity) risk?
Description	Detailed, clear, and thorough description of risk and what it means
Triple Constraint	Will this risk potentially impact Scope, Budget, and/or Schedule?
Trigger	The risk trigger should be described and documented
Responses	Document planned or implemented risk responses
Comments	Document the iterative process of risk management as the risks evolve
Risk Owner	Who is responsible to monitor risk and implement risk response?
Status	Is risk being monitored, in risk response, or closed?
Probability	The probability of occurrence
Impact	The project impact if the risk is realized
Risk Score	The numeric product of Probability and Impact from the Risk Matrix
Risk Exposure	Schedule risk exposure
Earned Value Management (EMV)	Budget risk exposure

Figure 6.4 Common risk register fields.

6.4 Risk Analysis

"Never test the depth of water with both feet."

– Warren Buffet

Risk Identification must be completed before the PM and team can advance to Risk Analysis. Risk Analysis is the evaluation and prioritization of identified risks. This is a two-step process. The risks must be qualitatively assessed, and then quantitatively evaluated. Qualitative Risk Assessment is a project management technique that subjectively analyzes the probability and impact of a risk occurrence. The risks are categorized on a probability and impact matrix, and those deemed significant may undergo a quantitative risk analysis. Quantitative Risk Analysis is the mathematical analysis of the probability and impact of a risk occurrence. This analysis should be conducted after qualitative analysis so it can assess risks that have been identified as significant.

6.4.1 Qualitative Risk Assessment

The purpose of Qualitative Risk Assessment is to prioritize the identified risks so you can strategically move forward. This subjective evaluation should be made of each risk identified on the Risk Register. Most qualitative risk evaluations utilize a Risk Matrix that is detailed in applicable organizational procedures, such as the one in Figure 6.5.

A Risk Matrix, or Probability and Impact Matrix, is a visual framework for categorizing risks based on the probability and impact of occurrence. Risk matrices may have increased granularity, but they should not be less than a 3 × 3 grid. The probability and impact ratings should be logged on the Risk Register, and updated as needed. If not already defined by your organization, below are general guidelines for delineated categories:

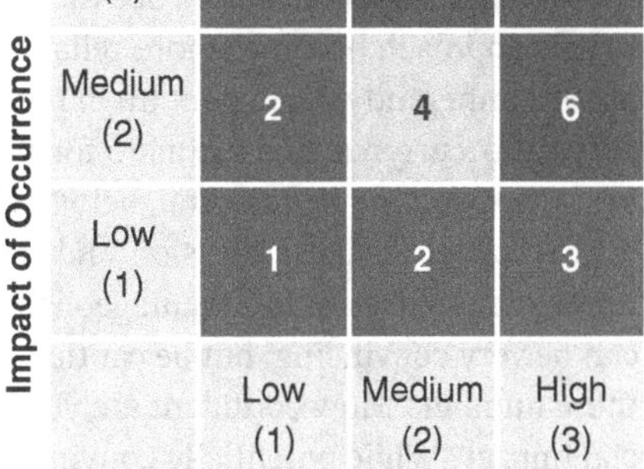

Figure 6.5 Risk assessment matrix.

Probability of Occurrence

- High = Risk is likely to happen regardless of your actions.
- Medium = You can direct or influence the likelihood of realizing the risk.
- Low = Risk is likely not to happen regardless of your actions.

Impact of Occurrence

- High = Catastrophic
- Medium = Critical
- Low = Marginal

Risk Score is the numeric product of the probability multiplied by the impact. This can provide clarity on prioritizing the urgency of the risk, and may also dictate organizational reporting requirements. This should be documented in the Risk Register that was initiated during Risk Identification. Risks should then be quantitatively analyzed.

There is a species of swans in Australia that happen to be black in color. For those who did not know this, the idea of a black swan seemed impossible, until it wasn't. A Black Swan event is a realized risk with a low probability and high impact. Nassim Nicholas Taleb described this phenomenon in detail in his book entitled, *The Black Swan: The Impact of the Highly Improbable*. These outlier results are beyond normal expectations. While rare, these Black Swan events are challenging to risk management. These can be difficult to anticipate, but require prompt attention when realized.

6.4.2 Quantitative Risk Analysis

A prediction is a best guess. A forecast is a prediction with an associated probability, or confidence level. Quantitative Risk Analysis allows you to integrate schedule and budget forecasts into your project management. This is especially critical during the Scoping process so that you correctly establish your Scoping schedule and budget baselines. The Risk Exposure is calculated by multiplying the risk probability of occurrence by its potential impact to budget or schedule. Adding Risk Exposures allows you to better plan and manage your project. The resulting schedule impact must be carefully evaluated as risks can impact the critical path in very different ways. The Expected Monetary Value (EMV) describes the project's financial Risk Exposure. The result, which should be in today's dollars, will be a positive number for negative risks (threats) and a negative number for positive risks (opportunities). See the example table in Figure 6.6.

This approach becomes more reliable when more risks are included. EMV calculations are also inherently dependent upon the accuracy and associated assumptions of the probability of occurrence and estimate for financial impact. Accordingly, this approach works best when you are risk neutral, being neither risk averse nor risk seeking.

Most people make a decision and then search for data to rationalize it. For business-minded audiences, it is advantageous to present the risks in financial terms. Precise data can be very convincing, but be cautious of its accuracy. There are other risks imbedded in these numbers. How confident are you of its likelihood? How certain are you of the financial impact? While potentially convincing, EVM calculations can be difficult to defend.

Pending the complexity of the project or specifics of certain risk(s), you may choose to utilize Probabilistic Analysis tools within your quantitative analysis. The most common of these are three-point estimates and the PERT (Program Evaluation and Review Technique) estimate. These both rely on estimate ranges. For each risk you prepare three different numbers:

- Pessimistic estimate (P) – this is the maximum estimate of impact
- Optimistic estimate (O) – this is the minimum estimate of impact
- Most likely (M) – this is the most likely estimate of impact

The Three-Point estimate adds P + M + O and divides by three. The PERT estimate is a weighted average using these same three estimates. You add P + O + (4 times M), all divided by six. These simple probabilistic estimates can be effective ways to increase accuracy of your EMV calculations. For more advanced requirements, a Monte Carlo statistical simulation can be used that analyzes all possible outcomes to create an even more accurate probabilistic estimate.

The cycle of risk identification, risk analysis, risk response, and risk monitoring should be repeated throughout the project. The PM is responsible to drive this process and document all relevant information on the Risk Register.

While you need to be vigilant, also be cognizant that it is easy to reach the point of diminishing returns; meaning beyond a certain point, efforts to accurately capture all possible aspects of a specific risk may not be worth your time. Organizational guidance on Risk Exposure can provide clarity on prioritization thresholds.

Risk Name	Risk Type	Probability		Impact		EMV
Risk A	Threat	75%	×	$40,000	=	$30,000
Risk B	Threat	50%	×	$10,000	=	$5,000
Risk C	Opportunity	30%	×	($50,000)	=	($15,000)
Risk D	Threat	10%	×	$200,000	=	$20,000
Total EMV						**$40,000**

Figure 6.6 Quantitative risk analysis example.

6.5 Risk Responses

"Risk is like fire: If controlled it will help you; if uncontrolled it will rise up and destroy you."
– Theodore Roosevelt

If you don't proactively manage risks, you are deciding to allow risks to manage you. Think of the PM as being the coach of a football team preparing for a game. You scout the other team and evaluate their strengths, weaknesses, and tendencies. You make your offensive and defensive game plans and contingency plans, and think about possible game scenarios. If you don't prepare, it is impossible to be proactive. You will spend the limited and valuable game time trying to figure out how to respond in real time, and likely fall further and further behind. You want to be the coach that controls the game. You want to win. Successful coaches are remarkable risk managers. They are constantly planning, identifying, analyzing, responding, and monitoring risks. They make assumptions, make decisions, and strive to be structured yet agile as they evolve to stay ahead of changing conditions in a dynamic environment.

You will either respond to risks proactively or reactively. The choice is yours. Incompetence, busyness, naiveté, laziness, or procrastination are not valid reasons to ignore risks. Successful PMs choose to be proactive, which positions them to better control their project. PMs who feel constantly behind and surprised by events that seem to overtake and overwhelm them are almost all passive risk managers. Being reactive severely limits the number and effectiveness of your available responses, and eliminates virtually all possibilities of exploiting positive risks (opportunities).

Risk Response Planning is conducted after risk analyses to determine the best response for each identified and prioritized risk. While the PM is ultimately responsible to manage all risks in order to successfully develop and deliver the project, the chosen risk responses should be consistent with the organization's risk tolerance and appetite. Risk Tolerance is the level of uncertainty an organization is willing to accept for each specific risk. This can be quantified in an acceptable variance in performance measures. Risk Appetite is the total amount and types of risk an organization is willing to accept in order to pursue its objectives. These represent the level of risk you are comfortable assuming. It doesn't mean you like it, but you can tolerate it. Most organizations have historic precedents of risk tolerance that reflect their culture and objectives. Your personal risk tolerance may be different than that of your organization, but be cautious of straying too far from accepted thresholds without appropriate concurrence. There may be no faster way to run to the far end of the limb by yourself than mishandling risks.

The PM is responsible to ensure the Risk Register is maintained, and its contents available to the team. This includes ensuring each assigned Risk Owner is updated on risks for which they are responsible, including full knowledge of their risk triggers. Should a trigger be passed, the PM is responsible to ensure the identified risk response is implemented by the Risk Owner.

6.5.1 Negative Risks (Threats)

Once a negative risk is identified and analysis has prioritized it for action, a risk response should then be identified and implemented should the associated risk trigger be passed.

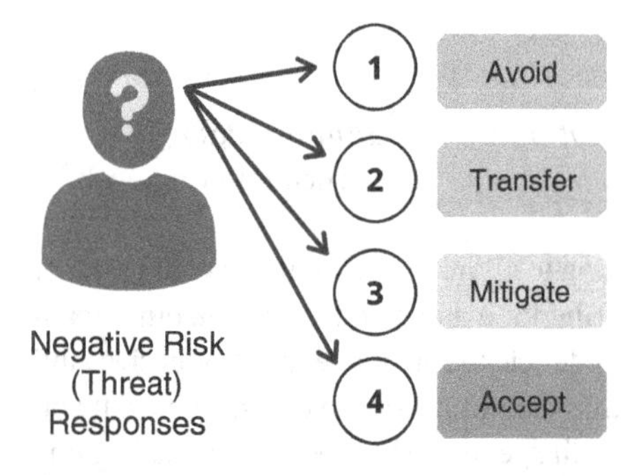

Figure 6.7 Negative risk (threat) responses.

For any negative risk (threat), there are four possible responses as shown in Figure 6.7: Avoid, Transfer, Mitigate, and Accept.

Avoiding the risk is changing the project so that the issue is no longer a risk. You exercise change controls or modify parameters so as to remove the risk from the equation. Examples may include avoiding a RW risk by modifying the design so you no longer need to access the property, or finding an innovative stream-bank reinforcement solution that will allow installation within the spatial constraints on a stream restoration. Risks that are identified early can often be avoided by clarifying requirements and operational parameters. Avoiding risks can involve two different actions. First, you can take action to eliminate the risk trigger, or cause of the risk. Second, you can administer or execute the project in a way that avoids the risk but still satisfies the project's objectives. Not all risks can be avoided. In some circumstances, avoidance may jeopardize the project's budget or schedule. However, generally, avoidance should be the first considered risk response strategy.

Transferring the risk means shifting the risk to a third party who is better positioned to mitigate or withstand it. This does not eliminate the risk, but rather transfers ownership and management of the risk to another. Risk transference is especially common with financial risk exposure. Organizations' procurement, risk management, and legal departments do this all the time. In public sector transportation projects, the system owner/operator is well positioned to strategically structure contracts, task orders, and change orders so your organization bears appropriate and reasonable risks, while transferring the potential liability of specific risks to a third party. While there is often a risk premium for this transference, it can still be a cost-effective way to manage certain risks.

Mitigating the risk means taking action to decrease the probability or impact of the threat, should it be realized. Risk mitigation can have intended and unintended consequences. In chess, you must think a few moves ahead or your initial move to mitigate a threat may put you in an even more precarious position. Identifying, assessing, and managing secondary risks should be a part of the initial risk management consideration. In risks, an ounce of prevention is often worth a pound of cure. Proactive, preventative responses can be dual-focused: first, reducing the probability of the risk trigger, the cause of the risk; second, reducing the budget or schedule impact, should the risk be realized.

Accepting the risk means taking no intentional actions to avoid, transfer, or mitigate the risk. Organizations and PMs that do not practice proactive and effective risk management often accept most all risks by design or default. Generally, this is not a good approach. However, there may be good reasons to accept a risk. There may be instances where no other response strategies are practical or cost effective. Acceptance may also be appropriate where the impacts are minimal, the probability of occurrence is low, or the risk is completely uncontrolled. If the probability of occurrence is high, you may include the risk as a project assumption. When you accept a risk, you have a choice. You can Passively Accept the risk, meaning you take no action other than documenting and providing awareness of the decision. Or you can Actively Accept the risk, meaning you take action to prepare for the consequences. This often takes the form of budget or schedule risk-specific contingencies.

Exceptional PMs understand when to escalate risks. Risk Escalation is elevating the risk to a higher authority for management of that risk. Highly sensitive or political risks may need to be escalated to gain concurrence or direction on how best to handle that risk. Elevating a risk can also be an effective way to manage project expectations while securing concurrence on leadership's preferred risk response. Especially with escalated risks, the PM must be diligent in monitoring the risks and implementing appropriate project change management measures if needed.

Implementing risk responses can often be difficult and unpopular, but it is the necessary and right thing to do. It may be beneficial to remind yourself and others that the purpose of risk responses is not to cast blame, or find an acceptable scapegoat should a negative risk be realized, but rather to proactively manage risk in such a way that the project may continue to advance within its established constraints of budget, scope, and schedule.

The selected risk response should be based upon the specifics associated with each risk. Figure 6.8 can be used as a guide as to when the four risk responses may be most appropriate, based upon the qualitative risk analysis.

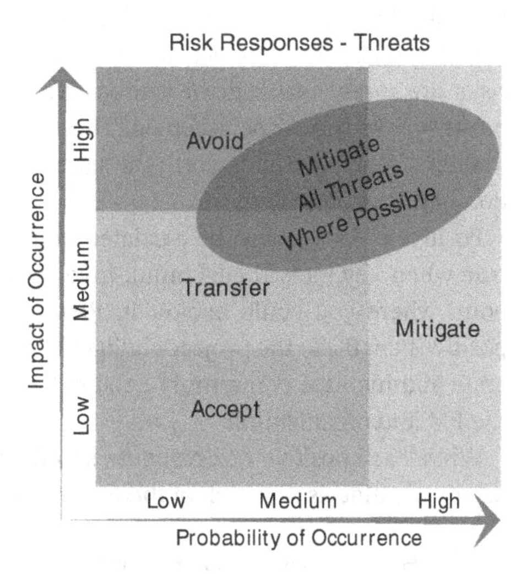

Figure 6.8 Negative (threat) risk response guidance.

6.5.2 Positive Risks (Opportunities)

Winston Churchill said, *"The pessimist sees difficulty in every opportunity. The optimist sees the opportunity in every difficulty."* Positive risks are present on most projects. However, opportunities are rarely realized if left to their own devices. They must be proactively identified and then actively pursued. For any positive risk, there are four possible responses as shown in Figure 6.9: Exploit, Share, Enhance, and Accept.

Exploiting the risk is taking action to ensure the positive risk occurs. This is the opposite of avoidance. Exploiting is deliberately modifying the project or approach so the opportunity is realized. Similar to Avoidance for negative risks, Exploitation is the most aggressive of the positive risk response strategies. This response is typically reserved for those positive risks with high probabilities of occurrence and anticipated impacts, or benefits.

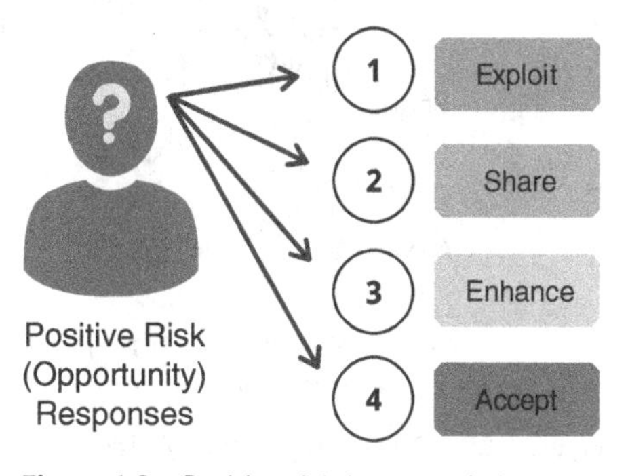

Figure 6.9 Positive risk (opportunity) responses.

Sharing the risk means partnering with a third party who is better positioned to realize the opportunity. You need to find those who are best able to capture the benefits for the project. This may be accomplished by pushing the risk trigger, increasing the probability of occurrence, and increasing the positive impact to the project. As Transference often distributes the negative risk liability, so Sharing often distributes the realized benefits. Examples may include shared cost savings in design innovations or construction means and methods.

Enhancing the risk is to take intentional action to increase the probability of occurrence and or the impact of the risk. Resulting modifications in design or approach may increase the likelihood of capturing identified benefits, or increase the scale of the beneficial impacts. One common tactic is to focus on facilitating the risk triggers that enable the risk to be realized. If an organization can enhance the probability of occurrence to 100%, then they have exploited the positive risk.

Accepting the risk means you taking no intentional actions to exploit, share, or enhance the risk. If you do nothing, by default you are accepting all positive risks, even if they are not identified. You may get lucky every now and again; but generally, accepted positive risks are rarely realized. In a proactive risk management approach, only insignificant positive risks should be accepted. These should be documented and tracked, but the bottom line is the juice isn't worth the squeeze to commit resources to actively pursue realization of this opportunity.

Positive risks can also be escalated in an effort to best manage them. This is especially true when key individuals' input, involvement, or influence may enhance the risk to the point where you could exploit it. Realizing opportunities can do more than providing positive benefits to the project's budget, scope, and schedule; it can boost team morale, generate and increase community goodwill, and strengthen the reputation and credibility of the PM and organization.

While each positive risk response should be based upon the analysis, Figure 6.10 can be used as a guide as to which of the four response strategies to select for each risk.

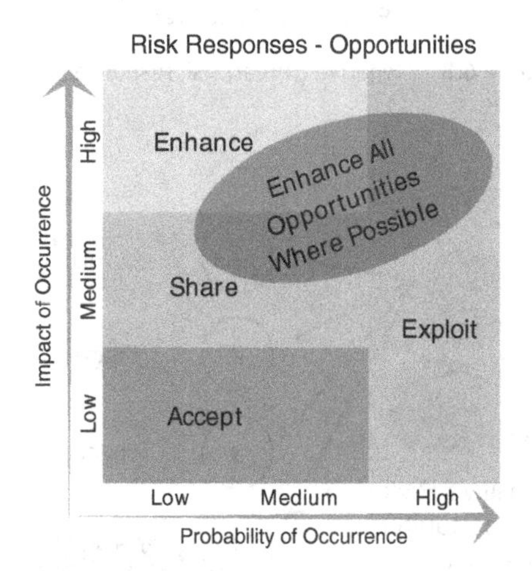

Figure 6.10 Positive (opportunity) risk response guidance.

6.5.3 Change Control

Successful risk response implementation to a negative or positive risk is almost always accomplished through change control. This is because most risk response strategies directly or indirectly impact the project's budget, scope, or schedule. Change management and risk management are both the responsibility of the PM.

Mitigating risks can create intended and unintended consequences. Identifying, assessing, and managing secondary risks should be a part of the initial risk mitigation consideration. Regardless of your risk response, some residual risk may remain.

6.5.4 Contingency Plans

Contingency Plans are similar to Risk Responses in that they strive to reduce the threat impacts and increase opportunity benefits. The difference is in the timing. Risk Responses are planned and implemented while the risk is not yet realized. Contingency Plans are planned while the threats and opportunities are still risks, but implemented only after the event has occurred. A contingency plan outlines the intentional steps to be taken after the risk is realized. Risk responses focus on prevention of loss, while contingency plans focus on surviving the impacts once they are occurring.

Contingency plans should be created for significant, identified risks. As with any good plan, it should answer the basic questions. Who will be involved? What will they do? When should they do it? Where will this happen? How will the plan be executed? Monitoring and reporting expectations should also be included. The plans should be brainstormed and vetted with critical team members. As with many aspects of project management, engineering judgment should be exercised to appropriately tailor and scale the effort to the significance of the risk.

Common contingency plan challenges seem to center on the fact that it may never be needed. As such, contingency planning is often viewed as a lower priority than other urgent matters of the day. Team members may question the necessity of the effort, or become unreasonably attached to their preferred approach. Similarly, securing leadership or stakeholder buy-in on hypothetical realities can present its own challenges while unintentionally influencing project expectations. Regardless, contingency plans are an integral part to any successful risk management program. Everyone needs a Plan B.

6.6 Monitoring Risks

> *"There is no doubt that Formula 1 has the best risk management of any sport and any industry in the world."*
>
> – Jackie Stewart

Risks on a public transportation project often develop over time, are challenging to identify, are difficult to quantify, and are even more difficult to control. Public Sector projects frequently present very high levels of largely undefined, unique, and occasionally unprecedented risks. They are a high-stakes game characterized by significant and irreversible commitments that aim to satisfy a wide range of success criterion that are initially set and then tend to evolve from a diverse group of stakeholders. To further complicate matters, public sector project solutions may not be selected from an optimized list of alternatives. More commonly, project boundaries and critical design features are often dictated and are not easily changed. The sanitary sewer extension will start at point A and end at point B at predetermined elevations, following the natural creek line between the two drainage basins. The roadway improvement will begin at this intersection and end at that intersection complying with applicable design speed geometry in-between. The bridge replacement will begin at this bank and end at that bank and tie into existing approach grades. Because these designs are not selected or scoped to intentionally minimize risk resolution, the PM must play the hand he is dealt. He must actively seek to proactively identify and evaluate potential risks, diligently prioritize each based upon their likelihood of occurrence and possible impact to the project, and successfully manage and respond to these risks until the project reaches completion.

Following Risk Identification, Analysis, and Response, the fourth essential process in the Risk Management Project Workflow is Risk Monitoring. This is a continuous effort that is ongoing until project completion. Risk Planning is essential, but the reality is that risks will change and evolve. It is not uncommon for a PM to find themselves steering the project ship through a very different sea than that upon which they set sail. Throughout the duration of a public transportation project, risks will continue to be added and dropped from the risk register. Risk monitoring is an essential component that empowers the PM to be positioned to proactively respond to threats while seizing opportunities. The knowledge it produces grants team flexibility and enables innovation.

The Project Management Institute's (PMI) *Project Management Body of Knowledge (PMBOK)* defines Monitoring Risks as, "*The process of monitoring the implementation of agreed-upon risk response plans, tracking identified risks, identifying and analyzing new risks, and evaluating risk process effectiveness throughout the project.*" Inputs for this process are primarily the risk register, change management documentation, and utilized risk management metrics.

At least at every major design milestone (30%, 50%, 60%, 90%, and 100% design meetings), the Risk Register should be updated, and both Qualitative and Quantitative Risk Analysis should be conducted. Risk Monitoring should be a rolling process throughout the life of the project. However, Risk monitoring is a task that often gets buried under the proverbial piles of urgent papers. Some of these simple steps may make it more manageable to achieve. Be fanatical about collecting status information on risks. Be equally fanatical that it only be the status you need, not extraneous information. Document all essential information. As Clifford Stoll said, "*If you didn't write it down, it didn't happen.*" Monitor the status and trends as a procedural matter of routine. Make the effort meaningful by promptly and intentionally acting on gained insights.

Think of risk monitoring as exercising. A haphazard and undisciplined approach will yield marginal results, at best; whereas, a commitment to an intentional workout plan will produce meaningful results. Similarly, all risks may not need to be monitored with the same frequency, or with the same intensity. Significant or volatile threats and opportunities may warrant special attention.

6.6.1 Risk Triggers

Perhaps just as important as identifying potential risks, is carefully and accurately identifying the risk triggers. A Risk Trigger is an indicator that a risk has occurred, or is about to occur. These triggers are symptoms of an impending threat or opportunity, serve as precursors to preempt its occurrence or minimize its impact, and are critical to a timely response that can most effectively mitigate or enhance the risk. Risk Triggers should be carefully chosen to accurately reflect when a risk crosses the right risk threshold. The Risk Threshold is the level at which the likelihood or impact of a risk becomes significant enough that th PM or Risk Manager deems a risk response necessary.

Risk triggers are tricky business. While some are obvious and straightforward, many are less so. Some risks may have multiple triggers that are in very different forms. Typically, the more advance warning the trigger provides, the less reliable it is. For instance, you may be able to identify a worrisome trend, but the longer the trendline of your projection, the more assumptions you must make to substantiate your conclusion.

It is important that you strive to identify triggers that accurately warn of the associated risk. If you are having difficulty identifying the trigger, speak to those most likely to cause the risk or feel its impact. Ask them how they would know if the problem occurred; then work backward. Many soft, or qualitative, triggers can be effective at warning of a problem without clearly identifying the core issues. For instance, you may receive forewarning that a Home Owners Association is planning to attend an upcoming public hearing in mass to protest a roadway expansion due to traffic concerns. While stakeholder discontent may be a risk, that doesn't truly prepare you for the possible breadth of core issues that may arise. Many hard, or quantitative, triggers can provide precise data, but often not soon enough to take corrective actions to avoid the issue. For instance, a PM might be reviewing the monthly consultant bill and notice one work activity has spent 80% of its budget while only 40% of the work is complete. It is better to discover this situation before the entire budget line item is spent, but it may be too late to take corrective actions that will not end up impacting other aspects of the project. Some triggers are secondary in nature, meaning you observe the secondary risk trigger after the original risk is realized.

Many useful triggers are not able to be predicted and are completely unexpected. For example, you might overhear a conversation in the elevator that a contractor on one of your jobs just won a big project with a short fuse that will demand most of their resources once underway. Or you are at your child's PTA meeting and overhear some grumblings from other parents whose neighborhoods are adjacent to an ongoing construction project. In both of these examples, you can proactively act on this knowledge before the consequences of that risk continue to escalate. As PM, you never know from where you will pick up useful bits of information. Consequently, it is important to keep your ear to the ground and be ever vigilant and observant.

It is imperative that you associate a Risk Owner with each risk on the Risk Register. The Risk Register should also list the Risk Trigger. The Risk Owner should be carefully chosen as they are the ones best positioned to monitor the risk trigger and implement any risk response. If the PM is not the risk owner, they are responsible to ensure the Risk Owner is aware of the risk trigger and monitor updates.

The goal of triggers is to keep your finger on the pulse of the correct spheres of influence so you have advance warning and can more effectively respond to risks if they are partially or fully realized. Experience teaches the best triggers are personal warnings that are officially, or unofficially, provided by those with whom you have an established relationship based on trust and mutual respect. An afternoon of clever analysis of all the project metrics cannot replace a single phone call from a colleague in a permitting organization giving you a heads up that you may be receiving a letter later that week that says thus and so. Likewise, a casual conversation in the hall from a coworker telling you this issue is going to be a problem is worth more than any weekly progress report. It is invaluable that you actively establish a network of such connections within your own organization, within your consultants, and within the many federal, state, and local entities with which you must interact to build your project. This takes time and deliberate effort, but is an investment that can pay back immeasurable benefits. As with everything you do, be ethical, be honest, be smart, and maintain appropriate confidentiality and discretion.

6.6.2 Monitoring and Controlling Risks

Risk Monitoring includes the two distinct PM efforts of Monitoring and Controlling the risks. Monitoring is the data collection. Controlling is doing something with the data that is collected. Accomplishing these two initiatives encompasses a number of separate and important tasks. New risks should be identified and processed through risk identification, analysis, and response. Existing risks should be tracked. This includes updating the status of the risk and the qualitative and quantitative evaluations. Risk triggers should be examined to determine if any established thresholds have been breached. Risk metrics should be evaluated, and appropriate communication should be made both upward, downward, laterally, and externally with all who need to know, as defined in the project's communication plan.

Pending the length and complexity of the project, risk audits may be appropriate. A Risk Audit is a dedicated effort to evaluate the Risk Management process. This can be a separate meeting or incorporated into monthly status meetings. The purpose is to ensure the risk management processes are being followed as per the risk management plan. The emphasis is not to evaluate individual risks or responses (e.g., the product of the process), but rather to evaluate if the process is working and adding value to the project. If not, then appropriate adjustments should be made.

6.6.3 Risk Metrics

Like all performance measures, risk metrics should answer a fundamental organizational question, and their results should add value to the project, program, or portfolio by providing insight that increases efficiency and productivity. They should not be burdensome to staff, and should strive to reflect, rather than influence, behavior. Ideally, information from a metric should lead to improved performance, not just improved metrics.

Organizations may track a wide variety of risk metrics, some of which are summarized in Figure 6.11. These can be analyzed in combination to provide additional insights and trendlines for projects, programs, and portfolios.

The analysis results should be used to evaluate the health of the project, program, or portfolio. Trend lines can be identified. Lessons learned should be updated and shared throughout the organization. At the very least, Risk Monitoring should allow you to answer the simple question...is the project better positioned now than before to be developed and delivered within the established budget, scope, and schedule.

Results can be distributed in a report format and documented in the Project Management Plan. More advanced organizations may have a risk dashboard they maintain to summarize key risk performance metrics.

A Risk Burndown Chart is another extremely useful tool to help PMs monitor and manage risks. This shows trendlines of risks over time. Each risk is a separate line on a graph where the y-axis is time (typically weeks, sprints, months, or quarters), and the x-axis is either the risk exposure (probability of occurrence times anticipated impact in units of time or money)

Risk Metric	Key Question Answered
Number of Identified Threats	How many threats were identified?
Number of Threats realized	How thorough, accurate, and effective is risk analysis and response?
Negative Schedule Impacts	What are the negative schedule impacts to the project from identified threats?
Negative Budget Impacts	What are the negative budget impacts to the project from the identified threats?
Predicted risk severity compared to realized impact	How accurate were the estimated risk impacts to the project compared to what was realized?
Number of risks that occurred more than once	How effective are your risk responses?
Number of unidentified and unanticipated risks	How many risks did the risk management plan and framework fail to identify?
Speed and effectiveness of risk responses	How nimble and responsive is your project team?
Number of closed threats that were never realized	Are we correctly analyzing and prioritizing the right risks?
Number of Identified Opportunities	How many opportunities were identified?
Number of Opportunities realized	How thorough, accurate, and effective is risk analysis and response?
Positive Schedule Impacts	What are the positive schedule impacts to the project from identified opportunities?
Positive Budget Impacts	What are the positive budget impacts to the project from identified opportunities?
Number of closed opportunities that were never realized	How effective was the team in analyzing and prioritizing opportunities?
Risk Management Cost (combining threats, opportunities, and Admin costs)	What is your ROI for risk management activities?

Figure 6.11 Risk metrics and questions they answer.

or probability of occurrence (in units of percentage). This tool can be especially effective in communicating with the team. A Risk Burndown Chart, as shown in Figure 6.12, can be a powerful visual to focus responses and mobilize momentum toward solutions. This can also provide a historic record of the effectiveness of the team's risk response. As threats are identified and risk responses are implemented, the risk exposure and probability of occurrence should diminish over time.

6.6.4 Risk Communication

Effective risk communication is critical to maintaining team efficiency and productivity while guiding the project to success within the established budget, scope, and schedule. This is true for team members, stakeholders, and leadership. It reduces confusion and brings clarity to project expectations.

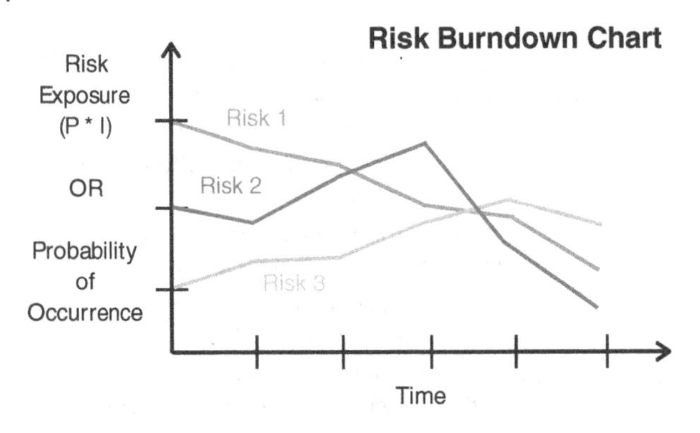

Figure 6.12 Risk burndown chart.

Less confident or secure PMs are often reluctant to pass potentially bad news up the chain of command for a number of reasons. Perhaps they don't want leadership to feel compelled to act and unintentionally make things worse. Or perhaps they don't want to damage the impression that all is under control, gain the reputation of one who cries wolf, or reinforce the impression that they cannot handle the responsibility. Perhaps they feel overworked or overwhelmed already and don't want to invite additional questions and requests for status reports from leadership. Perhaps they don't know how to manage the risk and are choosing denial, desperately hoping it goes away. Some may even have a hero-complex and consciously or subconsciously want the risk to be realized so they can save the day.

While these may be understandable or defendable, at least in the PM's own mind, it is their responsibility to keep leadership informed of project progress, which includes risks. Remember that perhaps the greatest risk to a PM is for leadership to be blindsided by a complication that could have been avoided. Exceptional PMs have already spent time cultivating trusted working relationships with leadership that are based upon consistent performance and reliable counsels, so that notifying them of potential risks does not initiate an inquisition. There are benefits to notifying leadership of risks in a timely manner. If the risk is realized, leadership may need to know, and the sooner the better. Regardless of the resulting impacts, foreknowledge can help protect them, and you, from the fallout. Leadership also may approach the issue from a wider perspective, and perhaps offer insight on how to better manage the risk. Exceptional PMs understand that more than knowing all the right answers, knowing the right questions to ask makes them valuable. And most importantly, it's the right thing to do. Part of your job is to present good, and bad, news. Delaying the inevitable almost always makes it worse.

Having said this, wise PMs understand how to exercise judgment and discretion in what risk information is shared, with whom, and when. Pending your organization's culture and expectations, risks that have a high anticipated impact on the project should be discussed with leadership. As with many aspects of project management, there is no substitute for wisdom and experience.

7

Managing Resources

"The strength of the team is each individual member. The strength of each member is the team."

– Phil Jackson

7.1 Developing and Managing a Team

"It is amazing what you can accomplish if you do not care who gets the credit."

– Harry S. Truman

Roughly translated, the second law of thermodynamics states that a system left to its own devices will naturally devolve into increasing states of disorder. This *Entropy* is self-evident in public transportation project teams.

Culture matters. The culture, tone, and productivity of a team is the PM's responsibility. Yes, the corporate culture is incredibly important, but a PM can directly impact the team's culture. Harmonious, high functioning teams do not happen by chance. They take time, but this is a critical and powerful investment. The PM must be intentional to this end, strategically fostering trust, acceptance, and commitment to shared common values and goals.

Developing and managing a team are two very different functions. Developing a team means establishing and improving the competencies, team interactions, and overall team environment to enhance project performance. This is concentrating on the beehive, not the bees. Managing a team means focusing on the individuals. This involves providing feedback, resolving issues, and managing team changes to optimize individual team member's performance. Two very different foci that require two very different skill sets.

Successful team development requires intentional planning and consistent follow-through. First, the PM must accept that they run the team. While there are likely no project schedule milestones related to team culture, it is the undercurrent that can carry your objectives toward shore or pull you back out to sea in spite of your best efforts to the contrary. Push decisions to the lowest possible level. Be alert and aware of the pace and pulse of progress and relationships. Emphasize collaborative decision making. Be humble. Actively listen. Be kind. Praise in public; criticize in private. Lead with empathy. Truly care about your people. Become vested in them as individuals and their individual success,

Transportation Project Management, First Edition. Rob Tieman.
© 2023 John Wiley & Sons Ltd. Published 2023 by John Wiley & Sons Ltd.

which will be a part of, and extend beyond, this project. Develop a safe, supportive space that builds trust, empowers innovation, and energizes greatness. The culture is the framework and momentum that positions your team for success.

William Wrigley, Jr., said, *"When two people in business always agree, one of them is unnecessary."* Conflict is inevitable. In a vibrant team, the absence of conflict is a danger sign. As Mark Sanborn said, *"In teamwork, silence isn't golden, it's deadly."* When conflict occurs, it is critical that you manage conflict in constructive manner, meaning people arguing about the problem, as opposed to a destructive manner, meaning people arguing about people. Team members should resolve their own conflict, with the PM stepping in only when necessary. The team's culture you develop can establish a structure that promotes healthy debate and constructive conflict resolution.

Differences of opinion can challenge assumptions and spark innovation, provided the conflict does not turn contentious. There are five basic strategies of conflict resolution. *Withdrawing* or *Avoiding* is retreating from conflict. While cooling off may have some temporary benefit, eventually you still need to find a solution. *Smoothing* or *Accommodating* emphasizes areas of agreement while ignoring the true problem. This is another lose-lose situation. *Forcing* or *Directing* pushes one's viewpoint at the expense of others. *Compromising* and *Reconciling* is where both parties give something up. While all of these may be appropriate given a specific situation, the preferred method is *Collaborative problem-solving*, which examines alternatives and involves cooperative give and take.

True leadership extends beyond positional power that is based on one's formal position within the organization or team. Pending your organizational structure, some or all of the team members may not report directly to the PM. Especially in these circumstances, it is imperative you develop other forms of power. Expert power is based upon your skill or knowledge. Referent power relies on respect or admiration others may have for you based on your personality, actions, and decisions. Informational power controls the gathering or flow of information. Persuasive power is the ability to convince others to embrace a preferred decision or course of action. Personal power is sourced form one's charisma. Ingratiating power relies on the application of flattery. Reward power is the ability to give praise or bonuses. Coercive power is the ability to discipline. Pressure-based power is the ability to limit another's freedom, options, or movement. Find the combination that fits your style and then refine it.

Assume positive intent. This is good practice for most all personal and professional relationships, but it also holds true for teams. Assume positive intent. This can be a challenge when personalities clash or messages may seem overtly confrontational, but it is a benefit to the entire team when the leader consistently and predictably assumes positive intent in both the message and the messenger. Sometimes, this trust can be compromised and you may desire to adjust your approach for specific situations or individuals. And be aware that while many may have their own angle or agenda, that is not inherently inconsistent with assuming positive intent.

For anyone who played team sports, you know you don't have to like all of your teammates to win. It can help, but it is not required to achieve success. By contrast, each team member needs to feel included and committed to each other and the common goal. The *"you mess with one, you've got us all"* mentality is extremely powerful and effective. It can be a challenge to find this delicate balance of collective and individual accountability, but when you do it is a powerful force for progress.

Some teams create team charters. This is a document that is created by the entire team, and is signed off by each team member. It contains the ground rules, describing team values, agreements, and operational guidelines. It may also include collective expectations for communications, meetings, decision making, conflict resolution, and anything else important to the group members. Even if not formalized, this exercise can be a powerful team-building experience, particularly when a new group is forming or you are onboarding new members. One short exercise is to give everyone a pad of sticky notes and then write down answers to a few questions targeting values, priorities, past times, hobbies, expectations, shared objectives, and goals. Share your answers, and then group the notes on a wall to show commonalities. This is an effective way to accelerate relationships and quickly begin to lay a strong foundation of trust. It may be the most valuable hour or two of your entire project.

Teams can be like children in that they usually live up or down to expectations. Few things will encourage a team more than the expectation of success. Make team building a priority. Explain why it is important, invest in your people, and have fun. Be it on-site or off-site, during work hours or on the weekend, doing something meaningful or something silly, in your comfort zone or something completely new. Use your imagination and be all in. Team building activities are vital to getting to truly know each other and building trusting relationships. Additionally, they can create powerful opportunities to set the ground rules of behavior and performance expectations. These should address how the team members work with each other, how the team will handle obstacles, and how the team will meet deadlines. Some examples may include a commitment to promptly return communications, to being honest and listening to fresh ideas, to the quality and timeliness of project deliverables. A team with defined boundaries will be happier and more productive. Once expectations are clarified, then the stage is set for goals.

Set SMART (Specific, Measurable, Achievable, Realistic, and Timely) individual and team goals. Goals can give each employee a united purpose and a cohesive mission. Achieving realistic goals can create a sense of unity and trust and this can be self-motivating, where unrealistic goals will generate apathy and erode morale. Clear goals should be paired with motivation that is consistent with the team culture. The results can be cumulative and astounding. Some of us who played sports were fortunate enough to have a coach that created a cohesive team with personalized motivations that pushed us to achieve more as an individual and a team than we ever thought possible. Those life lessons endure and are directly transferable to the business world.

Mature teams invest in their team and team members' development. One way to accomplish this, while insulating against SME loss, is cross-training. Another method, often conducted concurrently, is mentoring. An experienced coach or advisor can provide needed support and guidance and perspective to team members at all levels, including the PM.

Know when to lead and when to take a step back. Delegation is the action of trust. No one wants to work for a micromanager. Tell your team what needs to be done, and then grant them the respect and freedom to best determine how to accomplish the task. When delegating, define the *Who*, *What*, *When*, and *Why*, while providing some freedom in the *How*. Remember you want to control the outcome, not necessarily the process. Experience will help you discern when to be a dominant leader driving progress with lasered focus toward the goals, and when to give the team the creative space to innovate and lead themselves. Successful PMs find a way to do both within the team's established cultural fabric.

Recognize team and individual successes. Everyone likes to be appreciated. And there is something to appreciate in each of your team members. Highlight conquered challenges, collaborative innovations, supportive actions, and specialized skill sets.

Successfully managing a team requires tact, interpersonal skills, emotional intelligence, and courage. This is not a spectator sport. The ability to identify, assess, and manage your own and others' personal emotions is critical. There are many effective ways to lead and inspire your team, you need to find the style that fits you and your unique skill set.

7.1.1 Theory of Constraints

The Theory of Constraints is a management methodology that asserts any system is inherently limited by a small number of constraints. In his book, *The Goal*, Eliyahu M. Goldratt described the Theory of Constraints using the analogy of boy scout troop on a hike. The goal was for the troop to hike together to their destination. While there may be many factors that impact their speed, their progress was limited by the speed of the slowest scout, "Herbie." If the troop wanted to get to the camp site sooner, they should put Herbie at the front to set the pace while they lighten his load so he can walk faster.

A PM must run the project within the established triple constraint of budget, scope, and schedule. However, they must also work the realities of their teams and resources. Therefore, in order to increase efficiency and productivity, the system (or team) should be analyzed, constraints should be identified, and everything should be subservient to the purpose of exploiting that constraint toward the point where it is no longer the limiting factor for progress. On teams, the constraint is often a critical Subject Matter Expert (SME) that is pulled in many directions. At any time during your project the PM should know, who is your Herbie.

7.1.2 Tuckman Ladder

Projects have a clear beginning and end. So do project teams. It is advantageous to plan how to define, staff, manage, and eventually release team members. Mature teams will go through the following five development stages, also known as the Tuckman Ladder, as shown in Figure 7.1.

7.1.2.1 Forming – Learning about Each Other

Forming is the first stage of team development, when the team comes together. While there may be clarity on purpose, most are approaching the team in an independent manner. There can be apprehension and uncertainty about roles, responsibilities, and processes. This can cause team members to feel anxious. Most will look to the PM for guidance and direction. Team members are getting to know each other and figuring out how to function at a team. Whether reserved or eager, most will be well-mannered, tentative, and compliant.

Figure 7.1 Tuckman ladder.

Interactions may be cordial, with conversations concentrating on work and safe topics. The PM should take the lead while providing clear expectations, consistent guidance, and prompt responses.

7.1.2.2 Storming – Challenging Each Other

Storming is the second development phase, where the purpose is clear but relationships are blurry. Team members may still be uncertain in trying to express their individuality by positioning on how best to accomplish goals. Concerns can arise about the team hierarchy. This phase can be characterized by uncertainty and conflict, which can promote deeper understanding of each other's expectations. Teams are starting to sort themselves out and build trust. The team can work through the resistance, lack of participation, competition, power struggles, and high emotions to provide increased clarity of purpose and lead the group toward establishing norms. Members start to open up and communicate feelings, but still view themselves and a collection of individuals and not yet a team. This phase should be passed through as soon as possible. Successful PMs focus on encouragement and coaching, while managing relationships and task-oriented issues.

7.1.2.3 Norming – Working with Each Other

Team efficiency increases as you enter the Norming, the third development phase. Team relationships are well understood, and there is a shared commitment to goals. Ground rules are established and accepted. Roles and responsibilities are clearly defined and understood. Conflict is being replaced with reconciliation and an open exchange of ideas. Members are engaged and supportive of each other. The team is developing an identity, and real cohesion and confidence. Social relationships may develop among the team as interactions and trust deepens. Differences are now discussed and negotiated constructively. Successful PMs facilitate, providing meaningful feedback and recognition, while closely monitoring the energy of the team.

7.1.2.4 Performing – Working as One

Team efficiency skyrockets during the Performing development phase. Now the team is fully committed and running like a well-oiled machine. Members have embraced the team interdependence and are effectively producing as a team. Healthy, balanced communications, and relationships based on trust and respect are the bedrock of substantial progress, even when stressful situations arise. Cooperation and collaboration enable the team to focus on performance as there is a clear vision and purpose. The team naturally becomes more efficient and flexible as it collectively focuses on achieving the results. During this phase, the PM may step back a bit and delegate as the team may need minimal oversight. Celebrate successes, encourage group decision-making and problem-solving, and keep your pulse on the team dynamics.

7.1.2.5 Adjourning – Celebrating Together and Individually Transitioning

By definition, all projects will come to an end, thus Adjourning the project team. Productivity naturally declines as individuals begin to prepare for their next assignments. Sadness and other emotions can permeate especially close teams, which can also reduce efficiencies. Successful PMs recognize this change, and strive to acknowledge team and individual efforts and achievements. Meanwhile, the PM needs to ensure the project finishes, and finishes strong. Always take time to celebrate with your team. Experienced PMs understand the public transportation world can be small; there may be a good chance you will work with these team members again on future projects, so treat them well from the beginning to the end.

7.1.2.6 Tuckman Ladder Challenge

Teams can accelerate or regress to different stages throughout their life. Some experts say that adding new team members brings the team back to the Norming phase. One challenge in public transportation projects is the extended project duration. As such, you as PM can be frequently onboarding new team members due to staff turnover or other resourcing realities. Intentionally integrating them into the team's culture is important for the productivity and efficiency of the entire team.

7.1.3 Motivators

Employees are as loyal to an organization as their options allow. Fredrick Herzberg established his Hygiene Theory of motivation that asserts one's work environment does not cause satisfaction, but if missing, can cause dissatisfaction. Thus, one's compensation, benefits, and work conditions are only preventative in measure. In 1943, Abraham Maslow proposed a hierarchy of human motivations that now bears his name, as shown in Figure 7.2. He asserted that people can only ascend to higher levels after achieving security in all previous levels. Knowing where your team members are on Maslow's Hierarchy can produce insight as to their engagement and commitment to your team.

For any individual or team to reach its full potential, culture matters. When a team environment is one that is inclusive, promotes safety, makes its members feel valued and respected, recognizes contributions, and challenges our creativity, the team does more than just survives, it thrives. It is the PM's responsibility to develop and manage the team in ways that makes this happen.

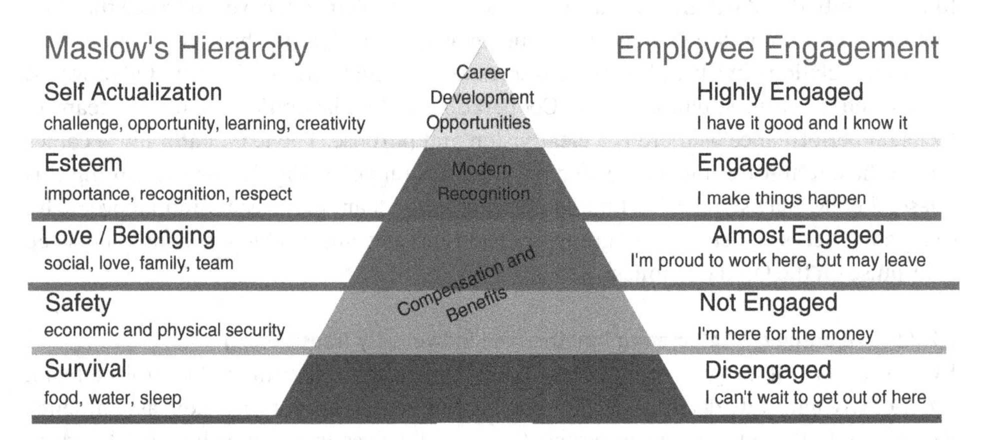

Figure 7.2 Maslow's hierarchy.

7.1.4 Some Additional Thoughts

True leadership is a long-term commitment. Great PMs focus on positive and productive, long-term relationships, not transactional exchanges with team members or stakeholders. Like exercising or a nutritious diet or investing, evaluating progress on a day-to-day frequency is too small of a window to notice and appreciate enduring trends. Whereas, disciplined behavior pays off over time. Consistently leading with integrity, empathy, and respect will build trust and earn a reputation that will empower your team to accomplish amazing results. As Johann Wolfgang von Goethe said, "*You can easily judge the character of a man by how he treats those who can do nothing for him.*" Most outstanding leaders consistently live their life in a way that makes others want to follow them, and feel good for doing so.

A quick reality check; if your team is coming in early, leaving late, working through lunch, and sending work emails after hours or on the weekend, there are two possible reasons. Perhaps you have an unusually passionate, dedicated, and committed team that lives, eats, and breathes your project? More likely you are burning them out. While perhaps necessary for certain pushes at critical times, this is not a sustainable reality.

In his book *Leaders Eat Last*, Simon Sinek describes how the group expects the leader to mitigate all threats, even at the expense of their own well-being. Understanding this fundamental expectation can help you transition from an authority to a leader. Successful PMs are effective buffers, strategically filtering expectations and frustrations from executive leadership down to their team, and tactically advancing team challenges and successes up the chain of command. This approach will foster a supportive team culture in which you can quickly build trust. Take care of your team, and they will take care of you!

7.2 Personality Assessments

"Diversity: the art of thinking independently together."

– Malcolm Forbes

Many are promoted to PMs because they are exceptional engineers. These can be very different roles, requiring very different skill sets. This may be most evident in how one assembles and leads a team.

Successful transitions from engineer to PM are rarely by accident. Some typical engineering traits are easily transferrable. Focus and follow-through. Accurately defining the problem, risks, and constraints. Creative problem solving. Attention to detail. Others can be far more challenging. Delegating. Managing stakeholder expectations. Managing change. Resolving conflict. Making decisions amidst ambiguity. Operating in an unpredictable environment where influence management is key to success. Most of these soft skills center on relationship building and effective communication.

Many engineers make the mistake of assuming everyone thinks like they do, or discounting those that differ in opinion, approach, or priority. Exceptional PMs know that this is a recipe for disaster. When broadening one's focus from a task or engineering discipline to an entire project, it is beneficial to have the maturity to recognize there can be strength in diversity of thought. Sometimes an outlier of thought or personality can be a magical force multiplier, while other times they can almost instantaneously subterfuge momentum and catapult the best laid plans to ruin before you even realize what is happening. Successful PMs have the wisdom to identify and embrace the former, and quickly jettison the latter. Successful projects have PMs that invest in the team and foster an environment that builds and maintains healthy relationships. When you hire an engineer, you are not buying a widget, or other interchangeable commodity. You are renting access to their minds, hoping to benefit from their thoughts, creativity, and problem-solving skills. In the best of cases, underlying all of this is a foundation of trust, agreement of ethical principles, and complimentary personalities.

The reality is we don't often have the opportunity to hand-pick our team. More typically, we must play the hand we are dealt. Perhaps we are able to choose some key team members, but others will almost assuredly be assigned. A few may be superstars. Some may be cooperative. Some may be functional. Some may be distracted and unfocused. Others may be openly disruptive and a daily challenge. An effective PM identifies, and then leverages, the strengths of the team members to remain focused on the project objectives. Navigating personnel issues and fostering positive and productive team dynamics are not skills typically taught in engineering education. Nor are most engineers particularly good at it.

Fortunately, there are well-established tools that provide incredibly useful insights into the psychological undercurrents and motivations of ourselves and others that can increase one's emotional intelligence. Understanding some of these foundational principles will enable you to be a better leader and facilitator in how you connect with others to build and maintain high performing teams. As Charles Maurice said, "*I am more afraid of an army of one hundred sheep led by a lion than an army of one hundred lions led by a sheep.*"

There are a number of outstanding available personality resources. Each approaches the topic with a slightly different perspective and purpose. Any can provide unique insights into yourself and others. If combined, you can capture a more complete picture of your team and yourself that will enable you to more effectively work, communicate, and lead in a way that builds trust and increases productivity. Five of the most common and trusted personality assessments for business are:

1) Myers-Briggs Type Indicator (MBTI) looks at how you process information and make decisions.
2) DISC (Dominance, Influence, Steadiness, and Compliance) examines your preference in communication and behavior.
3) Strength Deployment Inventory (SDI) evaluates the motives of what drives behavior when things are going well and when in conflict.
4) Enneagram determines your core motivations.
5) Clifton Strengths identifies what you naturally do well.

7.2.1 Myers-Briggs Type Indicator (MBTI)

The *Myers-Briggs Type Indicator Handbook* was first published in 1944, and has been a staple of personality testing ever since. The MBTI personality inventory catalogs differing psychological preferences in how one perceives the world and make decisions. This personal inventory asks of series of introspective questions designed to focus on answering the four fundamental questions in Figure 7.3:

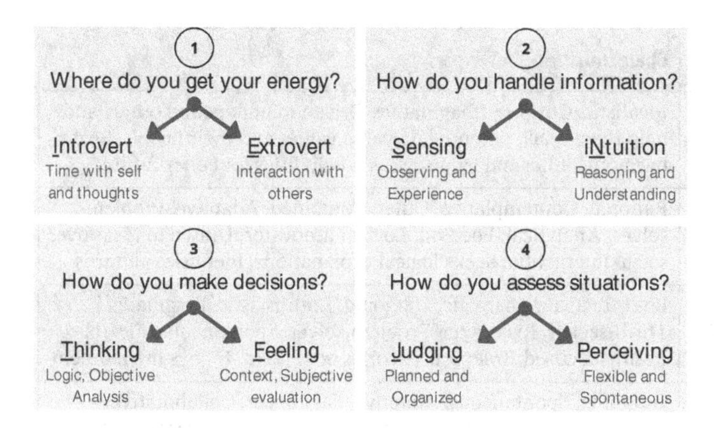

Figure 7.3 MBTI overview.

The underlined letters in your answer to the each of these questions combine to form your four-character personality type. There are sixteen uniquely identified personalities. Results detail a host of insights and cascading work implications. In a time of change, MBTI can suggest what people will likely do to manage the change. MBTI shows how team members approach similar tasks and situations, as well as how we go about getting what we want. The results show what information we value, and our differing approaches to creativity, problem-solving, and decision making. Figure 7.4 is a summary of the sixteen personality types:

Type	Description	Characteristics
ISTJ	The Inspector (Responsible Executors)	Serious, Quiet, Diligent, Dependable, Realistic, Responsible, Steadfast, Methodical, Logical, Thorough, Practical, Pragmatic, Seeks order and organization, Values tradition and loyalty
ISFJ	The Protector (Dedicated Stewards)	Quiet, Friendly, Responsible, Conscientious, Careful, Considerate, Compassionate, Steadfast, Thorough, Accurate, Loyal, Committed to obligations, Seeks order and harmony, Remembers details
INFJ	The Counselor (Insightful Motivators)	Insightful, Future-oriented, Conscientious, Creative, Nurturing, Emotional intelligence, Seeks meaning and connections, Wants to understand what motivates others, Committed to values
INTJ	The Mastermind (Visionary Strategists)	Logical, Analytical, Problem-solver, Independent, Ambitious, Driven, Original thinker, Visionary, Quickly identifies patterns, High standards of competence and performance, Can be skeptical
ISTP	The Craftsman (Nimble Pragmatics)	Artisans, Methodical, Organized, Efficient, Tolerant, Candid, Silent observer, Flexible, Analyzes cause and effect, Quick with solutions, Processes information without emotion, Logically organizes facts
ISFP	The Composer (Practical Custodians)	Passionate, Quiet, Empathetic, Compassionate, Friendly, Flexible, Kind, Spontaneous, Sensitive, Open-minded, Loyal to values and people, Prefers to work alone, Avoids conflict, Enjoys the moment

(Continued)

Type	Description	Characteristics
INFP	The Healer (Inspired Crusaders)	Idealistic, Curious, Imaginative, Driven to understand others and help them reach potential, Loyal to values and their people, Seeks harmony of life and values, Sees possibilities for better future
INTP	The Architect (Expansive Analyzers)	Rational, Contemplative, Quiet, Contained, Adaptive, Problem-solver, Analytical, Focused, Logical innovator, Drawn to ideas over social interaction, Seeks logical explanations, Identifies patterns
ESTP	The Dynamo (Dynamic Mavericks)	Bold, Tactical, Energetic, Outgoing, Enthusiastic, Pragmatic, Thrill-seeker, Risk-taker, Problem-solver, Spontaneous, Flexible, Results focused, Tolerant, Theories bore them, Enjoys the moment
ESFP	The Performer (Enthusiastic Improvisers)	Outgoing, Spontaneous, Adaptive, Passionate, Collaborative, Friendly, Accepting, Spirited, Common-sensed, Loves people and life, Entertainer who enjoys the stage, Makes work fun
ENFP	The Champion (Impassioned Catalysts)	Charismatic, Enthusiastic, Imaginative, Spontaneous, Flexible, Improviser, Kind, Supportive, People-focused creator, Loves helping others explore creative talents and reaching potential
ENTP	The Visionary (Innovative Explorers)	Clever, Entrepreneurial, Quick, Stimulating, Alert, Outspoken, Resourceful, Emotional intelligence, Ambitious, Bored by routines, Embraces intellectual challenges
ESTJ	The Supervisor (Efficient Drivers)	Decisive, Efficient, Practical, Realistic, Systematic organizer, Hard worker, Dependable, Focused on achieving results, Likes logical standards, Finds solutions, Eager to lead groups
ESFJ	The Provider (Committed Builders)	Outgoing, Loyal, Follow-through, Cooperative, Conscientious, Kind, Pragmatic, Devoted, Responsible, Seeks harmony in their environment, Enjoys deep relationships, Notices others' needs
ENFJ	The Teacher (Engaging Mobilizers)	Goal-oriented, Empathetic, Facilitator, Responsive, Sociable, Kind, Caring, Responsible, Charismatic organizer, Inspiring leadership, Highly attuned to others' emotions, needs, and motivations
ENTJ	The Commander (Strategic Directors)	Candid, Decisive, Strategic leader, Organizational problem-solver, Well-informed, Planner, Goal-setter, Shares knowledge, Skilled at coordinating people and activities, Sees underlying relationships

Figure 7.4 MBTI type description.

A high-level understanding of your own MTBI personality type, as well as your coworkers, will grant you valuable insight in how to best to relate and communicate to others on your team as you will better understand how they think and what they value.

7.2.2 DISC (Dominance, Influence, Steadiness, and Compliance)

DISC is a behavior assessment focused on our preferred behaviors to others and everyday tasks we do. DISC can be especially applicable to work environments. The assessment relies on four distinct areas: Dominant, Influential, Steady, and Compliant. These four areas describe how we prefer to work, and interact with others. Figure 7.5 visually depicts the DISC approach, and Figure 7.6 provides more detailed descriptions of each focus area.

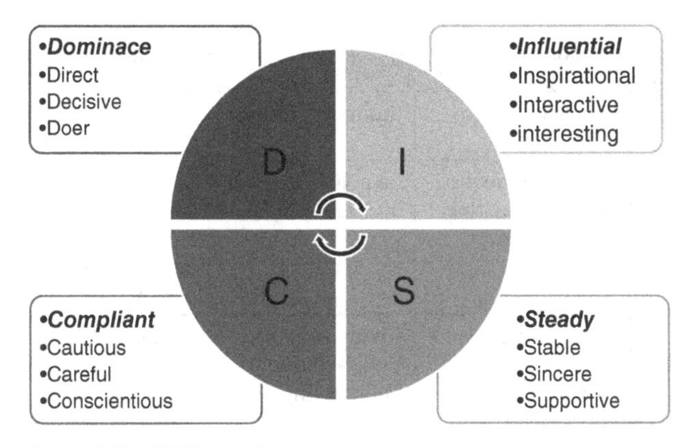

Figure 7.5 DISC overview.

	Dominant (D)	Influential (I)	Steady (S)	Compliant (C)
Focus	Tasks	People	People	Tasks
Personality	Outgoing	Outgoing	Reserved	Reserved
Preferred pace	Faster	Faster	Slower	Slower
Motivated by	Winning, Success, Competition	Having fun	Teamwork	Playing by the rules
Desires/Values	Productivity, Challenges	Recognition, Relationships	Sincerity, Loyalty	Accuracy, Quality
Strive to shape their environment by	Overcoming opposition to achieve results	Influencing and persuading others	Cooperating with others within existing system	Working contentiously within system
Measures/Values contributions by	Results	Recognition, Friends	Compatibility, Relationships	Precision, Accuracy, Activity
Gets security from	Control	Spontaneity	Relationships	Preparation
To be more effective needs	Challenging assignments	Freedom to express self	Reassurance of self-worth	Opportunity for careful planning
Wants to be	The boss	Admired	Liked	Right
Wants you to be	To the point	Interesting	Pleasant	Precise
Under stress can be	Domineering, Demanding	Impulsive, Irritating	Slow, Sensitive	Calculating, Condescending
Bugged by	Inefficiency	Routine, Boredom	Impatience, Insensitivity	Mistakes, Surprises
Probably should	Relax and pace self	Listen and be more realistic	Not take things personally	Not overly rely on data and details
Fears	Loss of control	Loss of Prestige	Confrontation	Embarrassment

Figure 7.6 DISC descriptions.

(Continued)

	Dominant (D)	Influential (I)	Steady (S)	Compliant (C)
Communication tendency	Direct and guarded	Direct and open	Indirect and open	Indirect and guarded
Keys to communicating with this type	Focused brevity, Avoid repeating self and making generalizations	Share experiences, Encourage dialog, Don't interrupt	Be polite, Express interest in them, Clarify expectations	Focus on facts and details, Minimize emotions, Be patient
Prioritizes	Accepting challenge, Taking action, Achieving results	Taking action, Collaboration, Expressing enthusiasm	Giving support, Collaboration, Maintaining stability	Ensuring accuracy, Maintaining stability
Challenges/May be limited by	Lack of concern for others, Insensitivity, Impatience, Skepticism, Getting into the details	Being impulsive Disorganization, Lack of follow-through, Staying focused, Speaking directly, Researching facts	Indecision, Multitasking, Working in ambiguity, Overly accommodating, Avoiding change	Delegation, Compromising, Social events, Quick decisions Being critical, Overanalyzing, Isolating self

Figure 7.6 (Cont'd)

When you take the assessment, your results (profile) are reported as a dot within the DISC circle, along with a highlighted area of your zoned preferences. The more intensely you align with a specific personality type, the further your dot is from the center of the circle. The highlighted area can overlay one or more other quadrants. Most have a dominant type that leans toward an adjacent quadrant with whom they share characteristics. The twelve DISC identified personality subtypes are:

- DC: The Challenger
- D: The Winner
- DI: The Seeker
- ID: The Risk Taker
- I: The Enthusiast
- IS: The Buddy
- SI: The Collaborator
- S: The Peacekeeper
- SC: The Technician
- CS: The Bedrock
- C: The Analyst
- CD: The Perfectionist

Your results profile may also include advice on how best to communicate and work with other personality types. As PM, you can plot the dots of everyone on your team on the same DISC circle to ascertain the makeup of your team. This can quickly provide insights on your team strengths and challenges, which can be extremely helpful in determining how best to communicate in ways that builds trust, and motivate your team toward productivity.

7.2.3 Strength Deployment Inventory (SDI)

SDI evaluates one's motivations that drive behavior by balancing one's concern for people, performance, and process. Uniquely, SDI evaluates individuals at two points; first when things are going well and then in situations of conflict. Comparing these two states provides insight into the degree and nature by which your and others' reactions change during challenging times. SDI focuses on why we do what we do within seven motivational value systems.

Your SDI profile is graphically reported on a color-coded, inverted triangle, as shown in Figure 7.7. There are three colors with associated scales ranging from 1 to 100. The blue scale focuses on our concern for people, and our nature to actively seek opportunities to help others. The red scale focuses on our concern for performance, and our nature to actively seek opportunities to achieve results. The green scale focuses on our concern for process, and our nature to actively seek logical orderliness and self-reliance. The assessment provides two results. A dot is placed at the intersection of the three scores which represents your Motivational Value System (MVS), how you are when things are going well. A vector, or arrow, is sourced at the dot and ends at the point of the Conflict Scenario (CS), how you are when things are not going well.

Results can have significant work implications. In times of change, SDI focuses on values and what matters most to us. SDI shows what people need during change to fully engage. SDI shows our motivations, what makes a task satisfying, and what role we prefer to play. SDI shows what we want. This knowledge can be invaluable in knowing, understanding, and managing team members' actions during calm and stressful situations. Figure 7.8 provides some more detailed insights.

It has been said our greatest weakness is our greatest strength taken to extreme. The SDI analysis also provides insight into one's characteristic strengths, which if overdone can become significant negatives. When someone is struggling to succeed with their strengths, they can push harder and cross a contextual line in terms of intensity, frequency, or duration. Knowledge of your own, and your team members', natural tendencies can be invaluable when guiding successful responses and driving effective results, especially in stressful situation. Although perhaps well-intentioned, these can trigger conflict in a team which can adversely impact harmony, efficiency, and productivity. Recognition of this tendency can start conversations to return the extreme to a strength for the individual and the team. SDI identifies seven strengths for each unique section of the triangle. Figure 7.9 lists these strengths, and the unwanted results if taken to extreme.

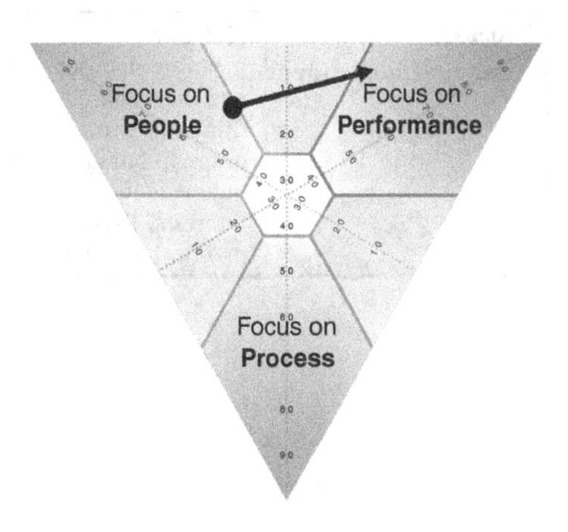

Figure 7.7 SDI overview.

Focus Area Value System	Description	Characteristics	Preferred Work Environment
People Altruistic/Nurturing	Motivated by protection, growth, and welfare of others, Strong desire to help others who can genuinely benefit	Seeking ways to bring help to others, Trying to make life easier for others, Avoids being a burden, Defending others' rights	Open, Friendly, Loyal, Trusting, Supportive, Respectful, Socially sensitive, Being accepted and appreciated
People/Performance Assertive/Nurturing	Motivated by growth and development of others, Strong desire to direct, persuade, or lead others for the benefit of others	Actively seeking to help others, Generating enthusiasm and support in overcoming obstacles, Challenging others to be their best	Enthusiastic, Open, Friendly, Sincere, Trusting, Respectful, Compassionate, Growth opportunities, Mentoring
Performance Assertive/Directing	Motivated by accomplishing tasks and achieving results, Strong desire to set goals, take action, and claim earned rewards	Alert to opportunities, Strives for immediate action, Accepts challenges, Persuasive, Competitive, Risk-taker, Responsible	Competitive, Creative, Innovative, Fast-paced, Challenging, Stimulating, Potential to advance, Material rewards available
Performance/Process Judicious/Competing	Motivated by smart assertiveness and fairness in competition, Strong desire to assess risks and develop opportunistic strategy	Seeks efficiency, Assesses risks, Decisive and proactive when facts are known, Challenges opposition with logic and strategy	Complex challenges requiring expertise, Recognition for achievement, Develops winning strategies, Technical resources
Process Analytic/Independent	Motivated by meaningful order and thorough thinking, Strong desire to pursue independent interests, To be practical and fair	Seeks clarity, accuracy and correctness, Cautious and thorough, Keeps emotions under control, Plans ahead and follows the plan	Organized, Clear, Precise, Appreciation for utility, efficiency, and reliability, Time to develop ideas, Logical decision-making

(Continued)

Focus Area Value System	Description	Characteristics	Preferred Work Environment
Process/People **Cautious/Supportive**	Motivated by developing self-sufficiency in self and others, Strong desire to analyze the others' needs and help them	Builds effective processes and resources to protect or enhance others' welfare, Fighting for principles that are fair	Conscientious, Patient, Congenial, Respectful, Fair, Encouraging, Tasks that require thoughtful analysis to aid those in need
Hub (all three) **Flexible/Cohering**	Motivated by adapting to others or situations, Strong desire to collaborate and remain open to different options and viewpoints	Considers other perspectives, Open-minded, Adaptable, Maintains balance, Brings people together and makes connections	Cooperative, Sociable, Interactive, Playful, Spontaneous, Being heard and listening, Consensus building, Open-minded

Figure 7.8 SDI descriptions.

SDI Focus Area	Strength	When taken to extreme
Focus on People Altruistic/Nurturing Color: Blue	Supportive	Self-Sacrificing
	Caring	Submissive
	Devoted	Subservient
	Modest	Self-Effacing
	Helpful	Smothering
	Loyal	Blind Spots
	Trusting	Gullible
Focus on Performance Assertive/Directive Color: Red	Risk-Taking	Reckless
	Competitive	Aggressive
	Quick-to-Act	Rash
	Forceful	Domineering
	Persuasive	Abrasive
	Ambitious	Ruthless
	Self-Confident	Arrogant

Figure 7.9 SDI focus areas.

(Continued)

SDI Focus Area	Strength	When taken to extreme
Focus on Process Analytic/Automizing Color: Green	Persevering	Stubborn
	Fair	Cold
	Principled	Unbending
	Analytical	Obsessed
	Methodical	Rigid
	Reserved	Distant
	Cautious	Suspicious
Hub Flexible Color: All three	Option-Oriented	Indecisive
	Tolerant	Indifferent
	Adaptable	Compliant
	Inclusive	Indiscriminate
	Sociable	Intrusive
	Open-to-Change	Inconsistent
	Flexible	Unpredictable

Figure 7.9 (Cont'd)

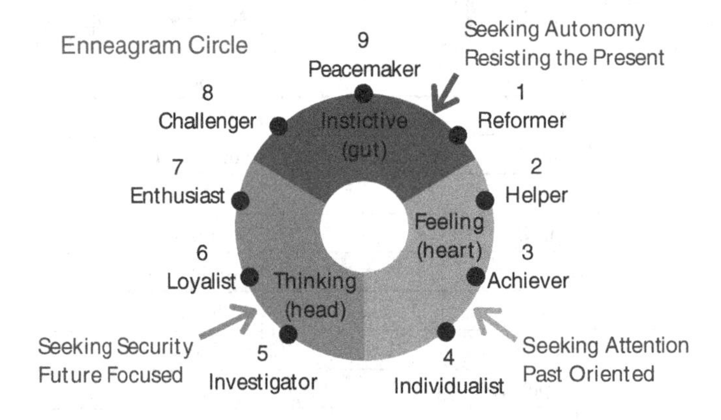

Figure 7.10 Enneagram overview.

7.2.4 Enneagram

Enneagram is a modern version of older, ancient wisdom traditions that differentiates those that rely on instincts, thinking, and feeling. This personality assessment tool then divides humanity into nine (9) different personality types that are graphically displayed around the circumference of a circle, as shown in Figure 7.10. Figures 7.11, 7.12, and 7.13 provide some detailed insights for each of the nine personality types, divided by focus area. The position on the circle becomes important relative to adjacent personality types. Lines can also be drawn between the personality types that add additional understandings. Enneagram can provide some deeper insights on individual's mindsets, as well as communication and productivity strengths and challenges.

Instinctive Center (Gut)	8. Challenger (The Powerful Person)	9. Peacemaker (The Peaceful Person)	1. Reformer (The Good Person)
Premise	Love and respect are gained by being strong and just	Love and respect are earned by blending in with others' agendas	Only perfect people are worthy of love and respect
Summary	Protective, Leader, Maverick, Intimidator, Self-assured, Assertive, Decisive, Direct, Willful, Confrontational	Mediator, Reassuring, Accommodating, Easy-going, Preservationist, Patient, Complacent, Receptive, Adaptive	Moral, Strict Perfectionist, Judge, Critic, Crusader, Contentious, Self-controlled, Principled
Focus of Attention	Power and control, Protecting those that need it	Others' wants and needs, Conflict, Wholeness, Peace of mind	What is right or wrong, correct or incorrect, Integrity, Balance
Focus of Energy	Being powerful and protective, Fighting, Self-protection	Avoiding conflict, Other people, Separation, Loss	Trying to improve
Fear/Avoidance	Vulnerability, Being controlled, Weakness	Conflict	Corruptness, Imbalance
General Strengths	Bold, Assertive, Action-oriented, Strength under pressure	Accepting, Calming, Strong relationships, Steady	Honest, Responsible, Improvement-oriented, Conviction, Serenity
General Challenges	Domineering, Defines own version of truth, All or nothing style, Quick tempered	Ambivalent, Passive-aggressive, Surrenders agenda, Lazy, Avoids conflict, Disengaging	Critical, Judgmental, Rigid, Uncomfortable with emotions, Anger, Guilt, Hypocrisy
Communications Strengths	Protective and concerned with providing for the group	Diplomatic and considerate	Once committed becomes extremely loyal
Communications Challenges	Compromise can feel like surrender, Low tolerance for ambiguity	Avoids conflict, Hard to say "no," Goes along to keep the peace	Focuses on perfecting the flaws, "Scorched earth" policy if angry
Job Performance Strengths	Strong leadership, Great at overcoming obstacles to get it done	Imaginative	Consistent effort, Dependability, Thrifty
Job Performance Challenges	Can fall prey to "My way or the highway" approach	Distracted in priorities and productivity, Ambivalence	Perfectionism leads to procrastination, Self-imposed pressure

Figure 7.11 Enneagram type descriptions – instinctive.

Feeling Center (Heart)	2. Helper (The Loving Person)	3. Achiever (The Effective Person)	4. Individualist (The Original Person)
Premise	Love and survival depend on giving to get	Love and recognition are only for winners	Others enjoy the happiness I deserve
Summary	Supportive, Advisor, Nurturer, Manipulator, Caring, Generous, Friendly, Considerate, People-pleasing	Successful, Motivator, Producer, Chameleon, Ambitious, Focused, Shrewd, Adaptable, Competitive, Driven	Romantic, Artist, Deep, Melodramatic, Intense, Creative, Emotional, Temperamental, Self-absorbed, Expressive
Focus of Attention	The wants and needs of others, Being appreciated	To feel valuable, Roles, Image, What brings success and approval	Sees the best in what's missing, and the worst of what's here
Focus of Energy	Giving and helping, To feel loved	Achieving goals, tasks, and prestige	Searching for the special, To be uniquely themselves
Fear / Avoidance	Own needs, Being unloved	Failure, Worthlessness	The Commonplace, Having no unique identity or significance
General Strengths	Helpful, Empathetic, Emotionally intelligent, Relationship-oriented	Achievement-oriented, Image conscious, Authenticity	Creative, Empathic, Idealistic, Emotionally balanced
General Challenges	Intrusive, Dependent on others' approval, Pride, Distracted in others' emotions	Out of touch with own emotions, Intolerance for negativity, Always pushing to be "the best"	Envy, Moodiness, Impatience with reality, Feeling like an actor in your own life
Communications Strengths	Helpful, Caring, Relationship-oriented	Socially aware, Social skills	Emotionally engaged
Communications Challenges	Easily distracted, Losing self in others' need, Mind goes blank	Status conscious, Need to produce something of value to prove worth	Alienation, Sense of being different from the people around you
Job Performance Strengths	Collaborative leadership style	Energetic, Adaptable	Creative
Job Performance Challenges	Conceding to others can interfere with autonomy of leading	Overworked, Competitive, Impatient	Unrealistic, Unpleasant emotions can ruin productivity

Figure 7.12 Enneagram type descriptions – feeling.

7.2.4.1 Wings

Everyone has a dominant personality type. However, most everyone also has aspects of one or both of the adjacent personality types on the circumference of the Enneagram circle. These "*wings*" compliment your dominate type in important, and sometime contradictory, ways. Wings don't change your core motivations, but they do influence them. Some have one wing. Those with two wings typically have a more dominant wing.

Thinking Center (Head)	5. Investigator (The Wise Person)	6. Loyalist (The Loyal Person)	7. Enthusiast (The Joyful Person)
Premise	Love and respect are gained by practicing self-sufficiency	Love and protection are gained by vigilance and endurance	Frustration can be avoided by attending to positive options
Summary	Thinker, Innovative, Sage, Reductionist, Cerebral, Private, Curious, Detached, Quiet, Perceptive	Guardian, Devil's advocate, Skeptic, Pessimist, Cautious, Anxious, Alert, Responsible	Entertaining, Optimist, Generalist, Adventurer, Escapist, Spontaneous, Uninhibited, Visionary, Versatile, Scattered
Focus of Attention	What others expect	Being without support or guidance, Evaluating risks and worst-cases	Positive possibilities in all things, Limits and constraints
Focus of Energy	Becoming self-sufficient, Mastery, Understanding	Becoming vigilant or questioning, To have support and guidance	Interesting ideas and activities, To be satisfied and content
Fear/Avoidance	Uselessness, Helplessness, Incompetence	Uncertainty	Discomfort and Pain, Being unfulfilled, trapped, and deprived
General Strengths	Scholarly, Analytical, Non-attachment, Self-Reliant	Strong relationships, Attentive, Perceptive, Courageous	Optimistic, Fun-loving, Positive visioning
General Challenges	Emotionally detached, Isolated, Replacing direct experience with concepts, Greed	Procrastination, Doubt, Reactive, Indecision, Seeking reassurance, Fear	Scattered, Impulsive, Avoids self-reflection, Boredom masks emotional confusion
Communications Strengths	Independent, Self-reliant, Brings clarity to confusion, Loyal friend	Enduring loyalty and support	Life of the party, Enthusiastic, Fun, Adventurous
Communications Challenges	Detach to observe, Can become invisible, Emotions are intrusive	Can attribute their own feelings to others	Inwardly-focused, Sensitive to criticism, Dismissive of limits
Job Performance Strengths	Systemic and well thought out	Strong under adversity, Asks hard questions to eliminate skepticism	Strong Generalist, Great at starting new projects
Job Performance Challenges	Useless specialization, Emphasizes thinking rather than doing	Anxiety peaks near the point of success, which may make you a target	Scattered thinking, Lacks follow-through, Struggles with details

Figure 7.13 Enneagram type descriptions – thinking.

7.2.4.2 Lines

The Enneagram also provides insight on how we can travel, or be pulled, to different personality types. Our dominant personality type is our home base. The enneagram lines indicate other personality types with which you have special relationships, as shown in Figure 7.14.

Figure 7.14 Enneagram lines.

These are other personality types you can easily travel toward. Some consider these two lines to be "*growth*" and "*stress*" lines. This is important as these can influence your thinking and provide a significant perspective shift. These lines are the powerful building blocks of enneagram that can be strategically targeted for growth, development, and balance.

7.2.5 Clifton Strengths

Clifton Strengths, or StrengthFinder, is an assessment tool that provides a basic understanding of what one naturally does well. This tool was created by Gallup to discern one's strengths to maximize their potential at work and in life. This well-established online survey defines 34 unique strengths that are divided into four main categories: Executing, Influencing, Relationship Building, and Strategic Thinking. Figure 7.15 shows the 34 strengths divided by category.

When you take this assessment, you receive your top five strengths. As you dig into the results, it is important to keep the following three principles in mind. First, focus on what is right. This positive assessment is informing you of your strength inventory so you can leverage it to be a better version of yourself. Second, interpret the results as descriptive, not prescriptive. The results should reflect those things at which you naturally excel. Don't stress that these results are a new directive, rather seize the power that comes with self-awareness. Thirdly, pair your strengths to amplify their effectiveness. Your top five

	Executing	Influencing	Relationship Building	Strategic Thinking
Focus	Knowing how to make things happen	Knowing how to take charge, speak up, and make sure the team is heard	The ability to build strong relationships that can hold a team together and make the team greater than the sum of its parts	Helps teams consider what could be, the ability to absorb and analyze information that can better inform decisions
Defined Clifton Strengths	Achiever Arranger Belief Consistency Deliberate Discipline Focus Responsibility Restorative	Activator Command Communication Competition Maximizer Self-Assurance Significance Woo	Adaptability Connectedness Developer Empathy Harmony Includer Individualization Positivity Relator	Analytical Context Futuristic Ideation Input Intellection Learner Strategic

Figure 7.15 Clifton strengths overview.

strengths are woven together in a fabric of your unique experiences and perspectives. Use two or more seemingly contradictory or unrelated strengths to support and strengthen each other.

Below is a quick summary of each defined Clifton Strength under the Executing Dominant Strength:

- Achiever – tireless, hard worker who finds satisfaction in being busy and productive
- Arranger – flexible organizer who arranges resources to optimize productivity
- Belief – unwavering core values that produce a clear life purpose
- Consistency – fair-minded individuals who ensure equality by following clear rules
- Deliberate – serious and cautious decision makers who anticipate obstacles
- Discipline – embrace processes and routines to create structure order
- Focus – prioritize directions with consistent follow-through to stay on track
- Responsibility – a natural obligation to honestly and loyally keep their commitments
- Restorative – easily discern what is wrong and enjoy reaching resolution

Below is a quick summary of each defined Clifton Strength under the Influencing Dominant Strength:

- Activator – adept at transforming thoughts or conceptual ideas into action
- Command – decisive individuals who take control with a strong presence
- Communication – effectively convey thoughts through natural conversations
- Competition – motivated to win as they naturally compare progress to others
- Maximizer – driven to achieve individual and group excellence through improvement
- Self-Assurance – confident in their values, identity, and their course of action
- Significance – seek validation through other's eyes and admiration
- Woo – charismatic social butterflies who can quickly connect with others

Below is a quick summary of each defined Clifton Strength under the Relationship Building Dominant Strength:

- Adaptability – those who live in the moment and accept and embrace life as it comes
- Connectedness – recognize and value links and connections between people and events
- Developer – coaches who spot and develop others' potential
- Empathy – understanding others by fully considering their perspectives and feelings
- Harmony – avoids conflict by constantly seeking consensus and areas of agreement
- Includer – identifies and intentionally involves everyone to ensure no one is excluded
- Individualization – embraces uniqueness and appreciates strength through diversity
- Positivity – optimists with infectious enthusiasm
- Relator – values and prioritizes close and meaningful relationships

Below is a quick summary of each defined Clifton Strength under the Strategic Thinking Dominant Strength:

- Analytical – logic-driven thinkers who search for reason and causes
- Context – understand and frame the present based upon the history and traditions
- Futuristic – visionaries who are fascinated with what can be
- Ideation – brainstormers who revel in ideas and find trends and connections

- Input – curious minds that yearn for more knowledge
- Intellection – introspective thinkers who enjoy intellectual pursuits
- Learner – loves the process of learning more knowledge than its application
- Strategic – clever problem solver who optimizes situational options

It is worth noting that the likelihood of another sharing your top five ordered strengths is 1 in 33 million. And each of these strengths may manifest itself in different ways. One true benefit of this assessment is to recognize that you are uniquely you. Celebrate and embrace your strengths!

7.2.6 Personality Assessments Summary

Any one of these assessments will grant you valuable insights into yourself and your team members. Since they all approach one's personality, motivations, and behaviors from differing perspective, completing more than one of these will provide an even clearer picture that will enable you to better leverage strengths, protect weaknesses, communicate effectively, and foster a team environment that values contributions and drives results. These insights can bring a common perspective to shared common goals.

There are many online or workshop opportunities to facilitate these assessments. Regardless of the maturity of your team, this can be a fun and powerful team-building activity that will promote productive discussions and accelerate trust. This time investment can pay surprising dividends in helping you learn and better lead your team to drive productivity. Perhaps even more importantly, you will know yourself better, which enables you to be more productive and effective in your own work and in leading others.

7.3 Resource Planning

> *"Even the best team, without a sound plan, can't score."*
>
> – Woody Hayes

Leaders motivate, inspire, and create buy-in with big-picture, strategic vision. Managers are task-oriented achievers that can dive head-first into the minutia. PMs must be able to shift between the CEO (strategist) and COO (operations) roles on a project. This is perhaps nowhere more evident than in resource planning.

The PM should pursue efficiency. This means intentionally planning to complete the tasks using as few of resources as possible. However, not enough resources can impact schedules, which can impact costs. Generally, resources should be negotiated at the functional manager level. Finding the right fit and attitude is essential to success.

Tasks in the Work Breakdown Structure (WBS) should be assigned to an individual who bears responsibility of completion. If multiple people are responsible, then no one is responsible. Someone needs to be responsible.

An assignment matrix can be especially effective in showing and tracking the relationship between work packages and team members. An RACI chart is typically created in initiation and identifies key roles and responsibilities for major tasks and decisions. It lists who is Responsible (does the work), Accountable (ensures work is done), Consults (input required by SME), and Informs (update status). An RACI chart can also provide valuable insight in balancing workloads.

Unfortunately, there are times when a team member just doesn't fit and you need to enable them to seek other opportunities where their skills may be better utilized and appreciated. In public transportation projects, you don't always have the option to readily replace team members. But one bad apple can spoil the batch. Consequently, the PM must proactively identify and address toxic behavior that undermines the team culture and jeopardizes project goals. If there are performance issues with a team member, they should be dealt with straight away. Be sure to include Human Resources and the team members functional hierarchy in any formal discipline proceedings.

Sometimes team members become too tasked focused, acting as if under the assumption their responsibility is of primary importance and all others should accommodate their preferences or constraints. Sometimes that is the case. Typically, though, it is not. The PM should understand the big picture and leverage the critical path to provide a perspective and clarify expectations.

Throughout the course of administering a project, a PM may be exposed to a number of resource terms and ideas. *Resource Allocation* is the assigning and scheduling of resources for project-related activities. Ideally this is done by the PM, but may be overridden by a program manager if resources are to be shared between multiple projects. *Resource Availability* is whether a specific resource is available for use at a given time. The *Resource Breakdown Structure* is a hierarchical list of resources needed for the project, classified by type and function. A *Resource Calendar* shows resource availability over time.

Resource Optimization Techniques can be used by a PM to balance supplies and demands for resources. *Resource Leveling* is done when resource availability is the constraint. This involves adjusting task start or finish dates when resources have been over-allocated or become scarce. This technique is also used when a limit is imposed on a particular resource for some reason. This often allows schedule slippage that can adversely impact the critical path. *Resource Smoothing* is done when time constraints are the priority. The objective is to leverage tasks floats to balance resources across tasks to they are completed on time while avoiding the peaks and troughs of resource demand. The schedule and critical path cannot slip with smoothing, so tasks can be delayed no longer than their free or total float.

7.3.1 Resource Management Plan

Resources are people, capital, materials, and tools used to execute tasks and complete a project. In short, they are the lifeblood of a project. The Resource Management Plan details how resources and teams will be developed, managed, and controlled throughout the life of the project.

Every Resource Management Plan should include an organization chart that details the names and roles of critical personnel in the execution and completion of this project. Additionally, each plan shall include a list of roles and responsibilities. This should be formatted in a logical and reasonable manner. Some examples may be delineating individual roles/responsibilities (what they are supposed to do), authorities (what they are allowed to do), and competencies (required skills). Some use a Responsibility Assignment Matrix (RAM) to clarify roles and responsibilities. Other projects may warrant a full RACI Matrix to define task owners and dependencies.

Additional resource management aspects may be added to the plan, that should be tailored to the size, complexity, and specifics of the project. Some examples may include training needs and personnel development plans. Team member recognition and award plans may be included, where applicable. There may also be a host of other Team Management issues you may wish to include. Some examples may include staff acquisition plans, staff onboarding and transition protocols, resource and work calendars, compliance, and safety. Remember the point of this plan is not to create work for anyone, but rather to establish an easily referenced foundation upon which you build your resource management processes.

7.4 Managing Consultants

"I only know one way to communicate, openly and honestly."

– Bob Gallagher

Civil Engineering consultants play a critical role in the development and delivery of public transportation projects. From survey to subsurface utility designation to geotechnical borings to traffic studies to hydraulic evaluations to environmental permitting to road layout to bridge design, and so much more. Most all transportation system owners and operators rely on consultants to successfully develop and deliver their transportation program. Forming and maintaining productive partnerships is essential.

This becomes increasingly important as owner-operators rely more heavily on consultant expertise. There can come a point when transportation organizations stop approving consultant plans, and realize they are instead accepting them. This transition can be favorable if the owner-operator acknowledges this change, and accommodates it by appropriately adjusting required workflows and documentations such that project risk is properly realigned with this approach. Other organizations can stumble if they practically are accepting plans, but still act as if they are approving plans. Each transportation organization should make a conscious decision if they are accepting or approving plans, project applications, scopes, schedules, and estimates, and then have workflows that align with this decision.

7.4.1 Professional Service Procurement

Most consultants are procured via Professional Services. This means it is not sealed, low-bid selection; rather fees are negotiated after a firm is selected based on their professional qualifications. States and localities may have their own procurement standards and procedures, but most include the workflow in Figure 7.16 to bring a consultant on board.

Consultant procurements for project-specific designs are relatively straightforward. The RFP would define the problem to be solved with enough detail to solicit interest and meaningful responses. Your organization's procurement office should carefully review the RFP to ensure all legal requirements are included, and that the RFP is reviewed by all who need to approve it prior to release. Pending the complexity of the project, some organizations choose to have a Pre-Proposal meeting where they provide a brief overview of the project

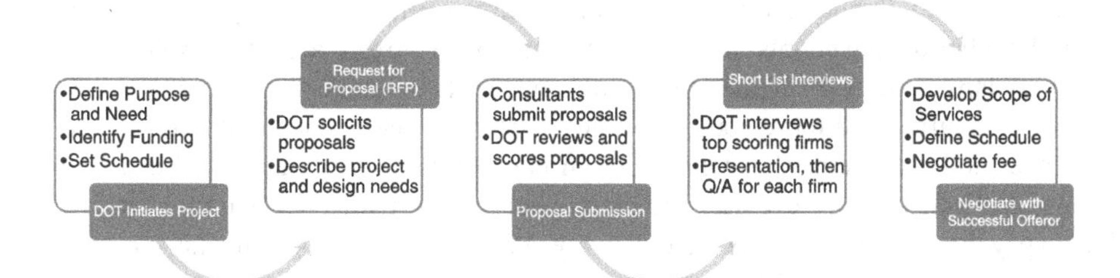

Figure 7.16 Professional service procurement.

and can respond to consultant questions. Submitted proposals typically include a proposed project team organizational chart, their approach to project design, demonstrated success in similar projects, and applicable resumes. The proposals should be scored based upon the criteria detailed in the RFP. Remember throughout this process that you are not judging their proposed solution, but rather their professional capabilities to successfully complete the design should they be selected. Once a firm is selected after the short list interviews, their contract is negotiated that details the schedule, budget, and scope of their work. For many projects, this consultant contract is a critical component for the scoping document.

On-call Annual contracts are where the transportation organization selects a firm to provide specific services for as yet undefined needs. Task orders are then assigned to the consultant under the Annual Contract for each project or effort. The task orders serve as the Scope of Service, subject to the terms of the Annual Contract. Once Annual Contracts are established, they are the quickest way to bring a consultant onboard. Most Annual Contracts have legislative or organizational limits on the initial term, the number of renewals, and dollar limits on each task order and for each contract term. These follow the same general workflow shown above except the RFP is tailored to the needs and specifics of the annual contract.

Pending applicable state and local procurement laws and regulations, there are different contractual methods to hire an engineering consultant under professional services contract. The three most common are time-and-materials, fixed fee or lump sum, and negotiated fee. Time and materials are when the owner pays the consultant based on their actual time and expenses. This arrangement can bring a firm onboard when the scope is not well defined. However, the realized schedule benefits should be weighed against the budget risks as this arrangement can quickly resemble a blank check if not carefully managed. A fixed fee, or lump sum, contract sets a firm price for the agreed services. This arrangement can be beneficial if the budget is fixed, as it prioritizes the fee over scope and schedule. A negotiated fee typically most values the scope. Consultant hours are assigned to accomplish the agreed Scope of Services. The hourly rates are totaled for a not-to exceed fee for the specified work. Consultant's billable hourly rates should be calculated per applicable federal, state, and local regulations. Direct labor rates are multiplied by approved overhead rates that can vary widely. Agreed expenses and profit margin are then added to the total, which may include escalating inflation rates on multiyear projects. Most project-specific transportation designs are of this fee structure. Each of these arrangements assigns different risks to different parties. Be intentional and choose the fee structure that works best for you given the situation.

Design-Build and Public-Private-Partnership arrangements are entirely different. In these, and similar arrangements, the transportation owner hires a team that collectively designs the plans, acquire all rights-of-way, and constructs the project. These approaches are common in larger and more complex projects where the close collaboration between the engineer and contractor throughout design can stimulate innovation and expedite construction. These can be extremely successful ventures, particularly when the owner is willing to pay the premium for expedited delivery, desires a true turn-key arrangement, or prefers to transfer significant risks on highly complex endeavors. It is important to understand that in these arrangements the consultant works for the contractor. This can and does shift the paradigm of the consultant's prioritized responsibilities. Intentional and cooperative communications are essential to establishing and maintaining an effective and productive working relationship.

Despite best efforts, some relationships are strained and difficult. Both the DOT and consultant PM bear responsibilities to ensure their partnership is positive and productive. Negotiation and compromise can be essential tools. The driving consideration should be ensuring efficient use of public funds to develop and deliver the project. However, sometimes due to personalities or other factors, it just isn't a workable fit. In these rare cases, the consultant bears the responsibility to find resource replacements that are acceptable to the DOT. In rare cases, the DOT–consultant relationship should be severed. Obviously, this is to be avoided, if possible, for a variety of reasons.

7.4.2 Roles and Responsibilities

Everyone should share the same goal of executing a successful project. However, in order for the DOT PM and consultant PM to function as an effective team, they must understand each may define success slightly differently. Both should strive to understand and be respectful of the others' unique organizational responsibilities, priorities, and constraints.

DOTs most commonly define project success by completing a project whose scope satisfies the stated purpose and need, on-time and on-budget. That seems simple, but there may also be other measures of success. Key stakeholders may need to be included in the process and satisfied with the final design. Influential businesses may have preferences for hours of operation and traffic flow during construction. This project may need to be coordinated with other projects to satisfy larger, regional goals. And so forth. Consultants define success by completing their project Scope of Services on-time and under-budget. But there may also be other considerations. Profit targets are established. Staff may have billable hour and cross-discipline development targets. Resource sharing across multiple projects to stabilized workloads. Perhaps they want to submit this project to win a prestigious award.

It is helpful for the DOT and consultant PMs to openly discuss what success looks like to each of them on each project. Find and focus on the common ground. Emphasize one goal to be a long-term relationship between the DOT and consultant. Outside of field work equipment, the DOT is paying for access to the consultants' thinking, skills, and abilities. It is in the DOT's best interest to develop long-term relationships with smart and principled engineers that are based on mutual trust and respect. Find conscientious firms that push you to innovate, and challenge you to be better stewards of public funds. Have the courage to form true partnerships over transactional commodities. In long-term relationship, you aren't focused on counting nickels to maximize profit on this one project. Instead, you can collectively focus on the project goals, knowing the symbiotic, win-win, long-term relationship will be positive and advantageous for both parties.

In general, the DOT PM drives the project, pays the bills, removes roadblocks, and secures decisions, while the consultant PM guides and provides the technical solutions that satisfy the project's purpose and need. Of course, it is more complicated than that. Below are some key responsibilities of both the DOT PM and consultant PM in key project areas.

7.4.3 Scope (Scope of Services)

The project scope defines the project objectives. The Consultant's Scope of Services (SoS) defines the consultant's contractual requirements, project deliverables, schedule, budget, payment terms, and project objectives. The project scope and SoS may not necessarily be exactly the same. When procuring Professional Services, ideally the SoS should be finalized before fees are negotiated. The SoS should be based upon information in the Request For Proposals (RFP), additional conversations between the consultant and owner, any project funding application, other owner project scope information, and professional judgment. Adjustments can be made to streamline the SoS to adhere to budget constraints, if needed. Figure 7.17 details the DOT and Consultant PM responsibilities regarding scope.

7.4.4 Budget (Cost)

The project budget includes all costs to complete the project. These include consultant design fees, environmental evaluation and mitigation, surveying, geotechnical investigations, materials testing, right-of-way acquisition, utility relocation, construction, CEI, administrative oversight, and any other legitimate project-related expense. The consultant's budget is that which was negotiated in the contract in order to satisfy their SoS. This is a typically a portion of the project's PE budget. Figure 7.18 details the DOT and Consultant PM responsibilities regarding budget.

DOT PM	Consultant PM
Develop the project Scope	Fully understand the project Scope
Develop the consultant's SoS and ensure it satisfies the project scope	Fully understand the SoS
Ensure schedule and budget properly adjusted before Scoping baselines are established	Satisfy all requirements in the SoS within the agreed schedule and budget
Officially close scoping	Perform no work outside of the SoS without prior consent
Work with Consultant PM to proactively practice established Change Management Procedures, if needed	Work with DOT PM to proactively practice established Change Management Procedures, if needed

Figure 7.17 DOT compared to consultant PM – scope.

DOT PM	Consultant PM
Complete the project within the project budget	Complete project SoS within the budget specified within the consultant contract
Promptly review consultant invoices to ensure charges are "reasonable and expected" and reflect design progress, per payment terms in SoS	Promptly submit accurate invoices (typically monthly)
Confirm invoice line-item percentages reflect actual work progress	Meet the profit target established by firm
Ensure prompt payment of approved invoices	Proactively identify and discuss with DOT PM any budgetary concerns or issues
Work with Consultant PM to proactively practice established Change Management Procedures, if needed	Work with DOT PM to proactively practice established Change Management Procedures, if needed

Figure 7.18 DOT compared to consultant PM – budget.

7.4.5 Schedule

The consultant's schedule should be specified in their contract. Ideally it would be the same as the project schedule. PMs that attempt to juggle multiple schedules for the same project do so at their own risk. There should be one project schedule. Having said that, the consultant schedule may contain tasks or interim milestones not on the project schedule. These should be consistent with, and not conflict with, the project schedule. Figure 7.19 details the DOT and Consultant PM responsibilities regarding schedule.

DOT PM	Consultant PM
Meet project schedule	Meet all schedule requirements in the SoS
Understand project schedule and aggressively focus on critical path tasks to protect float	Understand project schedule and aggressively focus on critical path tasks to protect float
Manage project schedule, and update it as needed	Regularly review SoS project schedule, and update it when needed
Coordinate this project schedule with other related active projects in program	Be aware of local and project-specific issues that may impact SoS project schedule
Ensure DOT reviews comply with SoS expectations and do not delay project	Ensure quality deliverables provided to DOT in SOS timelines to allow adequate DOT review
Work with Consultant PM to proactively practice established Change Management Procedures, if needed	Work with DOT PM to proactively practice established Change Management Procedures, if needed

Figure 7.19 DOT compared to consultant PM – schedule.

7.4.6 Contract

The consultant contract dictates what the consultant will do and when. Just as important, it often dictates what the consultant won't do. Great care should be taken in the formation of the SoS and Consultant Agreement to ensure both parties are satisfied with the intended deliverables, when they will be delivered, and the quality of what is provided. Once the contract is executed, this becomes the foundational document for all future performance and change management discussions between the owner-operator and consultant. Figure 7.20 details the DOT and Consultant PM responsibilities regarding the contract.

DOT PM	Consultant PM
Fully understand contract (consultant agreement)	Fully understand contract (consultant agreement)
Ensure all contractual obligations are fulfilled on-time and on-budget with acceptable quality	Ensure all contractual obligations are fulfilled on-time and on-budget with acceptable quality
Be responsible stewards of public funds, ensure expenditures and deliverables as expected	Ensure deliverables meet all contractual expectations
Monitor project and contract to proactively identify issues that should be addressed	Ensure advice, recommendations, actions, and design are ethical and in client's best interests
Take timely and decisive action to correct budget, scope, schedule, or quality issues	Proactively identify and promptly advise DOT PM of contractual challenges and proposed solutions
Work with Consultant PM to proactively practice established Change Management Procedures, if needed	Work with DOT PM to proactively practice established Change Management Procedures, if needed

Figure 7.20 DOT compared to consultant PM – contract.

7.4.7 Risk

Successful PMs are exceptional at managing risk. The DOT PM is responsible to complete the entire project. The consultant PM is responsible to complete the SoS. Most project risks impact both the DOT and consultants as they work collaboratively to push forward. However, it is important to realize that some situations may pose risks to one party, but not the other. There are also times when risks may be at conflict between the two. Proactive and open communication is critical to navigating these difficult situations. If done right, the partnership is strengthened and you cooperatively progress toward mutual success. Figure 7.21 details the DOT and Consultant PM responsibilities regarding risk.

DOT PM	Consultant PM
Proactively identify and evaluate project and SoS risks	Proactively identify and evaluate project and SoS risks
Communicate risks to Consultant PM and determine how best to proceed	Provide risk mitigation solutions to DOT PM for consideration
Promptly decide risk response approach based upon DOT's risk appetite and project specifics	Support and enact DOT chosen risk responses
Work with Consultant PM to proactively practice established Change Management Procedures, if needed	Work with DOT PM to proactively practice established Change Management Procedures, if needed

Figure 7.21 DOT compared to consultant PM – risks.

7.4.8 Resources

Like many other areas discussed here, there are two separate but intertwined arenas. The DOT PM is responsible for the resources to complete the project. This includes the consultant, typically the consultant PM or contract administrator. The consultant PM is responsible for all consultant and subconsultant resources required or utilized to accomplish the SoS. Figure 7.22 details the DOT and Consultant PM responsibilities regarding resources.

DOT PM	Consultant PM
Ensure DOT SMEs are available to assist and provide guidance, as expected	Ensure proper resources conduct work to satisfy SoS within contract schedule and budget
Ensure reviewing resources conduct prompt reviews and provide meaningful comments	Ensure key team members and SMEs participate in project, as promised during procurement
Ensure correct resources are engaged to secure prompt decisions so consultant may advance	Balance delegation with coaching and team member development
Ensure project Civil Rights obligations are being met	Ensure all contractual Civil Rights obligations are being met

Figure 7.22 DOT compared to consultant PM – resources.

7.4.9 Quality

The DOT PM is responsible to provide a project solution that satisfies the project scope and meets or exceeds all applicable local, state, and federal standards. The consultant PM is responsible to satisfy the contractual SoS. The DOT PM should ensure these two quality efforts are consistent. Figure 7.23 details the DOT and Consultant PM responsibilities regarding quality.

DOT PM	Consultant PM
Ensure consultant's quality QC/QA processes comply with established plans	Conduct contracted services in compliance with established QA/QC standards and procedures
Ensure consultant work product and deliverables meet established quality standards	Produce deliverables that comply with established QA/QC standards and procedures
Promptly address and resolve all quality issues	Promptly address and resolve all quality issues

Figure 7.23 DOT compared to consultant PM – quality.

7.4.10 Communications

Communications are critical to any successful partnership. The DOT PM and the consultant PM have different communication responsibilities. The vital exception is the shared responsibility to communicate with each other. The relationship between the two PMs is critical to project success. Figure 7.24 details the DOT and Consultant PM responsibilities regarding communications.

DOT PM	Consultant PM
Appropriately escalate identified issues in a timely way to secure decisions, direction, or responses so as not to delay the project	Promptly inform DOT PM of issues that are limiting or inhibiting progress
Ensure consultant PM knows and understands all applicable DOT policies and procedures	Provide enough technical detail to the DOT PM so they fully understand key technical issues
Communicate project updates to stakeholders, as defined in the Stakeholder Engagement Plan	Support DOT PM in stakeholder engagement, as requested and specified in SoS
Provide timely and accurate project updates to leadership	Provide timely and accurate project status updates to DOT PM
Coordinate all public-facing project communications, including media, websites, and social media	Support DOT PM in all public-facing communications, as requested and specified in SoS
Coordinate all related agency interactions, unless previously agreed Consultant would direct these efforts	Support DOT in all related agency interactions, unless previously agreed Consultant would direct these efforts

Figure 7.24 DOT compared to consultant PM – communications.

7.4.11 Tips for Managing Consultants

- Stay involved. Do not hand over the keys to the project and expect the result to satisfy all of your objectives and expectations.
- Appreciate their expertise. They may be the SME and know more than you on certain issues. That is why you hired them. Listen. However, as PM you should know the pulse of your organization's and stakeholders' needs, expectations, and constraints. You need to guide their expertise to ensure project objectives are being met.
- Establish communication expectations. Have consultant provide weekly status updates that summarize what they accomplished this week, what they intend to accomplish next week, outstanding risks and challenges, a snapshot of schedule and budget milestones, and any decisions, directions, or actions they need to proceed.
- Meet in person. Have in-person status update meetings on a biweekly or monthly schedule.
- Build real relationships. Get to know their key team members, the SMEs, and those doing the grunt work. Visit their office. See their work product as it develops.
- Realize transportation development can be small world. As your career progresses, you may work with these individuals again, be it with the same company in the same role, or with other firms in very different capacities. In many cases, it is reasonable to expect your paths will cross again in some way. As such, approach each project with the personal goal to build or strengthen long-term professional relationships.
- You are hiring people. Companies may bring resources and processes and institutional knowledge, but individuals still will design your project. Choose wisely.
- Consultants are not all created equal. Consultants are not a commodity, that being the product is essentially equivalent regardless of who produced it. When your organization

hires them, you are renting their mind, buying their ideas and advice. Build long-term relationships with those that do exceptional work.

- When negotiating fees, consider the big financial picture. While some consultants charge more than others, consider the total project cost. Smart, sound, and innovative designs will save you construction money, as well as long-term operations and maintenance costs. Solutions designed with practical and cost-efficient MOT and SOC approaches will save you construction dollars. A complete and thorough set of plans and specs with minimal errors or omissions will save you construction money.
- Respect their chain of command. Just because you can go directly to their key personnel to get answers or action, don't do it unless those direct lines of communication have been discussed and are acceptable to both parties. Follow the agreed upon communication protocol that should be detailed in the Project Management Plan.
- Be open and honest. Strive to create an environment where the consultants truly are an extension of your staff. The relationship should not be contentious or confrontational, but rather a collaborative cooperation built on trust and mutual respect.
- Address issues promptly. Be it performance, personality, expectations, budget, billing, schedule, scope, or quality related. Ignoring the problem usually only makes it worse. As PM, you need to tackle difficult situations promptly and professionally. Not only is the smart thing to do for your project, it is the right thing to do.
- Especially when conflict arises, remind yourself you are on the same team. You need each other in order for you both to be successful. Search for win-win solutions. Find a way to make it work.
- Celebrate successes. Although in different ways, the consultants may be vested as much or more in the project as you are.
- Say *Thank You*!

8

Managing Quality

"Quality is never an accident. It is always the result of intelligent effort."

– John Ruskin

8.1 Defining Quality

"Quality means doing it right when no one is looking."

– Henry Ford

8.1.1 Overview

I was in a portfolio performance meeting once when someone wisely asked, "What is the point? We can push a lot of crappy projects out the door on-time and on-budget. But are we making good decisions?" This is the essence of why quality matters.

At its most basic level, the goal of most transportation projects is to design and build an improvement that increases safety and positively impacts commerce and the community. Ideally this means developing and delivering a fiscally responsible final product that serves its intended purpose and satisfies stakeholder expectations.

Meanwhile, the old adage remains true, "*Good, Fast, and Cheap...pick any two.*" Public transportation organizations should have a deeper perspective when answering this question as they typically own and maintain the asset. As such, life cycle costs should often be considered. Even on well-constructed projects, the long-term maintenance costs may be significant. Quality matters.

So how do you as the owner effectively ensure a quality product, especially in those circumstances where the construction contractor is selected by low-bid? The textbook answer is to create a set of approved plans and specifications that satisfies the project scope while being designed in accordance with all applicable standards and regulations, and then build it in accordance with these contract documents. This straightforward objective can be challenging to achieve.

If you concentrate too little on plan quality, you will pay for it in construction change orders, operation, maintenance, and life cycle costs. If you concentrate too much on plan quality, you may never get the plan set approved for construction as error- and

Transportation Project Management, First Edition. Rob Tieman.
© 2023 John Wiley & Sons Ltd. Published 2023 by John Wiley & Sons Ltd.

omission-free plans and specifications are truly rare...like spotting Bigfoot rare. Striking this delicate balance within the organizational framework and established boundaries is often the responsibility of the PM.

8.1.2 Competing Definitions of Quality

Max Depree said, "*The first responsibility of a leader is to define reality.*" It is important to determine how your organization does, and perhaps should, define quality. Accomplished PMs acknowledge and understand that different team members and stakeholders may hold and operate under very different definitions of quality. As such, the waters may quickly become muddied when these different success criteria mix. It is the PM's responsibility to bring clarity and focus to these discussions.

Stakeholders may judge quality from a single or subjective issue that they extend to the entirety of the project (e.g., how competently a presenter answered questions at citizens information meeting, how smoothly the right-of-way acquisition process went, how business's access concerns were addressed, how efficiently school bus routes were detoured during construction, and so forth). Design consultants may judge quality by the clarity of interaction or direction with the project owner (e.g., number of scope supplements due to project changes, number of comments by plan reviewers, number of non-billable hours to correct errors or omissions, number of questions by contractors at pre-bid meeting, and so forth). Contractors may judge quality by the inherent ambiguity, completeness, and constructability of the plans (e.g., are quantities accurate, are there adequate staging areas, does the Maintenance of Traffic plan include enough temporary pavement to stage the necessary equipment, and so forth). The owner-operator may judge quality by the project's triple constraint and leadership expectations (e.g., on-time, on-budget, number and severity of plan review comment, did addressing plan review comments add additional time or cost, number and cost of consultant scope supplements and construction change orders, ongoing operations and maintenance costs of the assets, degree to which process and product met or exceeded executive and political leadership and key stakeholder expectations, and so forth).

It can sometimes seem like consensus on defining quality will never be reached. This may be especially true for PMs as they float between the different project partners and stakeholders. It is possible that some of these various definitions may not only be inconsistent with each other, but may be in outright conflict. It is often the PM who navigates resolution in these situations.

Acknowledgment of the differences can often lay the framework for a productive discussion. The reference point from there should be how the project's owner-operator define success, as they are the entity that is paying for, and administering, the project. The definition of quality may be explicitly explained in their guidance, or may be implicitly implied in their workflows and performance metrics. Regardless, start there and then work to incorporate other definitions as you are able.

8.1.3 Qualitative vs. Quantitative Quality

At its core, most evaluate quality based upon the degree to which the perceived product or process exceeds our expectations. These can take two distinct forms: qualitative (subjective) or quantitative (objective).

Qualitative quality can be difficult to capture, or even ascertain. A new road widening may look fresh and new and impress some citizens, while others notice slightly off-parallel lane lines, or poorly timed traffic signals, and conclude it was a sloppy job. It is not uncommon that many form their quality judgment based upon a small number of aesthetic elements, and then the conclusion is extended to the entire project. As such, finishing touches matter.

Quantitative quality is able to be measured and has direct and indirect costs associated with a low-quality product. There are two main reasons for quantitative quality on a public sector project. First is to ensure the product is built according to the plans, specifications, contract documents, and applicable standards and regulations. It should be noted that demonstrating this as evidenced through consistent execution of an approved quality control plan can be an essential requirement of some public funding sources. Secondly, transportation projects are often owned and maintained indefinitely by the project's owner-operator; therefore, you have a vested interest in ensuring you "get what you paid for." The only reasonable and reliable way to make certain of this is to plan, implement, and execute an effective quality control plan. Quantitative quality is where you and your project team should and will spend most of your quality management time.

Impressions of perceived quality can also be formed over time, or be extended to your project based on something completely unrelated. Imagine a driver annoyed by an unrepaired pothole on a road that he recalls just being built a few years ago. He may easily extend his opinions of quality and service regarding that pothole to the administration of your nearby road project. Your organization's reputation and credibility matter when it comes to the public's perception of quality. They will often approach your project from a benevolent or skeptical viewpoint, and evaluate your project through that lens.

8.1.4 Quality Control (Product) vs. Quality Assurance (Process)

For some, Quality Control (QC) and Quality Assurance (QA) are interchangeable ideas. In reality, there are fundamental differences between these two concepts that have an inherent and interdependent duality, as illustrated in Figure 8.1. Both are important, and serve very different functions. However, it is when they are combined that they work together to dramatically improve quality. Their intentional and strategic union should be the foundation of every project's, program's, portfolio's, and organization's QA/QC plan.

At its core, Quality Assurance focuses on the process, while Quality Control focuses on the product. Figure 8.2 compares these two complementing concepts. Understanding the differences between these two will strengthen your ability to effectively manage and control project quality during development.

QC is inherently harder to ensure and evaluate. As such, most firms and organizations concentrate their effort on QA. QA processes can be standardized, documented, and audited. It is important to

Figure 8.1 QA/QC relationship.

	Quality Assurance (QA)	Quality Control (QC)
Primary focus	Process of product creation	End product (deliverable)
Approach	Proactive	Reactive
Purpose	Prevents defects	Identifies defects
How to monitor	Audit workflows	Plan reviews and field tests
Function	Managing quality	Correcting quality
Most useful to	Managers	Designers
End goal	Enables better QC by ensuring workflows followed	Produce better product (PS&E) – Plans, Specifications, and Estimates
Priority to bottom line	Less Important – supports behavior to encourage better product	More Important – directly impacts product
Difficulty in evaluating	Easier to measure	Harder to measure
When done	Performed in parallel with product development	Performed at milestone plan submissions and reviews

Figure 8.2 QA/QC comparison.

realize that this is one of those areas where good intentions can unwittingly create a bureaucratic beast that requires constant feeding.

QA is meant to complement and empower QC. QC rarely occurs on its own without QA. But successful QA does not inherently ensure successful QC. Organizations and project managers should beware the point of diminishing returns. Make sure the juice is worth the squeeze.

8.1.5 Layers of QC/QA

Pending the size and complexity of the project, there may be many layers or QC and QA both in development and delivery of the project. These layers may be within a single organization, or shared across divisional or organizational lines.

While not all projects will have all of the following components, every project should have some of them, both during development and in delivery.

- Quality Control (QC) focuses on the quality of the product. Examples during development may include the compliance of the plans with applicable standards and regulations, and the amount and severity of errors and omissions on the plans and specifications. Examples during delivery may include compliance of installed materials with applicable standards, and ensuring approved products are installed per manufacturer's recommendations. QC efforts often involve the designers, engineering discipline leads, plan reviewers, construction inspectors, and others with designated QC roles and responsibilities.
- Quality Assurance (QA) focuses on the quality of the process. This effort can be summarized as checking the checkers; the checkers being those who conduct the QC effort to check those who produce the plans or build the end product. QA should be done by

someone who did not do the QC. In design, this is often others within the same engineering firm. In construction, it is often personnel within a different company, organization, or department who complete this effort. Examples during development may include documenting QC reviews or cataloging quantitative statistics on plan comments. Examples during delivery may include random material testing to verify the accuracy of the QC tests, and ensuring compliance of QC testing frequency and results with the approved QA/QC plan.

- Independent Assurance (IA) is an unbiased monitoring of the QC and QA efforts. IA is checking the checkers (QA) who check the checkers (QC). To ensure an unbiased evaluation, this should be an independent effort by those not associated in any way with the QC or QA activities. Any testing or sampling that is done under IA should be conducted on different equipment. IA is not typically performed on design, and used in construction when specified in requirements.
- Independent Verification (IV) is an unbiased monitoring of the IA effort. This becomes the fourth layer of quality management – checking the checkers (IA) who check the checkers (QA) who check the checkers (QC). IV is not typically performed on design, and is used in construction when specified in requirements.
- Owner Independent Assessment (OIA) is similar to IA, but is conducted by the project owner. This is perhaps most common during Design-Build. If the Design-Build team is contractually obligated to adhere to the approved QA/QC plan by conducting QC, QA, and IA, the project owner may implement OIA to "spot-check" and monitor the Design-Build team's QA/QC efforts. OIA allows the project owner to judge for themselves if those responsible are correctly implementing and complying with the approved QA/QC plan.
- Owner Independent Validation (OIV) is similar to OIA, but different in that it is commonly focused more on ensuring QA/QC compliance in order to process payment. As such, OIV often concentrates on verifying the quality of products incorporated into the project meet acceptable specifications.
- Quality Assurance Manager (QAM) is the designated individual responsible for ensuring project adherence to, and compliance with, the approved QA/QC plan. This should be one individual, not a committee, who is defined in the approved QA/QC plan. Pending the size and complexity of the project, this is often a more senior management role with the authority to direct other personnel or contracted organizations' efforts.
- Construction Engineering Inspection (CEI) is an overarching umbrella under which can be construction administration, testing, inspections, surveying, and other construction oversight-related tasks. QC/QA often drives much of the CEI effort. On larger, more complex projects this is often a service that a third party provides for the owner. In this instance, the owner would oversee the CEI firm who is overseeing the contractor.

8.2 Monitoring and Controlling Quality

"Be a yardstick of quality. Some people aren't used to an environment where excellence is expected."

– Steve Jobs

8.2.1 Fundamental Approaches

Plan quality management is the process of identifying quality requirements and standards for the project, and documenting how the project will demonstrate compliance. Managing quality is the process of translating the quality management plan into executable quality activities that incorporate the organization's quality policies into the project. Controlling quality is the process of monitoring and recording results of executing the quality activities to assess performance and implement necessary changes.

There are a variety of quality management methodologies that fall in and out of favor over time. Their focus may vary from promoting uniformity while eliminating variability, to identifying and removing causes of defects, to eliminating wastes, to emphasizing compliance with applicable standards. Regardless of the current quality certification or mindset, they all generally include the steps in Figure 8.3 for existing or new processes.

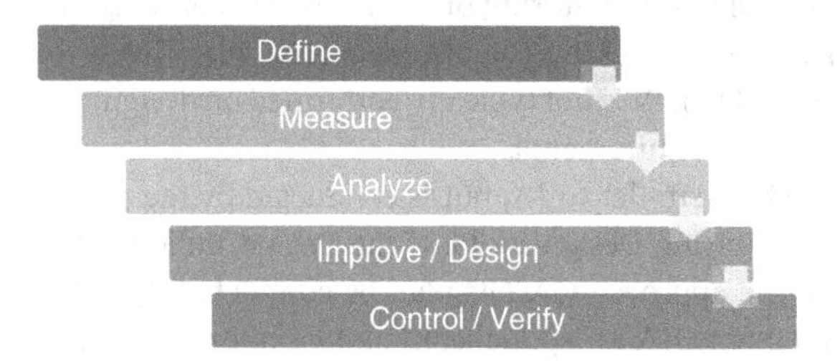

Figure 8.3 Approach to monitor and control quality.

8.2.2 Organizational Maturity

There are many ways to describe an organization's quality management maturity. While it is difficult to capture the many workflow and cultural aspects of an organization that inherently promote or inhibit effective quality management, it can be enlightening to reflect upon where your organization falls within the relative spectrum.

W. Edwards Deming said, "*85% of the reasons for failure are deficiencies in the systems and processes rather than the employee. The role of management is to change the process rather than badgering individual to do better.*" Your organization's maturity level will, in many ways, define and constrain your challenges to effectively manage quality at the project, program, and portfolio levels. Figure 8.4 depicts the spectrum of an organization's quality maturity. Figure 8.5 compares these organizational maturity levels.

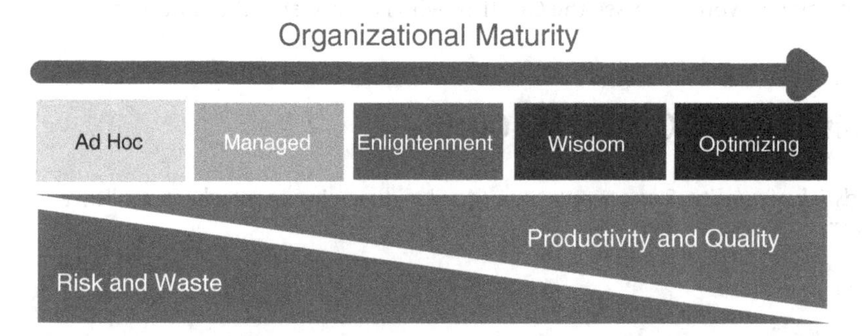

Figure 8.4 Quality and organizational maturity.

	Ad Hoc	Managed	Enlightenment	Wisdom	Optimizing
Project and Workflow Realities	Projects get completed, but are often delayed and over budget	Projects are planned, performed, managed, measured, and controlled at project level	Organization-wide standards provide guidance across project, programs, portfolios	Organization is data driven with quantitative performance improvement objectives	Organization is focused on continuing improvement, and quickly responds to seize opportunities
Quality Control Processes	Very few processes, unpredictable, poorly controlled, reactive, success based upon individual effort	Basic and consistent processes established; project managed reactively with approach that each project is unique	Processes are well defined, documented, standardized, and integrated across organization, consistent practices are proactive	Well-defined processes quantitatively measured and controlled with strategic analysis	Proactive process improvement implemented through qualitative feedback; Industry leader, quality values embedded in corporate culture
How quality problems are handled	Problems addressed as they occur, often without resolution and lots of finger-pointing	Teams formed to address major issues, often settling on expedient solutions	Quality issues are openly discussed and worked to resolution using constructive dialog	Quality issues identified early in development within a culture open to suggestion for improvement	Quality issue prevented
Leadership's attitude to quality management	Does not consider quality management a tool; Tends to blame others for quality problems	Recognizes quality management may be of value, but not willing to provide resources to make it happen	Grasps value of quality management as quality improving, becoming helpful and supportive	Recognize their personal role in continuing to emphasize critical importance of quality management	Consider quality management an essential part of company system and culture
Summary Quote regarding problems with quality	"We don't know or understand why we have quality issues."	"We accept there will always be some quality issues."	"We are actively identifying and resolving quality issues."	"Defect prevention is a routine part of our operation."	"We know why we do not have quality issues."

Figure 8.5 Organizational maturity comparison.

8.2.3 Product vs. Process Issue

When a defect occurs, mature organization and PMs seek to identify the root cause of the issue. Is this a product or process issue? Is the problem with what we are making, or how we are making it? Sometimes it can be both. Consider if an environmental scientist made a critical error on the permit application, and then the internal QA reviews missed it. Does this require an engineering solution, a process change, or both? Generally, one-off errors or omissions should be addressed with engineering solutions, while multiple or repeated errors in a discipline should be addressed with process changes.

In these circumstances, it is important to maintain a healthy perspective while striking a balance between vigilance and understanding. Most of these issues, particularly process changes, will likely be made at a program or portfolio management level. In order to be effective, organizations must have the courage not to dismiss issues for the sake of expediency, or to walk the path of least resistance. Likewise, they should embrace the wisdom not to overreact and to see outliers as what they are. Remember the old adage, "*hard cases make bad law*."

Kaizen is a Japanese business philosophy that focuses on continuously improving processes. The Kaizen approach is often framed in terms of continuous improvement of quality. What this really means is small changes will build incrementally over time. These changes can be the product and the process. Minor improvements compound over time to dramatically improve quality. Quality-focused organizations and PMs build momentum in pursuing excellence. Otherwise, they will struggle with apathy and scramble to react to the inevitable errors and omissions.

8.2.4 Quality Management Plan

The Quality Management Plan is the part of the Project Management Plan that details how QC and QA will be managed, monitored, and controlled throughout the life of the project. The purpose of the quality management plan is to ensure the project is being designed and constructed according to contract documents, specifications, relevant manuals, and applicable standards. Most projects have separate QA/QC plans for design and construction.

While some more complex projects may require a project-specific quality management plan, most projects will largely default to existing organizational protocol. An organization may require different levels of QC/QA pending the size and complexity of the project, the funding source, the procurement or delivery method, or other preestablished decision points or thresholds. Other owner-operators have little-to-no QA/QC plan, and rely on that which the consultant has defined within their corporate structure.

Given that many or most of the project's QC and QA requirements are often predefined by the organization, some PMs may not dedicate much effort to the quality management plan. This can be a costly mistake. The quality of a project is ultimately a PM responsibility. They may attempt to share the blame, but the PM is uniquely positioned to monitor the QC/QA effort and ensure appropriate corrective action is taken when needed.

The quality management plan is a powerful tool that experienced PMs leverage to assist them in this effort. This is the perfect place to document stakeholder quality expectations and detail QC/QA processes in such a way as to bring clarity and direction to the team. Tailoring this plan to your specific project can be enormously helpful in managing expectations and resolving conflict in a constructive way. Remember many quality judgments are comparisons of perceived reality to one's expectations. Plan the work, then work the plan.

8.2.5 Coordination Issues

On large or complex projects, there can be sizable intentional or unintentional ripples from coordination issues that directly impact project quality. Consider if the water resources team increases the slope of a drainage pipe to mitigate up-system hydraulic grade line concerns. This now creates a conflict with the utility group's planned sanitary sewer extension at the outfall end. The sanitary sewer designers can effectively shift the gravity sanitary sewer, but it means more temporary wetland impacts. This increased wetland disturbance sends the environmental group into panic mode because it was not in the submitted permit application and will cross the area threshold pushing them into an Individual Permit which will optimistically add six to nine months to the project schedule. And through all of this scenario, no one thought to bring the lead roadway designer or RW back into the discussion until it surfaces as an error in plan review that the easement areas are insufficient. Small problems can quickly mushroom into larger ones if left unchecked. Well-intentioned modifications can cast a particularly large net of ripple impacts when one makes changes or improvements to the Sequence of Construction, Maintenance of Traffic, or Erosion and Sediment Control plans without ensuring coordination with the other responsible team members.

Exceptional PMs have a diverse knowledge and experience base that provides an insightful perspective. They have the ability to quickly size up a situation and determine the fundamental risks and issues. They recommend solutions that will work and identify those that will not. I remember many occasions as an eager, junior engineer where I was with a senior engineer who demonstrated this omniscience. When I would ask, "*How do you know that won't work?*" or "*Why were you concerned about this and not that?*"; the response often was, "*I've seen it before*" or "*I just know.*" I think most better engineers have had a private moment of pride when realizing they have reached this level of experience. That is certainly not to say they cannot and do not learn new things every day, but there is a foundational base of knowledge and real-world experiences that can give you an advantageous perspective from which you can approach most any situation.

Generally, consulting engineers work with many clients and gain experience designing unique solutions for a variety of operational and situational circumstances. Generally, contracting or construction engineering grants the experience of seeing how plans are built, grasping equipment constraints and potential, understanding effective work sequencing, and the like. Generally, public sector engineers gain the appreciation of building quality products that last and satisfy broader success criteria as they can be involved with the traffic flow, citizen input, and long-term maintenance. You can be an exceptional PM if you have any of these backgrounds, but it is easier if you have done two, even easier if you have done all three. If you have done all three, you have a rare perspective that can be a powerful asset as you have first-hand insight others can't and won't offer. The broader your perspective, the better you will likely be at proactively identifying the intended and unintended consequences of plan modifications, and the more you will know who to bring into the resulting discussions to capture all that is needed so that ripple impacts will not become quality issues.

8.2.6 Is Value Engineering (VE) QA/QC?

Yes and no.

Value Engineering (VE) is an effort to increase design efficiency and excellence while reducing costs in a manner that adds value to the product or process. These seemingly contradictory goals are the mission of a multidisciplined VE team that usually consists of five and ten individuals who have had limited to no exposure to the project. The idea is that fresh eyes from experienced perspectives may see new previously unseen or missed opportunities. The VE objectives are rather straightforward.

- Increase product quality.
- Decrease construction costs.
- Decrease long-term facility ownership costs.
- Reduce construction time.
- Simplify the construction process.
- Minimize project risks.

The VE team typically prepares recommendations that are then evaluated by the design team. Examples of recommendations may include changing a bridge design to ease or accelerate construction, saving pavement sections as opposed to a complete rebuild, targeted narrowing of lane or shoulder widths, leveraging innovative E&S measures to eliminate a sediment basin and the associated easements, redesigning drainage to avoid temporary travel lanes or utility relocations, and the like.

VE can be traced back to the Federal-aid Act of 1970 which required VE on Federal-aid projects. Since that time, various regulations have defined the criteria when a VE study is required. The Federal Highway Administration (FHWA) requires VE on any project with a total estimated cost exceeding $50 million that is designated on the National Highway System (NHS), or any NHS bridge project exceeding $40 million. State DOTs often have significantly lower project cost thresholds. If required, the owner has no choice. If not required, the owner may still desire to conduct VE.

There are many VE strategies and methodologies, but they all generally contain the following steps:

- Information Gathering – VE team reviews the project history, information, and plans with project management and designers. During this step, the VE team strives to understand the project background, context, goals, and operational parameters.
- Brainstorm – VE team brainstorms on ways to build a better mousetrap. This step includes questioning of assumptions and searching for creative, outside-the-box resolutions. The ability to pursue tangential thinking free from initial judgment is crucial to success.
- Analysis – VE team evaluates brainstormed alternatives for materials, approaches, construction means and methods, sequencing, and the like. They then rank these ideas based upon viability, ease to enact, value added, and other project- or organization-specific criteria.
- Development – VE team investigates viable alternatives with more in-depth analysis to determine feasibility, risks, and rewards of options. They then summarize and present the findings and recommendations to the designers and decision makers. Effective presentation and marketing of these ideas can facilitate consideration and expedite acceptance.
- Implementation – This is often the most challenging of steps where the recommended and accepted ideas are integrated into the project. Effective project management leadership is essential to successfully incorporating these alternate solutions into the project design in a cooperative, productive manner that I matched with positive morale among the staff and designers.

As with all design QA/QC efforts, strategically choosing the responsible individuals who are conducting the study is much more critical to success than the process itself. If done correctly, VE can prove to be a powerful tool to reduce cost or construction time while ensuring the Scope still satisfies the project objectives.

VE is a process that should be leveraged early on in the design process, before the designed solutions get too much meat on the bones. VE should ask big questions first, such as is this the right solution for this situation? FHWA offers relevant guidance, which is often strengthened by DOT protocols. The goal is to be smart, practical, and innovative.

VE is a process.

VE is also a mindset.

Experienced PMs and designers capture the mindset from the process and apply it to their smaller projects throughout development.

8.3 Measuring Quality

> "Trust, but verify."
>
> – Ronald Reagan

8.3.1 Hard Truth

Most transportation organizations espouse that plan quality is of the utmost importance, and yet having trouble articulating exactly how plan quality is measured. What does that even look like?

One organization may gauge plan quality by the number and value of construction change orders. Does this capture needless gold-plating or construction issues that never surface because of exceptional construction inspectors? Another organization may gauge plan quality by the number of review comments at each submission. Does this capture the severity of comments, or the competence of the reviewers? Yet another organization may gauge plan quality by whether the project made it through a detailed project development process on-time and on-budget. Does this acknowledge that you can produce a bunch of crappy projects on-time and on-budget? And so it goes.

The hard truth is that plan quality is extremely difficult to quantitatively measure. This challenge is complicated when organizations focus on a single quality metric, which can easily provide a distorted view of reality.

During development, engineering design is professional service. It is not a widget. Consultants are hired for their expertise and professional judgment. Professional Engineers are legally recognized experts in their field. Many other industries measure quality by the number of defects or the financial impacts. These are valid approaches, but far more difficult to implement on public transportation projects.

The reality is QC is hard to measure. QA is far easier to measure and control. By intention or ignorance, many organizations silently accept this reality. That is why most QA/QC programs emphasize the QA, assuming the QC will follow suit.

QA is process oriented. As such, engineers can readily determine ways to ensure processes and procedures are followed, the correct reviews and signatures by the correct levels of employees are completed, and how to track and document all of this in ways that are defendable and auditable. It is important to understand that many transportation owner-operator organizations have their own QA/QC processes and procedures that often need to be combined with those that are established within consulting firms or other partner organizations. These requirements should be captured and discussed to reduce redundancy of effort, and to adequately accommodate all QA/QC efforts into the project's schedule and budget.

8.3.2 Construction Engineering Inspections (CEI)

Managing quality (QC) is much clearer during construction. Projects are built in the field. After project close-out, what matters is the product that was constructed. The vast majority of all project budgets are allocated for construction. And in most cases, it is being built by the low-bid contractor. To complicate matters, your organization or entity will likely own and maintain the facility in perpetuity. Imagine hiring a low-bid contractor through sealed bids for a fundamental home improvement that you will own, maintain, and rely on the use of every day for the foreseeable future. Throughout the home improvement, you would naturally want to check and ensure acceptable materials were being used and properly installed. It is the same for public sector projects.

CEI is a well-developed field, the purpose of which is to ensure you get what you paid for. This allows you make adjustments if the tests fail or to ensure compliance if the results are favorable. It is worth noting that CEI can be expensive. It is imperative you budget for this in your project development estimates. One may determine they can sacrifice CEI as a way to cut costs, but know that CEI performs an important function. If you are pondering this course of action, the question to ask yourself if you can't afford CEI is can you afford the alternative?

It is worth noting that there are different kinds of construction inspectors that I have divided into the following categories:

- Construction Inspectors are responsible to ensure the project is built according to the approved plans and bid specifications and that the project and the construction process complies with all applicable laws, standards, and regulations. They typically work to this end by ensuring the project is constructed according to the approved QA/QC plan. These critical individuals work for whoever is responsible for construction management, typically, the project owner, contractor, or third party hired by the project owner. This is an incredibly important position as it is where the proverbial rubber meets the road. An experienced, innovative, pragmatic, and professional construction inspector can make a bad set of plans look good and make a difficult project go smoothly; whereas a mediocre inspector will cost your organization time and money.
- Enforcement Inspectors are those individuals who inspect the specific aspects of the construction to ensure compliance with specific, applicable codes or regulations. These are typically from the governing locality or other governmental organization with jurisdiction over the project. Examples include building inspectors, erosion and sediment control inspectors, and inspections from permitting agencies.
- Governing Agency inspectors are from governing organizations that have some involvement or jurisdiction over the project, are from the organization not directly responsible

for construction management, and are not acting in the capacity of an Enforcement Inspector. An example would be a local roadway project that utilizes federal funds. At any time during this construction, inspectors from the state's Department of Transportation or Federal Highway Administration may show up to review any aspect of your project.

- Manufacturer inspectors are those who conduct inspections at businesses that manufacture materials for the project. These individuals typically work for the manufacturer. Examples include bridge beam inspections at the steel mill, concrete pipe fabrication inspections, and aggregate inspections at an asphalt plant.
- Specialty Inspectors are those individuals who inspect a specific aspect of the project for which they have a unique and uncommon knowledge. The most common example might be bridge inspectors who require specialized training to achieve bridge inspection-specific certifications or certified SCUBA inspectors for underwater inspections.
- Testing Inspectors are those individuals who conduct materials testing and work for geotechnical, environmental, civil, or laboratory testing service companies. Testing inspectors perform the predefined tests utilizing properly calibrated equipment and following the standardized procedure. Their role is critical to provide accurate and timely information as to what is being put in the ground, but they typically bear no decision-making responsibilities other than to report the results to the construction inspectors. I separate these from construction inspectors who are concerned with the materials, the process, and the product.

8.3.3 QA/QC Tips

Measuring quality is a difficult challenge for most transportation organizations. There is no silver bullet QC metric during plan development. Most organizations, by design or default, have settled on systems that work for them given their constraints and risk appetite. Below are ten best practices that can perhaps be incorporated pending your organization's culture and constraints:

1) If your state does not require it, encourage discipline leads to stamp and seal documents, not just the PM. Responsibility and initial quality rise with the signing of plans by SMEs.
2) Leverage consultants' and other agencies' QA/QC practices.
3) Be responsible by pursuing consulting firms with glaring errors and omissions.
4) Ensure you accommodate QA/QC efforts into the project budget and schedule.
5) Involve all disciplines, and the right SMEs, in milestone plan reviews.
6) Schedule adequate staff review time for milestone plan submissions.
7) Host a comprehensive final review of the Plans, Specifications, and Estimate (PSE).
8) Leverage over-the-shoulder reviews with gray-haired, experienced engineers who understand discipline interdependencies during development and know construction realities.
9) Establish QA/QC processes and procedures that serve your needs and then consistently follow them.
10) Remember the goal of QA/QC is to produce quality plans that enable smooth, surprise-free construction. Find the right balance where you are accomplishing this goal and not just feeding a created bureaucratic beast.

8.4 The Costs of Quality

"You may delay, but time will not."

– Benjamin Franklin

Most mature organizations embrace the notion that aggressively pursuing quality management provides a good return on their investment. In transportation, this becomes more evident as one widens their perspective.

Consider a plan design flaw, be it an error or omission. The design engineer may deem correcting it is not worth the required rework. The discipline lead may add that it can best be handled in the field as the benefit of fixing it not worth the time. The PM may realize this error will significantly impact other disciplines and have concerns over schedule delays to Advertisement. The Program or Portfolio Manager may be concerned pushing the error to the field will increase change order costs. Organizational leaders may be concerned field solutions will result in higher asset life cycle costs by increasing future operation and maintenance cost.

Obviously not all plan errors and omissions are created equal, nor do they result in equal impacts to the project's budget, scope, and schedule, and the asset's life cycle costs. As such, it can be a challenge to balance competing priorities. Add to this the reality that stakeholders don't often appreciate the cost of quality.

It is important to acknowledge that there has never been, and will never be, a perfect, error-free set of transportation plans. The PM should leverage organizational guidance, and the consultant's QC/QA protocols if applicable, to catch and correct significant issues. What is a significant issue? This should be defined by the organization, with more mature entities having tighter definitions. Generally, a significant issue is one that adversely impacts the project's budget, scope, or schedule, or introduces longer-term asset operation and maintenance costs, beyond the preestablished organizational thresholds.

It can be difficult for a PM to determine which issues are significant, especially when there are conflicting priorities. Imagine trying to reconcile conflicting organizational pressures to resolve issues on the plans so that they are not pushed into the field while still making the scheduled Advertisement date. In these instances, organizational protocol should provide guidance. Additionally, the PM and public transportation organization bears ethical obligations to the public. If you become aware of a quality issue, then you bear responsibility to take action. The decision may be to do nothing and accept the risk or resulting costs, but be intentional and take action. Denial is not an ethical choice.

It is important to remember that Quality Management should not be an afterthought or a series of selective decisions. To be effective, Quality Management must be woven into the fabric of project development such that significant issues are identified early and aggressively addressed to resolution. This requires transparency, trust, and cooperative collaboration toward a common goal.

8.4.1 Cost of Conformance and Nonconformance

Generally speaking, the cost of errors and omissions increase over time. The lowest cost of an error is never to make it in the first place. This is where experienced designers paired with a mature quality management culture provide a tremendous return on investment.

The cost of many errors and omissions are never realized because they don't occur, or are quickly identified and resolved. As the design progresses, rework to correct errors become more complicated. Discipline details are finalized. Permits are submitted or issued. Right-of-Way and easement lines are delineated and presented to the public. Maintenance of Traffic and Sequence of Construction plans take shape. The further the design progresses, the more the ripple impact of correcting errors increases. Costs typically increase even more if the error or omission is corrected in the field. Correcting errors or omissions on the plans before advertisement can be pennies-on-the-dollar compared to corrective action with a contractor's change order.

The different aspects of the quality costs can be divided into two different categories: Conformance and Nonconformance. Conformance costs are costs incurred in the effort to prevent errors or defects. Nonconformance costs are those incurred when a defect or error has occurred. Costs increase as a quality issue passes from conformance to nonconformance.

Conformance costs are frequently divided into two subsections: Prevention costs and Appraisal costs. Prevention costs are those efforts to provide and support a culture of error prevention. These may include training, process documentation, and adequate resources to do the job right the first time. Appraisal costs are costs incurred to measure, monitor, and control Quality.

Nonconformance costs are frequently separated into two subsections: Internal failure costs and External failure costs. Internal failure costs are rework costs to correct the error or omission before the product is delivered. This could be rework costs before the design plan are advertised for construction. External failure costs are costs incurred to correct the error and omissions after the product is delivered. This could be the change order costs to correct the errors or omissions in the field.

8.4.2 I Thought Engineers Were Supposed to Be Smart?

A question many PMs have pondered is how can so many bright, educated engineers be so wrong? This is especially frustrating considering most engineers are detail-oriented, and internally driven to produce quality work. How can accomplished firms of intelligent, conscientious professionals occasionally produce designs with such egregious errors and omissions, or make recommendations that so clearly defy common sense?

It could be a proposed directional island whose nose extends into a travel lane, or a proposed sediment basin atop an existing shallow sanitary sewer force main, or a sequence of construction that traps water behind temporary pavement, or not providing needed staging areas for a crane, or not sizing temporary easements to accommodate the construction equipment, or proposing a sound wall directly atop a natural gas pipeline vault, or countless other examples. Most experienced PMs have all looked across a conference table and politely paraphrased the unfiltered thought running through their head, "*Are you insane? What are you doing to me?*"

In these and other situations, it can be helpful to be mindful of Hanlon's Razor. This old adage states, "*Never attribute to malice that which is adequately explained by stupidity.*" In other words, bad things most often happen not because of bad intentions, but because of reasons not directly related to you. Maybe they are incompetent? Or maybe they are distracted, or unmotivated, or untrained, or inexperienced, or are burned out and thinking about their upcoming vacation, or going through a difficult medical situation with an aging parent, or stressing about childcare, or just having a bad day.

As PM, you are leading the team. When errors and omissions occur and are identified, they must be acknowledged and resolved. Part of your job is to ensure this happens. Consultants will often live up to, or slide down to, meet the owner-operator expectations. A PM should expect quality plans, and create a culture that promotes them. This often includes transparency of errors, which enables faster and more thoroughly capturing the intended and unintended consequences of the errors or omissions. Perhaps most important, this often includes the freedom to openly and honestly discuss errors and omissions. Work the issue; don't attack individuals. First examine the processes, their internal training, their internal QA/QC reviews, and so forth. If a pattern of errors or omissions develops, then the conversations need to change, and likely be quickly elevated so as not to jeopardize the project's budget, scope, or schedule.

It is important to acknowledge not all engineers are created equal. They have different strengths, different experiences, different level of commitment, different personalities, and different communication styles. Similarly, not all consultant firms are created equal. They have different cultures, different leadership, different profit models, different priorities, and different training and mentoring for their employees. Consultants are not interchangeable commodities. When you hire an engineer, you are renting their mind. You are leveraging their expertise in anticipation that they will solve your problems within given constraints and move your project forward.

Pending the firms you hire, the ones doing the design may not be the individuals who led the short-list interview presentation during the professional procurement process. Many firms use draftsman, technicians, and junior engineers to do much of the foundational design and modeling. In these instances, their work product should be closely reviewed before being presented to the owner-operator. This can be especially true for junior engineers whose perspectives, knowledge, and experience are commonly at vastly lower levels than a senior engineer. Many junior engineers calculate numbers with no real contextual understanding of what those numbers mean in the real world. Perhaps more important, most junior engineers have spent little to no time in the field, and truly have no idea how projects are actually built.

Plans are built in the field. Despite the obviousness of this statement, many designers do not design plans with this in mind. This quickly becomes apparent when in a meeting you ask the consultant how are you going to widen the road and maintain two-way traffic without temporary pavement, or what equipment will be used to set the stream embankment armoring solution you propose and how will you get it to the point of installation. The reality is that many designers don't have sufficient construction experience to draw from as they design the project. This can be especially true for those firms that specialize in large or complex public transportation projects where there can be some length of time between project origination, design, and construction. This is why I would ask designers in professional services short list interviews which of their sample projects have been built. It can be surprising how many projects engineers have designed that have never been built. Some lessons you best learn when your plan goes to the field. There is a huge difference between a design working on paper, and that which can be built. This is another case where you want those gray-haired engineers looking at the plans. Remember, the goal is to build things, not generate great looking plans.

8.4.3 Thoughts to Ponder

There are two spheres of QA/QC during design that should coordinate and be complementary of each other, those performed by the consultant and those performed by the project owner-operator or sponsoring organization. Most consulting firms have some internal QA/QC plan. Some take this far more seriously than others. This should be discussed and understood prior to retaining their services. For smaller projects where draftsman, technicians, or junior engineers are doing the bulk of the design, it is imperative that more experienced eyes review the work product before deliverables are submitted. It can be advantageous if reviewers have different areas of expertise. For instance, one may look for constructability issues with the design while another may review the accuracy of hydrologic and hydraulic numbers. These same issues are present on larger and more complex projects, along with increased coordination issues. In general, the larger and more complex the project, the more time and effort should be spent on design QA/QC. Often larger projects will have a project-specific design QA/QC plan, complete with hierarchical charts, defined roles, regular meetings, and the like.

The good news is that gross plan errors and omissions are relatively rare. In general, this seems to be truer for larger projects where more experienced firms are employed that typically have more rigorous QA/QC programs that are administered by more experienced and accomplished senior engineers. However, as the cliché goes, "*the devil is in the details.*" Budget and schedule consequences from an erroneous unit cost line-item specification or surveying error can become significant. Identifying and plugging these potential contractor change order holes in the plans and bid specifications are where many better engineering firms earn their fees and well-deserved reputations.

Irrespective of the consultant's QA/QC efforts, the product is purchased by the project owner. Organizations have vastly different opinions on how best to ensure they are receiving quality plans and bid specifications. Regardless of where your organization falls on this maturity spectrum, it is dangerous for any owner agency to do nothing. They may assume they are hiring professional engineers who will seal the plans and bid specifications, don't review the work, and don't institute contractual measures to mitigate their risks. You can only bury your head in the sand for so long before your foolishness is irrefutably confirmed.

It should be noted that one critical question many owner-operators never bother to ask or answer is whether they are approving or accepting consultant-designed plans. The answer to this question should directly impact your project's QA/QC approach and plans. If you are approving the plans, then you are inherently implying you have the expertise to adequately do so, and should take your organization's QA/QC responsibilities very seriously. If you are accepting the plans, then you should ensure your contracts adequately transfer appropriate risks to the consultant such that they protect the owner-operator in a way that best ensures quality plans and products.

So, what can your public organization do to manage plan quality? First you need to truly accept that managing quality is an investment worth making and not wasted effort. Make a concerted effort to advance your organizational maturity. Evaluate your staff resources and enact a pragmatic approach that adds value to the project. When developing and implementing a design QA/QC program, there is no substitute for having the right people involved in examining and reviewing the plans and bid specifications. A colorful hierarchical QA/QC team chart and detailed workflow document is worth far less than a few candid meetings with respected professionals who are experts in their fields. For example, if your organization is administering the construction, get your construction SMEs with decades of institutional knowledge to review the plans and specs. They may not like it, but it is an investment that will pay big dividends.

So, what can you as a public sector PM do to manage plan quality? This is an especially critical role on larger, more complex projects where the QA/QC risks are less based in lack-of-knowledge and more rooted in coordination. I recommend having regular project walk-throughs between the major milestone plan submissions. These can be in the field where you walk the site or in a large conference room where you can walk around a roll-plan. Have all engineering team disciplines present, start on one side of the project, and walk and talk your way through it. This helps ensure each of the engineering discipline trains that are concurrently running down their own tracks are still heading in the same direction. Equally important is to accept the role of the communications clearing house for your project. As PM, you are in more meetings on more aspects of this project than anyone else. As best you can, the PM needs to ensure information, improvements, concerns, and decisions are distributed to the right individuals to minimize unintended ripple effects that may eventually surface as plan error or omissions.

9

Communications

"Courage is what it takes to stand up and speak; courage is also what it takes to sit down and listen."

– Winston Churchill

9.1 Project Communications

"The single biggest problem in communication is the illusion that it has taken place."

– George Bernard Shaw

This may be the most important chapter in this book.

Project Management Institute (PMI) estimates PMs spend 85% of their time communicating with others. It is almost impossible to overemphasize the importance of consistent, effective communication between the PM, the project team, executive leadership, and internal and external stakeholders. The ability to consistently communicate effectively is foundational in order to implement much of what this book discusses. PMI studies have determined "*Ineffective communications is the primary contributor to project failure one third of the time, and had a negative impact on project success more than half the time.*" Given the obvious importance of communications, why do so many continue to struggle? As a PM, how should you effectively manage project communications?

Communication is not just a PM obligation or responsibility, it is fundamental to all you do. With the exceptions of perhaps risk management and operating within the triple constraint of budget, scope, and schedule, Communication is a priority that should permeate all we do. Effective project management relies on effectively controlling the flow of project information. It is imperative that you, as the PM, know the right information at the right time. Similarly, it is imperative your team and stakeholders know the right information at the right times. This is central to all that you do. The PM is in a unique position to receive, evaluate, and distribute project information. Discerning the importance and unintended consequences of new information, in addition to knowing who needs to know what and when, is foundational to a successful project.

Besides discussing and distributing new information, regular updates are an important technique to build trust and credibility with your team and stakeholders. Even if there is

Transportation Project Management, First Edition. Rob Tieman.
© 2023 John Wiley & Sons Ltd. Published 2023 by John Wiley & Sons Ltd.

nothing new to report, a polite check-in can provide assurance and validation. You may be working your tail off to track down the answer to someone else's question, but unless you communicate that you are actively working on it, how are they to know. In silence, their default assumptions regarding you and your priorities may be quite different than reality. And especially for stakeholders, their perception is their reality. Effective communication is the cornerstone for managing expectations, which is discussed in Chapter 10 of this book.

Successful PMs differentiate between communication and communications. Communication (singular) is the process of exchanging information. These are focused acts that heavily rely on our personal communication strengths and weaknesses. These are soft skills that we can individually work to improve. Communications (plural) is the means by which communication takes place. This includes the framework within which communication occurs. A project's communications can, and should, be managed, which involves three distinct processes: Establishing a Communications Plan, Managing Communications, and Monitoring Communications.

9.1.1 The Communications Plan

A project's Communications Plan is an intentional approach that documents when and how, who needs to know what. This effort would ideally be completed early in the project and included as part of the Project Management Plan. However, if one is not completed, better late than never. It is important, perhaps more so than you might think. PMI cites that, "*High-performing organizations (those completing greater than 80 percent of their projects on time, on budget and within goals) create formal communications plans for nearly twice as many projects as their lower performing counterparts (those completing less than 60 percent of their projects on time, on budget, and within goals).*" The essential components of a communications plan are shown in Figure 9.1.

The size and complexity of the project will often determine the granularity level of the Communications Plan. But in general, they all should have the same basic components. It starts with a listing of *who* needs to receive information and updates. This should include both internal and external stakeholders, and can be individuals or groups. The Stakeholder register should be consulted during this process. The plan should also specify who is responsible to prepare and distribute the specified communications. You then need to determine *what* informational content should be conveyed. The plan should define the anticipated frequency of *when* to communicate, and the method of *how* best to communicate. Pending specifics, it can also be useful to note *why* they should be contacted. Advanced Communications Plans also include clear guidelines on issue escalation, which is often in conjunction with other risk and issue resolution specifics. The premise and intention is to tailor the Communications Plan to satisfy the unique informational needs of all stakeholders.

Figure 9.1 Communications plan components.

For instance, your boss may want a project status update every Friday in email format. Your Program Manager may want an in-person briefing the third Wednesday every month. Your team may benefit from personalized task lists every Monday morning. You may decide to provide key community stakeholder groups a video update every quarter, while others who signed up for the project email list on your organization's website prefer email updates only at major milestones. Regardless of your specifics, the Communications Plan's goal is to establish an intentional, strategic, and workable framework by which project information will be shared. This is a remarkably effective tool to leverage when managing expectations.

The Communications Plan is also a great place to document expectations for project documentation. What will be saved, by whom, where, and in what format? Many organizations already have existing documentation protocols. Perhaps there is a centralized electronic document repository with established naming conventions, versioning nomenclature, file hierarchy, and associated project metadata. If so, this is easy to reference, and augment with any project-specific needs. If your organization is less structured, this plan should carefully detail all project documentation expectations and requirements.

There can be overlap between the Communications and Stakeholder Engagement plans. While they may seem very similar, the Communications Plan should focus on what to say, and to whom it should be said. The Stakeholder Engagement Plan should focus on who to listen to and what feedback you seek to elicit. The Communications Plan ensures the right project information is being distributed to the right people at the right time in the right way, all for the purpose of efficiently driving project progress. The Stakeholder Engagement Plan ensures the right stakeholders are being engaged in an effective way in reasonable frequencies for the purpose of efficiently managing expectations to promote a successful project delivery.

9.1.2 Manage Communications

Managing Communications is executing the Communications Plan that you have prepared. In the day-to-day PM rush of responsibilities, this is one that can easily slip through the cracks. While it rarely shows up in project performance metrics or personal evaluations, know it is foundational to your personal and professional success. There may be no faster way for a PM to build credibility with internal and external stakeholders than to humbly, reliably, and honestly convey progress in a proactive way that exceeds expectations. Credibility leads to trust. More than any budget, scope, or schedule, a PM must fiercely guard their credibility and stakeholder trust.

Throughout the life of the project, you will leverage other work documents as foundational inputs to what you communicate. Plans, engineering studies, issues logs, risk register, and other project-working documents will drive much of what is conveyed and discussed. For it all to work together smoothly, a PM should ensure all these project management tools are fully utilized with current information, and available to the right people at the right time.

Executing the Communications Plan is as much about discipline and persistence as it is communication skills. If you are to provide your boss status updates every Friday and your Program Manager an update the third Wednesday of each month, then make sure you do it. Experienced PMs realize these are not obligations, as much as they are opportunities to report project progress, proactively discuss risks and challenges, and praise team members' success. These are scheduled events designed for you to actively manage expectations. Seize these powerful opportunities.

9.1.3 Monitor Communications

As the project progresses, a strong PM will periodically review the Communications Plan to ensure it is still effectively satisfying the informational needs of all internal and external stakeholders. In your review, start with the *Who, What, When*, and *How* of the plan. Does the Communications framework still make sense? In order to manage the process and control the product, you need to periodically revisit and answer the following two questions: (1) Are we abiding by the plan? (2) Are we getting the desired result?

Throughout the life of a project, it is natural to adjust the communication frequency and method to accommodate changing informational urgencies and sensitivities. As such, the PM should also examine the efficacy of the communications. Perhaps a video call would be more effective than an email update for certain stakeholders. Maybe your weekly status update email to your boss would be more effective if it were a brief conversation instead. Be strategic in your modes of communications. Should the communication be formal or informal, written or spoken, in person or remote? Is this best shared individually, or in a group? Are the methods effectively pushing (one way communication to specific stakeholders), pulling (retrieving information you need), or conversationally interacting with the right information at the right time with the right people? Be smart and intentional.

9.1.4 Four Different Communication Styles

Ned Hermann pioneered the whole brain model, which took the science of the four major brain quadrants and applied it to business. He created the Hermann Brain Dominance Instrument (HBDI). This brain-based study describes how we process information, which can provide remarkable insights allowing us to more effectively communicate with all types of people.

The four major brain quadrants are right brain, left brain, cerebral cortex, and the limbic system. Those that are right-brain dominant tend to value and rely on emotions, intuition, ideas, and human interactions. Conversely, those who are left-brain dominant tend to value and rely on logic, reasoning, details, and facts. Those who are cerebral cortex dominant tend to value and rely on longer term, less tangible solutions. Conversely, those who are limbic system dominant tend to prefer shorter term, more concrete and practical solutions. These four quadrants can be graphically displayed. This creates four distinct communication styles: Process-Oriented, People-Oriented, Action-Oriented, and Idea-Oriented, as shown in Figure 9.2.

Process-Oriented communicators focus on details, processes, and implementation. They tend to be practical, logical, methodical, factual, cautious, patient, sequential, and organized planners. They talk about controlling, management, and analysis. Their assertions often rely on accepted references, proof, tradition, established protocols, and personal experience. This style often engenders reassurance and credibility as they tend to be reliable and

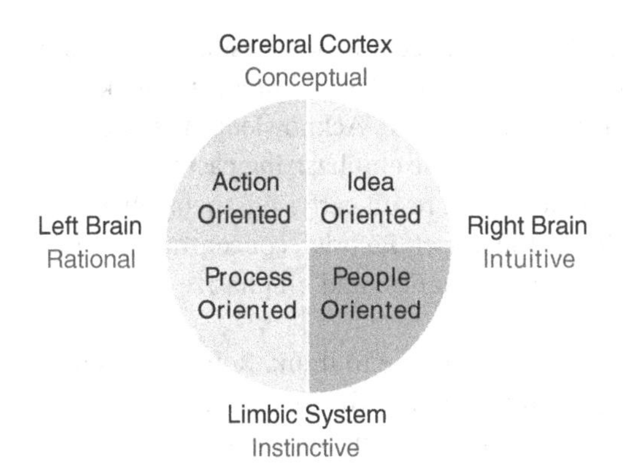

Figure 9.2 Four different communications styles.

task-driven individuals with a get-it-done personality. However, the pacing, levels of detail, and protracted thought process can undermine their points. These individuals can be drawn into proving minor points, to the detriment of the bigger picture.

People-Oriented communicators focus on relationships with the audience, while being keenly observant of the emotional environment. They tend to be kinesthetic, supportive, empathetic, perceptive, emotional, and subjective. They talk about people, needs, collaboration, self-development, sensitivity, values, and expectations. Their assertions may be more improvised, as they read the room and adjust accordingly. They like to talk and teach others. These verbal and nonverbal styles are often lively and expressive. Often, they interweave anecdotes and stories to make their message more personable and relatable. However, they can become too emotional and underestimate the importance of substance, which may undermine their credibility.

Action-Oriented communicators focus on rational, analytic reasoning. They tend to be direct, pragmatic, impatient, critical, quantitative, energetic, and quick to decide and challenge others. They talk about objectives, results, facts, productivity, efficiency, performance, and next steps. Their assertions are precise, structured, and concise. This neutral style is generally easy to understand, and relies on them knowing how things really work. The message is structured and clear. However, it can convey a clinical-like detachment between the speaker and listener as it can easily discount emotions and human aspects of the issue.

Idea-Oriented communicators focus on concepts, innovation, and creativity. They tend to be holistic, synthesizing, imaginative, flexible, curious, charismatic, thought-provoking, and sometimes unrealistic. They talk about possibilities, opportunities, interdependencies, problems, and potential solutions. Their assertions are passionate and engaging. This style can lead listeners to consider different perspectives and realties. It can stimulate thoughts, drive innovation, and win support. However, their enthusiasm and conceptualized thinking may alienate some with more practical concerns.

While everyone has some natural and environmental affinities for each quadrant, most tend to have one quadrant that is more dominant. The power in HDMI is understanding how to communicate with all styles in ways that resonate. When speaking with an individual, a group of people, your team, or stakeholders, there will undoubtedly be a variety of communications styles. The best way to increase the effectiveness of your communications is to make sure you address the foundational needs of each style. This doesn't mean you transform your style. It does however mean you should modify your style to include critical aspects of the others to best reach all listeners.

To reach a Process-Oriented communicator be precise. Present a logical, organized discussion. Tie the issues to practical, key elements. Emphasize past precedents, and logically present your recommendations. Include a clear path forward of who will do what and when.

To reach a People-Oriented communicator, don't be rushed. Encourage small talk before you dive into the discussion. Include anecdotes, stories, examples, or humor that make the issue human. Use first-person pronouns that make it personal. Acknowledge the relationship between the people and your proposals or issues, while emphasizing past successes.

To reach an Action-Oriented communicator, define the result early. Tell them where the discussion is headed by stating the conclusion at the outset. Be brief, concentrate on your best recommendation, include facts, figures, and schematics, and emphasize the practicality of your solution.

To reach an Idea-Oriented communicator, allow them time to think. Remain patient if they go off on a tangent. While emphasizing the uniqueness of the topic at hand, mention options and take time to relate it to broader concepts. Tie the future value or anticipated impact to the future reality.

Work to integrate aspects of each into your own style, and sprinkle them throughout the discussion to dramatically increase your connection and efficacy when communicating with others. To become an exceptional communicator, pair this approach with effective active listening skills.

Active listening builds trust and connections in that it conveys you value the other's thoughts. Active listening is hard as it requires your undivided attention, not thinking about what you want to say next. It means concentrating on both what is said and what is conveyed with nonverbal communications. It means being patient and focused, often restating their position to confirm your understanding of their concerns. There are six key steps to active listening:

1) Pay attention
2) Withhold judgment
3) Reflect
4) Clarify
5) Summarize
6) Share

Like many things in life, active listening is a skill that can be refined and become a habit.

9.1.5 Effective Communication

There are abundant resources focused on improving one's communication skills. Most of these emphasize time-tested truths. When writing, use correct grammar and spelling. Be concise and clearly relate to the reader's needs with logical paragraphs and flowing sentence structures. Speak clearly and enunciate. Practice active listening. Be polite, kind, and respectful. Remain professional and courteous. Be politically and culturally aware. Praise in public; criticize in private. Be honest and forthright. And many more truisms your parents taught you that can take a lifetime to master.

Successful PMs communicate well. While they find and embrace their own style, there are some commonalities. Great PMs truly listen. They excel at nonverbal communication. They bring clarity and maintain focus. They can read the room. They ask probing questions and then give people time to formulate meaningful answers. They provide perspective. They keep everyone in the loop. They promote their team and celebrate successes. They talk to their audience, tailoring the message to the technical and situational awareness of the listeners. They focus on the issues, and don't take dissent or criticism personally. They empathize while not letting their emotions cloud their judgment or responses. They have an accurate bullshit-meter, and aren't afraid to use it. They know when to be direct and when to let an issue simmer. They promptly address conflict and seek win-win solutions. They are inclusive and build relationships built on trust and mutual respect. They are consistent and reliable.

Effective communication is the basis of almost everything a PM does. Perhaps more so than any other area, this is one of the professional skills where we have trouble seeing our own blind spots. We naturally assume our message was clearly conveyed, and if confusion resulted it must be on their end. We naturally assume the listener heard our position and completely understands it, even our mindsets and assumptions that we didn't share because they are so obvious. We naturally assume the audience absorbs the content and ramifications expressed in our presentations and well-prepared slides. We naturally assume the language we use to explain technical issues is understood by all those listening. Some of these assumptions are assuredly not true. Seek candid, trusted feedback to help you see your own blind spots.

Learn to become a storyteller. Especially in our increasingly distracted existence, grabbing and holding a listener's attention is hard. Stories are a powerful tool to do exactly that. We love stories. We remember stories. And many people are often motivated by feelings and emotions over facts. Stories provide a powerful way to bind these all together. Stories also can provide a platform to convey to the listener that their concerns are being heard and considered. While we each have and will develop our own style of storytelling, the formula is often the same. There is a situation, followed by an action, that produces a result, which leads to a lesson learned.

President Ronald Reagan was one of America's foremost orators, earning him the moniker of "*The Great Communicator.*" Ken Khachigian, a former Reagan speechwriter wrote, "*What made him the Great Communicator was Ronald Reagan's determination and ability to educate his audience, to bring his ideas to life by using illustrations and word pictures to make his arguments vivid to the mind's eye. In short: he was America's Teacher.*" In describing his own style, Reagan said, "*What I said simply made sense to the [man] on the street.*" He also advised, "*Talk to your audience, not over their heads or through them. Don't try to talk in a special language of broadcasting or even of politics, just use normal everyday words.*" His advice, approach, and example are directly applicable to every PM who wants to improve his own communication skills.

9.2 Stakeholder Involvement and the Stakeholder Engagement Plan

"*The biggest communication problem is that we don't listen to understand. We listen to reply.*"

– Anonymous

The PM is 100% responsible for stakeholder involvement. Team members may assist. Public relations and graphic designers may be leveraged for their expertise. Leadership may interject themselves in planned or impromptu manners. But the ultimate responsibility for stakeholder relations falls squarely with the PM.

And the reason is simple. Effective stakeholder involvement is an incredibly important component to efficient project advancement. As the project progresses, there is an inverse relationship between the influence of stakeholders and the cost of change (both in terms of time and money), as shown in Figure 9.3. This does not necessarily mean stakeholder influence diminishes as the project progresses, but the impact to the project to implement requested changes increases the further you are in the development process. Ideally stakeholder input would be captured prior to scoping. Changes can become much more costly to implement after Design Approval. Early engagement is not only the right thing to do, but there are important business reasons to do it, and do it well.

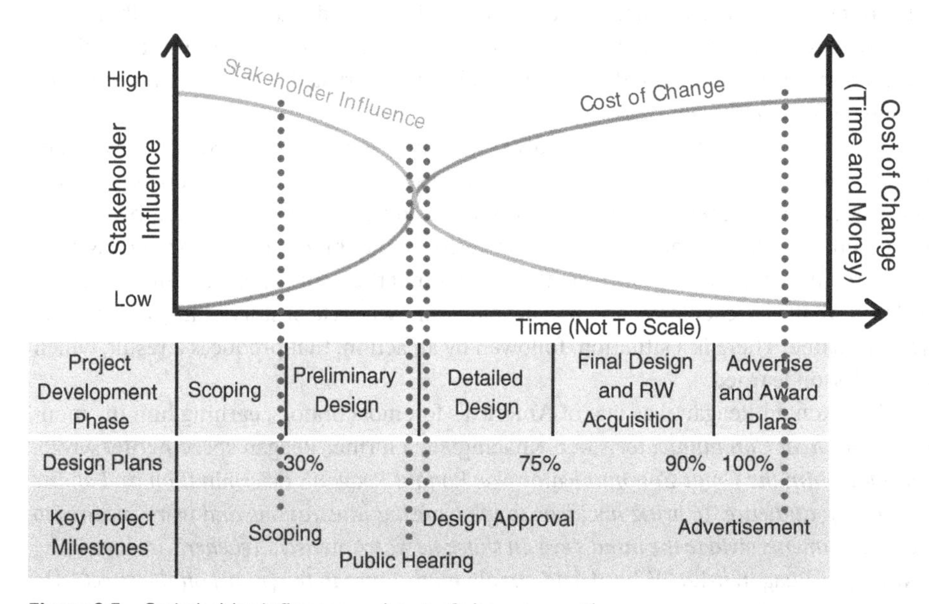

Figure 9.3 Stakeholder influence and cost of change over time.

9.2.1 The Stakeholder Engagement Plan

Stakeholder issues are best managed within the framework of a Stakeholder Engagement Plan. In this context, when you see "Plan," think "How." The Stakeholder Engagement Plan details how you will engage the stakeholders. It is a formal strategy detailing how you will communicate with stakeholders to build support for, and effectively drive, your project development. This plan will list the stakeholders and specify the frequency and type of anticipated communication to achieve the objectives. It may also include considerations for media and specific communication events. While preparing, managing, and monitoring the Stakeholder Engagement Plan may seem, and be, time intensive, this is a fundamental responsibility of the PM.

Figure 9.4 shows the four distinct processes that comprise the Stakeholder Engagement Plan: (1) Identify Stakeholders, (2) Plan Stakeholder Engagement, (3) Manage Stakeholder

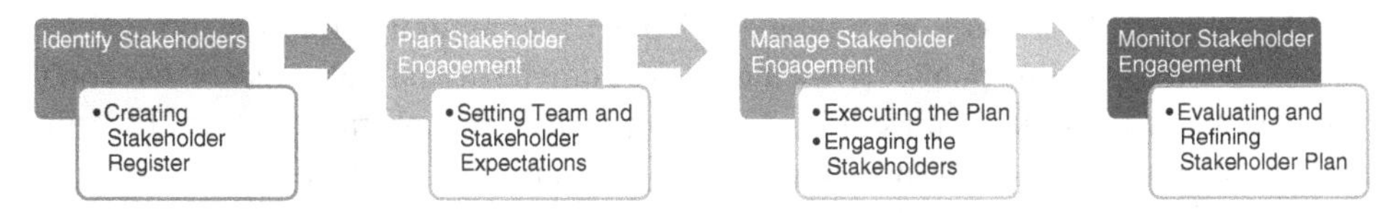

Figure 9.4 Stakeholder engagement plan.

Engagement, and (4) Monitor Stakeholder Engagement. These processes begin early in the planning stage, and extend through the life of the project.

Every project should have a Stakeholder Engagement Plan. Pending size, complexity, location, and design solution of your project, you may not implement all aspects, but a proactive and intentional communication mindset is critical to your project's success. The PM should tailor the Stakeholder Engagement Plan to be appropriate and reasonable for your specific project, while complying with your organization's guidance and requirements.

9.2.2 Identify Stakeholders

The process of Identifying Stakeholders involves four distinct steps: (1) identifying who is a stakeholder, (2) analyzing stakeholder's motivations and potential impacts on the project, (3) mapping stakeholder relationships and influences, and (4) listing and prioritizing stakeholders on the Stakeholder Register. More time and effort should be spent on these four tasks for larger and more complex projects, but all projects should go through the four steps in Figure 9.5.

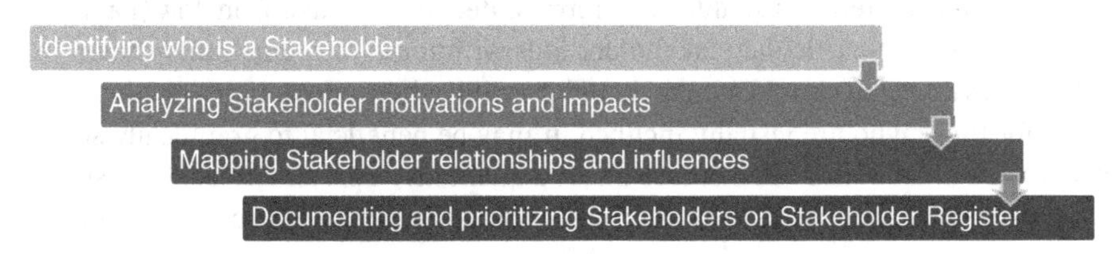

Figure 9.5 Identifying stakeholders.

9.2.2.1 Step 1 – Identifying Who Is a Stakeholder

A stakeholder is any person, group, organization, or entity that has a real or perceived stake in the development, delivery, or results of your project. Identifying potential stakeholders relies on wisdom, experience, and active brainstorming. The goal is to proactively identify all potential stakeholders by which you can be intentional in your communications. This doesn't mean you will contact all identified stakeholders. That will be determined in Step 4 of this process, after you analyze and map them.

Stakeholders can be internal or external to your organization. Internal stakeholders may include your boss, your project team, co-workers, financial support staff, consultant(s), elected officials, and executive leadership. External stakeholders can be much more difficult to proactively identify. These may include adjacent or nearby property owners, property renters, residential neighborhoods and their Home Owners Associations (HOAs), churches, schools, businesses, economic development, community groups, parks and its patrons, and the traveling public. There may also be other external stakeholders that don't readily come to mind, like Police, Fire, First Responders, school transportation (buses), local hospitals and other medical facilities, business organizations, organizations that meet in facilities that will be impacted (e.g., Scouting troops, Youth Basketball leagues, etc.), the press, trade organizations, and local, regional, or national special interest groups (e.g., bike and pedestrian advocates, environmental groups, etc.).

At the end of this step, you should have a list of stakeholders. This is the brainstorming section of this process, and the results may be a little chaotic. It is okay if it is lengthy and not as organized as you might prefer. This list is the start of your Stakeholder Register. Future steps will refine the list that you will finalize during Step 4.

9.2.2.2 Step 2 – Analyzing Stakeholder's Motivations and Potential Impacts on the Project

Once stakeholders have been identified, the next step is to perform stakeholder analysis. The purpose of this effort is to categorize the motivations and potential impacts each stakeholder may have upon your project. These are relatively easy exercises that can provide valuable individual insights and a comprehensive stakeholder picture that is crucial in shaping an effective engagement plan.

The two most common techniques for stakeholder analysis are the Power-Interest Matrix and the Salience Model. Both provide meaningful insights, but with slightly different twists.

The Power-Interest Matrix is a two-dimensional grid with four distinct quadrants. A stakeholder's Power, or Influence, is qualitatively measured from low to high on the Y-axis. A stakeholder's level of interest is qualitatively represented on the X-axis from low (passive) to high (active). Where a particular stakeholder falls within the four quadrants shown in Figure 9.6 dictates how you should approach and prioritize their interaction.

Especially for those who are visually inclined, it may be beneficial to graphically show key stakeholders on the Matrix. Remember this is a qualitative analysis. You should strive to correctly place the stakeholders in the correct quadrant, and then it becomes a relative

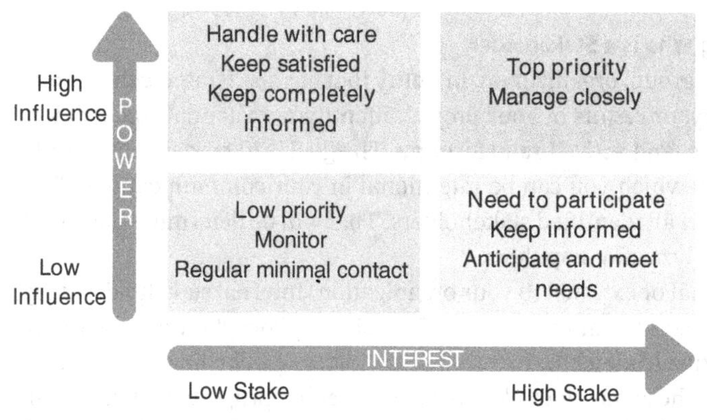

Figure 9.6 Stakeholder quadrant definitions.

scale. A simple way to also display the position of the stakeholders is to color code their names. In the example in Figure 9.7, green indicates advocates and supporters, red indicates critics or blockers, and blue indicates those that are neutral.

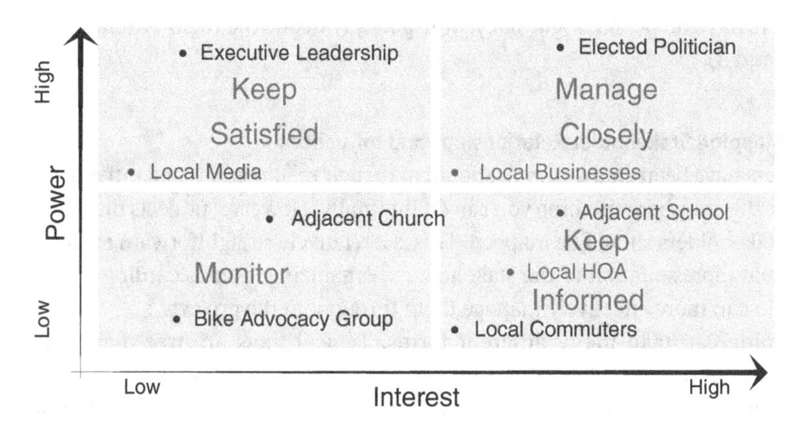

Figure 9.7 Stakeholder quadrant responses.

The Salience Model analyzes stakeholders in three dimensions: Power, Legitimacy, and Urgency. It is shown as a Venn diagram with eight specified regions, as illustrated in Figure 9.8. Each stakeholder should fall into one, and only one, category.

1) Dormant Stakeholders have lots of power but no legitimacy or urgency, and therefore not likely to become heavily involved (red region).
2) Discretionary Stakeholders have legitimate claims but little urgency or power, therefore not likely to exert much pressure (blue region).
3) Demanding Stakeholders want things to be immediately addressed. Although they have little power or legitimacy, they can make lots of "noise" (yellow region).
4) Dominant Stakeholders have both formal power and legitimacy, but little urgency. They often have expectations that must be met (purple region).
5) Dangerous Stakeholders have power and urgency but are not really relevant to the project (orange region).

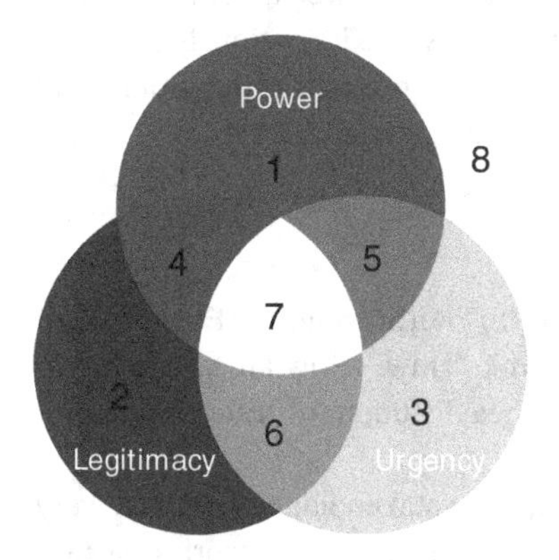

Figure 9.8 Salience model.

6) Dependent Stakeholders have urgent and legitimate stakes in the project but hold little power. These stakeholders often rely on others to amplify their voices (green region).
7) Definitive Stakeholders have power, legitimacy, and urgency, and therefore have the highest salience. (white region at the intersection of all other regions)
8) Non-stakeholders have no power, legitimacy, or urgency (outside the regions defined by the circles 1, 2, and 3).

9.2.2.3 Step 3 – Mapping Stakeholder Relationships and Influences

After the stakeholders have been analyzed, you should map their relationships and influences. On smaller projects, this may be something you can do internally. However, projects that have a large number of stakeholders should be mapped. This is a relatively straightforward exercise that produces a visual representation of the stakeholders, organizing them according to key criteria by which you can more effectively manage them throughout the projects.

The actual mapping can take many different forms. Typical tools are tree diagrams, spider diagrams, and mind maps. The format may depend upon the number and type of stakeholders, as well as our own preferences. Which format you select is less important than the results they produce. When complete, you should have grouped stakeholders in logical ways that provide clarity on formal and informal connections and influences.

9.2.2.4 Step 4 – Listing and Prioritizing Stakeholders on the Stakeholder Register

The Stakeholder Register is an incredibly important project document that relies on the results of your stakeholder analysis and mapping. There are an abundance of Stakeholder Register templates available to suit your particular needs and desires. They are generally formatted as a table, with stakeholders in the rows and columns for all relevant details. Typical columns may include contact information, preferred method of communication, frequency of anticipated communications, Power-Interest Matrix quadrant, Salience Model designation, priority, project attitude (e.g., supporter, detractor, neutral), role, critical relationships, and expectations.

The number of stakeholders and listed details should be tailored based on the size and complexity of the project. On simple efforts, this can be accomplished relatively quickly. Regardless of the time it takes to complete, it should be done. An organized and accurate Stakeholder Register is a PM time investment that will yield positive results.

The Stakeholder Register should be a living document that is reviewed and updated regularly. This document is for the PM use. Refer to it often, but share it cautiously so as not to inadvertently offend anyone on the list. This document should be a functional and valuable resource to the PM. If it isn't useful to you, then change it to make it so.

9.2.3 Plan Stakeholder Engagement

The second process in the Stakeholder Engagement Plan is to Plan Stakeholder Engagement. As a reminder, when you see "Plan" in this process, think "How." How are you going to effectively engage your stakeholders? This process relies heavily upon the information prepared and detailed in the Stakeholder Register.

As your stakeholder analysis and mapping identified, your stakeholders may have very different levels of power, legitimacy, urgency, and interest. These different categories warrant different approaches. As PM, your most valuable commodity is your time. Effective planning can help ensure you spend your time wisely, where it should be spent.

The plan can be tailored in a format that is most usable for you. Often the Risk Register can be expanded to suffice. Sometimes a separate and more formal document is preferred. Spreadsheets can be very effective. Regardless of the project size or complexity, the same basic components should be included: listed stakeholders, the content or deliverables, the intended frequency, who is responsible to prepare and deliver the communication, preferred method of communication, and the priority. Some general guidance is reflected in Figure 9.9.

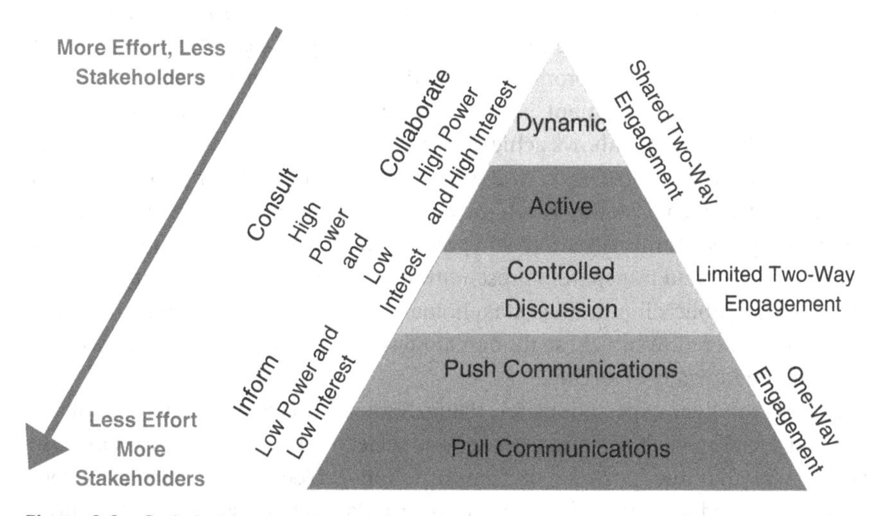

Figure 9.9 Stakeholder engagement.

This lays the foundation of what you should expect from yourself as PM in communicating with each stakeholder, a consistent and intentional approach. However, as with most plans, there should be some inherent flexibility. If you see a stakeholder in a store and they bring up the project, use that opportunity to determine if their expectations have changed and assure them you understand their positions. Ask the HOA President if you can attend the upcoming HOA social function and talk for ten minutes about the project, since social functions are almost always better attended than the business meetings. There will always be a place for face-to-face conversations and project meetings. But you and your organization may choose to explore websites, leverage social media and email lists, or even use radio ads, signage, or other advertising avenues. Be creative. Remember the goal is to establish trust, building credibility, and effectively managing stakeholder expectations.

It should be noted that there are fundamental differences between a Public Information and Public Involvement efforts. Public information is a one-sided dissemination of information from the organization to the public. Public involvement is a two-sided conversation. Examples of public information may include website or social media updates, and mass mailings. Public involvement seeks to interact with the public in such a way as to encourage input and include stakeholders in the decision-making process. Most successful stakeholder engagement plans include both of these. Problems often ensue when the public perceives an organizational public involvement initiative to really be a public information effort.

9.2.4 Virtual Stakeholder Engagement

Stakeholder engagement is a critical part of transportation project development. Federal and state codes require public engagement to acquire the needed easements and rights-of-way to build and maintain public transportation assets. Legally this establishes a public need and enables Eminent Domain, which is the government's right to expropriate private property for public use with fair compensation. But stakeholder engagement is so much more than a formality to check a legal box. It should not be discrete discussions at a Citizens Information Meeting or Public Hearing, but rather an ongoing conversation throughout the life of the project. Most organizations recognize the substantial benefits of more comprehensive and intentional efforts. These outreach opportunities are the basis for community engagement where you can mitigate risks, engender support, and promote cooperation during design and construction.

At its core, stakeholder engagement is where organizations build trust and manage community expectations. It also allows achieving inclusivity, encourages community participation, and provides a platform upon which consensus can be built. The goal should be to strive to creatively and consistently reach the entire impacted and effected community, regardless of race, age, economic situation, and political influence. Success relies on honest communications, timely and transparent presentation of accurate project information, and truly listening to the public's input. Most stakeholders want to be heard, understood, and have their ideas considered. While these fundamentals remain true, the methods by which we achieve them are radically changing.

The reality is stakeholder expectations are rising. Some of this is due to accessibility of information and social media's integral place in our societal fabric. Another reason is the general acceptance that the purpose of transportation projects now extends beyond connecting people, products, and places. Success criteria such as social justice, equity within the transportation space, responsible environmental stewardship, and community transformation are mobilizing passions and bringing new stakeholders into the conversation.

Transportation project development rarely experiences a watershed moment. Due to the nature of the business most changes are gradual, incrementally increasing efficiencies or adjusting workflows over time. This recently changed. While internet and social media have provided new pathways to connect to the public, it was the Covid pandemic that forced a rapid paradigm shift. Seemingly overnight, there were restrictions prohibiting in-person public meetings. Out of necessity, organizations scrambled to pivot their long-established practices and processes, hurriedly restructuring their outreach to utilize a wide spectrum of virtual platforms. As life returned to normal, the industry should take a deep breath and determine how best to move forward.

Virtual engagement brought many benefits, often including greater accessibility and increased participation. However, these gains came with some growing pains, and perhaps did not adequately replace the human connections from in-person interactions. So how do the industry and your organization strategically implement a coordinated, multiplatform approach that combines virtual and in-person components to all projects in your programs and portfolio? Generally speaking, challenges we face as an industry can be divided into three categories: technical, external, and internal.

Organizations will require technical solutions, including robust and dynamic systems to track individual and collective stakeholder engagement across platforms. This should capture and prioritize feedback across time based upon stakeholder metadata, such as contact information and other project specific categories (e.g. – adjacent land owners, HOA member, local business owner, etc.). Systems would ideally assist the organization in discerning the power, legitimacy, and urgency of each position. Forward thinking project managers and organizations want to leverage this data to prioritize feedback and more effectively tailor responses. There is also an organizational need for resources to stand up and support multi-platform initiatives. This would ideally enable an organization to seamlessly capture and leverage stakeholder input in all current social media platforms, as well as all traditional communications. Organizations may also pursue mobile applications to further integrate project digital visualizations, advanced mapping functionalities, and social data gathering trends such as crowdsourcing. Concurrent with these technical advances, the organization may need to review its workflows to ensure accurate and consistent messaging.

The big picture of the transportation game is to have the right money in the right place at the right time so the pipeline of projects can be predictably developed and delivered. Thinking externally, organizations need to balance project resources and constraints with increased stakeholder expectations. While some projects may warrant a fully interactive virtual public hearing, advanced simulations, dedicated website, and the like; others will not. What is the new baseline of effort? How is it decided, and by whom? Can it even yet be established, or is still a case-by-case decision? In the same state, the content, frequency, and quality of virtual engagement in affluent urban areas may significantly differ from that in rural areas where internet availability is limited and the community prefers in-person interactions.

Embracing these changes also represents significant internal workflow adjustments within transportation organizations. This goes beyond the immediate need for graphics and data specialists. There is an old engineering adage that says, "*We can do anything with enough time and money*." Increasing the quality and frequency of virtual stakeholder engagement is certainly possible, but only with real project schedule and budget considerations. This is particularly challenging for projects that are already active. It is common to underestimate the amount of time required by the Project Manager and development team to engage and respond to a comprehensive stakeholder engagement plan. Appropriate change management procedures should be followed that properly adjust the project's triple constraint of budget, scope, and schedule. In order to be successful, each organization should intentionally decide how to accommodate these project impacts.

It is clear we have crossed the Rubicon. Stakeholder engagement moving forward should include virtual and in-person components. What this hybrid approach looks like, and how it is implemented, may not be readily standardized in your organization or in the industry. As responsible public stewards, we must balance the benefits of a comprehensive, multiplatform approach with the realities of diminishing returns at the project, program, and portfolio levels. Schedule and budget constraints dictate organizations can't be everything to everyone; however, we should be intentional and strive to be effective. Organizations should focus on the noble purpose of stakeholder outreach, and then strategically tailor their approaches to the specific project and affected communities to best leverage the strengths of all available platforms and outlets to responsibly accomplish the overarching objectives.

9.2.5 Manage Stakeholder Engagement

The third process in the Stakeholder Engagement Plan is to Manage Stakeholder Engagement. In this context, when you see "Manage," think "Doing." This is executing the plan. This is the real work of day-to-day, week-to-week, month-to-month, quarter-to-quarter contact with the project's various stakeholders.

PMs should use the Stakeholder Engagement Plan and Communication Plan as references for the frequency and preferred method of communications with various individuals and groups. These will help synthesize expectations with practical realities. Use the plan to keep you on track with the regular updates, which are so important for others to feel connected, valued, and involved. While there will certainly be exceptions to the plan, think of it as a baseline of communications with stakeholders. There can be more, but there should not be less. Many PMs find it helpful to log and summarize each significant stakeholder interaction. The frequency and quality of these stakeholder touchpoints can build trust and credibility.

There may be times when you need to discover how your stakeholders currently feel about your project. This information will enable you to update your Stakeholder Register and provide you insight how best to engage them. A commonly overlooked technique is to directly ask them. *What financial or emotional interests do you have in this project? Who do you rely on for information? What do you think of our performance so far?* Polite and professional direct questions can save you time and jumpstart a respectful and productive relationship. In these circumstances, remember to strive to monopolize the listening. Stakeholders want to be heard. They want to intellectually know and emotionally feel that their concerns are being understood and considered.

When you are communicating with stakeholders, be intentional. Define the goals of the conversation before you begin. Is it to inform, brainstorm, decide, persuade, or something else? Let this guide how you strategically approach the conversation, how it proceeds, and how it ends. Be sure you leave with what you need, be it a decision, a plan forward, or acknowledgment of an update.

During the course of project development, it can be easy for stakeholder engagement to shift to a back-burner priority. The PM's job is to communicate and drive project development with the triple constraint of budget, scope, and schedule. In order to satisfy project success criteria, the PM must manage expectations. Managing expectations is foundational to stakeholder engagement. Stakeholder engagement is not something that gets in the way of your job; it is your job.

9.2.6 Monitor Stakeholder Engagement

The last phase of the Stakeholder Engagement Plan is to Monitor Stakeholder Engagement. This involves monitoring and controlling the Plan. In this context, when you see "Monitor," think "Collect performance data." When you see "control," think "Comparing actual performance data to planned expectations and then taking any preventative or corrective actions." This is all to evaluate the plan, the how.

This process is not monitoring the actual communications with the stakeholders, but rather the efficacy of the framework of the plan itself. It is time to look in the mirror and honestly see if you are doing what you said you were going to do. Are you sending out monthly email updates to those all who signed up for the update mailing list? Are you still sending your boss weekly project summaries? Is what you're doing still working? Is it still appropriate? Should stakeholders be added or removed from the Stakeholder Register? Should any of your stakeholders be re-analyzed or re-prioritized? This is your chance to adjust the plan to make it more efficient and effective.

The Stakeholder Engagement Plan should specify how frequently you will monitor the plan. Pending the length of the project, it could be quarterly or annually. You can certainly make changes whenever it is deemed prudent, but scheduled monitoring ensures you take a holistic look at specified intervals. This enables you to remain proactive in managing stakeholder expectations. All changes should be documented in the plan. This not only helps during this project, but also can be valuable information that is leveraged for Lessons Learned.

9.3 Media Relations

"It takes 20 years to build a reputation and five minutes to ruin it. If you think about that, you'll do things differently."

– Warren Buffet

The news media may play a pivotal role in shaping public opinion of your municipality or agency. As a PM, you may receive requests for media interviews, be interviewed during public participation, or hold a news conference. It is imperative you understand the opportunity this presents to convey a positive message for your organization, and you appreciate the damage it can cause if mishandled. As PM, you should be prepared should you pick up the phone and hear, *"This is John Smith from Channel 12 news. May I ask you a few questions?"*

9.3.1 Different Media Types

There are three types of media outlets: print, voice, and video. Each offers unique opportunities and challenges. A PM should be familiar and comfortable with all three of these very different medias.

Print media relies on written words. Examples may be newspapers, social media, websites, trade journals, regional magazines, homeowner association newsletters, online publications, internet articles, and blogs. Print is still a viable and reliable way to reach targeted or special interest audiences. The first thing to understand with Print is that the quotes and stories are permanent, and can be searched and retrieved indefinitely. Choose your words carefully. Print also may introduce a translation of your message, as the reporter may filter and edit the information for their readers. Their assumptions, interpretations, or agendas can subtly or overtly shape what is conveyed. It is prudent to attempt to discern the slant of their story before print so you can proactively correct any inaccuracies. The rapport you develop with the reporter is critical as you may have multiple conversations in preparation of a single article. As Yogi Berra said, *"It ain't over till it's over."* Regardless of how you feel the interview went, you never truly know how the article will represent you, your project, or your organization until it is printed or posted. So be sure to read it after publication. Unless there are truly egregious errors or omissions, it is usually not a good idea to demand corrections. In print, the reporter or publication has the last word.

Voice media relies on the spoken word. Examples are radio and podcasts. Voice is often considered the safest medium. This is in part because you can often use notes during the interview. Radio allows you to easily convey your message clearly and accurately, in your own words. Prior to the interview, prepare and practice sound-bite length quotes that concisely and effectively convey your message. Whether the station is airing answers to questions or pulling quotes from your interview, the message is in your words. Be thankful and humble, but exude control. Use a strong, relaxed, and comfortable voice and pace. Your tone can subtly promote confidence and authority while eliminating uncertainty and doubt.

Video relies on moving pictures. Examples are TV, video conferences, social media, webinars, and other internet video broadcasts and content distributions. Most citizens still rely on local TV news as the primary source for local information. Video is an emotional media where your message can have a powerful impact. In addition to all of the nonverbal communications, it is imperative that you accurately and succinctly respond to their questions while not providing a response that can be misconstrued. The biggest challenge in TV interviews is to be concise. This is especially challenging on complex transportation projects that take years to design and construct. Your conversation will likely not be aired in its entirety. Responses are generally edited to sound bites of three to six seconds, with the story narrated by the reporter. Plan ahead and be prepared with answers to anticipated questions that summarize your message in less than six seconds. That is not easy. However, like many things in life, there is no substitute for experience. Until you feel comfortable, practice, practice, practice. Role playing with peers or your organization's communications office can be a great time investment.

The advent and explosion of social media has transformed how organizations gather and distribute transportation project information. The ability to stay connected is constantly evolving and expanding. News cycles continue to shorten. The expectations of immediate access and updates are pervasive. While the applications and media vehicles may change over time, the basics of print, voice, and video media remain unchanged.

9.3.2 Who Gives the Interview?

Organizations have different relationships with, and postures toward, local media. The first question to answer is who gives the interview. Different organizations have different protocols. It may be senior leadership, or a designated Department spokesman, or an organizational Public Relations spokesman. Other organizations have the PM conduct the interviews. Many organizations don't have a specific protocol and handle it on a case-by-case basis. As a PM, it is your responsibility to know your organization's preferences before you get the interview request.

If you are not the one giving the interview, it the PM's job to brief the interviewee so they are able to convey the information in a manner that portrays they are in command of the facts and situation. This can prove especially challenging if the designated spokesman is not technically savvy or familiar with the history or details of the project. In these cases, you may prepare a brief, bulleted summary of the message and history for the spokesman. Be patient. Emphasize aspects of your message that are most important to convey. Based on experience or instincts, the spokesman may prepare for a certain line of questioning. Provide them what they need to be prepared. Perhaps most importantly, paint for them the big picture. The spokesman should have information to address specific details, but most stories are edited to sound bites of the highlights.

If your organization does clearly designate a spokesman, determine who is best suited to give the interview. Interview requests can be the proverbial hot potato that is quickly passed from one to another and ultimately up the chain of command. This may be due to fear or intimidation of being quoted, or simply being on camera. But letting this responsibility land on the desk of the one with the courage to do it is not always the wise choice.

There are times when it is important to have a senior leader give the interview as their position lends credibility to the seriousness with which your organization is responding. However, this is not always the case. Your organization should choose the individual that can best promote the message you want to send. Ideally, they would be competent, confident, and professional. They would convey a presence of control and authority while not seeming arrogant. They think before they speak and are careful, precise, and deliberate with their wording. They are concise, calm, and charismatic. There are very few people who can do all of this well, but many can convey an honest and straightforward message with confident credibility. For project-specific issues, this is often the PM.

As PM, it is imperative you are aware of who is speaking about what, with which media outlets. Public opinion can be a powerful force in project development, particularly during public involvement. You should keep a close eye on what is being reported about your project, as it is part of your job to keep your finger on the pulse of public opinion.

9.3.3 Setting up the Interview

When a reporter contacts you with a request, their goal is to schedule a time and place for the interview. Your goals for that initial conversation should also include discerning the nature of the story, the kinds of questions they may ask of you, and what aspects of the project interest them. What information are they hoping you provide to their audience? Most reporters will tell you: "*We haven't updated this project in a while and it's a slow news day,*" "*We have received some citizen complaints and want to hear your side of the story,*" "*We want to know how the recently passed budget cuts will impact project delivery,*" "*We want to know how this project will impact the new school,*" and so on. Finding out this information ahead of time will enable you to better prepare for the interview and dramatically increase the likelihood of a positive message being delivered. Sometimes it can also be helpful to ask the reporter if you can provide any background material, schematics, or maps to assist in the production of the story.

It is rare that you will answer the phone and they will want to put you live on air. If this occurs and you are not comfortable with the situation, it is acceptable to tell them you are busy now, but could call them back a specific time later that day.

9.3.4 Interview Location

Since you are the one granting the interview, you have control where the interview takes place. This is an important detail. There are advantages and disadvantages to each of the four most common settings.

TV reporters like on-site interviews because the viewers appreciate action. It is much more interesting to see a construction project than it is to hear someone talk about it. That is one reason why TV journalists report "*live from the scene.*" The advantage of an on-site interview is to present the picture of movement. A groundbreaking or ribbon cutting is a powerful picture of progress. However, there are also risks. Construction is messy. Visual pictures of you standing in front of bulldozers clearing trees can be unsettling to viewers and quickly erode hard-earned trust you have cultivated with citizens. Know that most news sources will shoot background footage of the ongoing construction anyway. So generally, there is no advantage to on-site interviews. Be especially wary of interviews on location when responding to a tragedy. It doesn't matter how well your words say the situation is under control; if the scene behind you leads people to believe otherwise, they won't believe you.

Interviews outside your organization's building entrance can lend credibility to your position and convey that you are "of the people." It can also be effective on sunny days in providing an optimistic feel. Conversely, avoid this setting on cloudy days as the background and setting can look gloomy, ominous, or disengaged. This can be a good location to provide an update on a particular situation. Generally, it is not a choice for a longer interview.

An office or conference room interview is a controlled environment where you can more easily control the setting, lighting, and sound quality. This is the best location for a longer interview where you can have a more in-depth discussion of the project or situation. It allows you to refer to plans, documents, or other visual materials that can explain and strengthen your positions. A conference room can make it seem more like a work session to the audience where you are approachable, cooperative, and productive. An office interview where you are sitting behind your desk can present an image of authority and power. Remember to ensure all background wall-hangings and desk displays present an image consistent with your, and your organization's, intentions and principles. It is important to thoughtfully determine where you want to host the interview based upon what image you want to project.

In studio interviews are not that common. When one is granted, perhaps for a local morning show, approach this opportunity with an increased commitment to preparation. Be especially cognizant of your body language and tone so you present a complete and consistent image that complements and underscores your message. Give some thought to what you will wear, and allow yourself plenty of time to prepare so you can effectively and succinctly deliver your message.

9.3.5 Preparing for the Interview

Never go into an interview without a clear plan. What questions may the reporters ask? How might they approach this topic from different perspectives to make the story more interesting to their audience? What is prompting them to do this story now? What have they already reported about this topic? What, if any, new developments have recently occurred. What is your message? Make notes and rehearse giving sound-bite length responses, less than six seconds, that emphasize your key points.

Preparation can also involve gathering background information to provide the interviewer. Some examples include project presentations or project schematics. Providing basic project information can answer many of the general and background questions. This enables reporters to prepare a more comprehensive report once they are back in the truck or at the studio editing for final production.

If you are the designated spokesman, it is important you understand what authority you possess. You may have authority to speak for all matters related to your project, or only over specific aspects of your project. You may or may not have authority to speak for other Departments, private entities, or other stakeholders involved in your project. You may or may not have authority to speak for related or unrelated organizational activities. During the interview remain focused on the project, and speak only to those issues upon which you can authoritatively comment.

9.3.6 During the Interview

Once the interview is underway, remember the reporter wants information from you. They are not interested in you personally. Reporters are doing their job. They need to get information from you so they can edit the story before the deadline, and then move on to the next assignment. While you have no control over the reporter or how they filter your information, you have complete control, and take full responsibility, for what you say and do in front of the camera. Be positive and remember this is a unique opportunity to provide information to your citizens that they may need or deserve.

Rule number one is never lie, ever. Answer their questions with accurate and reliable information. When possible, work in your key points. Throughout the interview, work in your practiced sound bites, perhaps even a few different times. Do this by utilizing the bridging technique, which answers the question and then segues to your main message. For instance, *"The planned shared-use pedestrian path is one example of the many benefit citizens will realize from this project. Another is..."* or *"When complete, this will be a six lane, divided roadway that will create a critical link between...."* If they ask a question you don't know, simply reply, *"I don't know, but I will find out and get back with you."*

Don't dodge unpleasant questions. You can only redirect questions so many times before it looks like you are actively avoiding the topic. It is better to address an unpleasant situation directly and honestly. Part of your purpose in giving the interview is to confront the truth of the topics and help the citizens understand what matters most. If you cannot answer the question, provide a reason. For instance, *"It would not be appropriate to comment on the ongoing negotiations."* Just as important as what to say is what not to say. Don't give responses that suggest you are hiding something. For instance, *"I'm not allowed to talk about that"* or *"no comment."* True or not, there is always a better response. Instead try, *"It wouldn't be proper to comment upon that at this time."* However, there may be times when you do need to deflect the question. Some possible responses are *"I'm not really the expert in that area," "I'll leave that speculation up to you," "My personal opinion is irrelevant on this issue,"* or *"You really need to direct that question to...."*

Start strong. Regardless of what the first question is the reporter asks, make sure your first answer includes the main message points you want to deliver. You do not want to dodge their initial question, but you can effectively bridge from that answer to your main message. Another point to work into your initial answer is gratitude for their interest and assistance in providing current and accurate information to your citizens.

Keep in mind the audience and use appropriate terms and vocabulary. Leave the "engineering-speak" behind. Strive to use nontechnical terms to describe the project. This avoids confusion, minimizes misunderstandings, and creates better sound bites. Speaking in layman terms can be challenging for engineers. Think about how you would describe it to a precocious second grader. For example, don't say, *"the project will require 300,000 cubic yards of borrow to achieve the necessary thirty feet of relief."* Instead, you could say, *"the citizens can expect to see a bunch of trucks bringing in dirt to create a thirty-foot-tall hill."* Similarly, limit the numbers and statistics you cite and try to create images and analogies that paint a mental picture for the audience.

9.3.7 Tricky Questions to Avoid

While some reporters intend to create controversy, most are focused on completing the story before their looming deadline. They are hoping you will make the story more interesting and compelling to their audience. Whether the reporter is intending to be deceptive or not, there are some common questioning tactics of which you should be aware.

The Hypothetical Question. *"What will be your response if a lawsuit is filed?" "If the necessary funding becomes available, will you proceed?"* These questions tempt you to comment on an assumed reality. Don't do it. You are the spokesman disseminating facts. Do not speculate on "what if" scenarios. Redirect the questions from the hypothetical back to reality, and end with a positive statement. For example, *"I do not think it is proper or appropriate to speculate at this time, but I would iterate..."*

The Other Guy's Opinion. *"How do you think the County attorney will respond?" "Will the Mayor agree?"* These questions prompt you to speak for someone else. Unless you have been given specific authority to speak on their behalf, do not speak for someone else, even if you are confidant of their position. It is best to suggest the reporter contact that person directly for his or her comments.

The Non-Question Question. *"Your department generally does a fantastic job, but it seems this one got away from you."* This is not really a question. The reporter is expressing an opinion to frame the story. In response you can acknowledge the compliment, if one is offered, then challenge the incorrect premise, closing with targeted points from your message. For instance, *"Thank you for recognizing our consistent performance, but I respectfully disagree that..., in fact ...."* Another approach is to turn their statement into a question. For instance, *"We followed all applicable State and Federal guidelines. Should we have responded differently?"*

The Off-the-Record Question. *"Just between us, did the contractor..." "So, tell me how does this really work..."* These questions may be misconstrued as being "off the record." A reporter will often seem very friendly as they try to build a rapport with you before or after the interview. As a public servant, nothing is off the record. Accordingly, be cautious when the interview seems to be over and the cameraman is packing up the equipment. Remain professional. Assume everything you say will be printed or aired.

The False Premise. *"Now that you're off the hook, what do you intend to do next?"* These questions are dangerous in that if you respond without challenging the inaccurate premise, you are silently accepting their assertion. The proverbial, *"So when did you stop beating your wife?"* is perhaps the most common example of an accusing question. When faced with a false premise, confront it, or the audience will assume it is true. Be tactful, but firm in your response.

The Open-Ended Question. *"So, tell me about this project?" "How will this project benefit the community?"* These softball questions you should hit out of the park if you have properly prepared. Use these opportunities to present key points of your message in previously rehearsed, sound-bite length responses. Don't ramble. When you have delivered your message, stop talking.

He-said-She-said Scenarios. *"I understand the Mayor said...." "According to Mrs. Smith she isn't being offered fair market value for her property...."* These questions rely on hearsay, what someone else said. Unless you know exactly what was said, how it was said, the context in which it was said, and what was intended; don't fall into this trap. Even if you say, *"Assuming what you cite is correct, I would recommend...,"* remember the editing room can easily remove your disclaimer from the final broadcast. Talk about what you know and keep on message. You can also acknowledge and respectfully decline their question before bridging to your talking points. For instance, *"I have no knowledge of that conversation and therefore cannot speak to it, but I would like to emphasize..."*

The Silent Pause. Many of us are naturally uncomfortable with silence. Reporters know this, and will often use this powerful technique. They will ask you a question, and when you are done talking, they will look at you with beckoning eyes as if to say, "*keep going*" or "*can you expand upon that thought*." It is a natural instinct to keep talking and fill this silent void. Don't do it. This can lead to providing more information that you intended to convey. Once you answer the question, stop talking and stare back at the reporter, patiently awaiting the next question. No matter how awkward it may seem at the time, be smart. They will edit out the silent pauses before broadcast.

The Bait-and-Switch Ambush Interview. While not common, some reporters will use this technique. They call to schedule an interview and describe the nature of their questions. At the interview, they will ask a few questions on this topic before completely switching gears and focusing on a completely different topic. Usually, the new subject is more controversial, and the questions are direct and confrontational. In the moment, it is easy to feel ambushed as they pursue a more sensationalized story. Their questions may be filled with false premises, he-said-she-said scenarios, or other tricky questions. When this occurs, it is imperative you remain composed, providing professional and diplomatic responses. Remember, your body language, facial expressions, and everything you say may be aired. Provide measured and authoritative responses that are securely rooted in facts. If questions begin to repeat themselves, you may say, *"I think we have covered that already. Is there anything I can further clarify for you? If not, I appreciate your time."*

9.3.8 After the Interview

You should be in contact with your chain of command immediately after the interview. Provide brief details on what was discussed, the angle of the story you suspect the reporter will present, and when the interview will likely air or be printed. No one wants to be surprised. It is your responsibility to let leadership know what information will be presented so they are not surprised by any follow-up or fallout.

In dealing with the media, experience is the best teacher. So, after your interviews, think about what you said and how you said it. Think about what you did well and what you can improve next time. Critically read the story or watch the interview and evaluate if you effectively presented your message. Each interview is an opportunity to further develop these important skills. And as with most things in life, say *Thank You*!

9.3.9 The News Conference (How to Prepare and Execute a Successful News Conference)

There are many reasons your organization may hold a news conference. Be it to make a major announcement that will profoundly impact your citizens, provide an update for an ongoing crisis, or to stay ahead of a breaking story. Regardless of the focus, the purpose for a news conference is the same. You want to distribute a single, consistent factual message to many media outlets at the same time. While not common for public organizations, these can be a powerful tool when used correctly.

News conference rule number one, it's your show. You control the timing, the setting, and what is addressed. It is likely your organization's Communications or Public Relations office has expertise that should be utilized. The timing is important so that you make, or miss, the various news submission deadlines. The setting is important because you want the backdrop to match the intended tone. Whatever you decide, deliberately choose the timing and setting. Plan the event to ensure there is adequate access, parking, seating, lighting, and so on. Once the news conference is scheduled, your Communications or Public Relations office can likely assist in notifying the local media of the event through their established connections.

Another decision is whether you will distribute media packets immediately after the news conference. These should be carefully crafted in advance and can effectively emphasize your message. Typically, these may include a fact sheet, written remarks or statements, relevant photos or schematics, and background information to help them complete their story. Include the full names and titles of those speaking at the event, as well as contact information for follow-up questions.

Once the event is underway, remember this is your show. Tell the reporters the agenda and format of the event, and stick to it. Read a prepared, and rehearsed, opening statement that incorporates key points of your message. Set the proper tone with your words and nonverbal actions. Remain poised, confident, and in control. Smile, if appropriate. Keep your remarks brief and on point. If you are taking questions, point to one reporter and recognize them as having the floor. When you are ready to end the news conference, provide advance warning, such as, "*We have time for two more questions.*" And remember to end by thanking everyone for their time.

It takes the coordinated contributions of many individuals to pull together a successful news conference. As with all team efforts, be sure to recognize and thank the team members whose efforts made it happen.

9.3.10 When Tragedy Strikes – Crisis Interviews

Accidents happen and occasionally tragedy strikes. If a disaster occurs, it may be critical that your organization approach or respond to the media. The task of emergency response media communications is a serious responsibility.

First, it is essential that you understand the citizens want, and may deserve, to know situational information, provided it does not compromise anyone's safety or privacy. Within the moment and throughout the duration of the event, keeping the media informed is a priority function. The public's impression of your response may be largely shaped by your cooperation and forthrightness with the media.

Most organizations have established emergency response plans. While no accident or tragedy ever exactly fits the practice scenarios, contingency plans can bring some semblance of order to chaos. All emergency action plans include notifying a long list of individuals and entities. These often include police, fire, third-party emergency response units, hospitals, schools, critical businesses, civic organizations, and the media. When in the midst of a crisis, make sure there is a single, authoritative voice through which all information flows to the media. This individual can change, but it is crucial to have only one spokesman talking to the reporters at a time. This will better ensure consistency of message and minimize misconceptions. During a crisis, it is imperative to remind staff that no one talks to reporters other than the designated spokesman.

If you are your organization's spokesman, know what you are going to say and how you are going to say it. Remain calm. Be compassionate. Be reassuring and controlled. Start your remarks by focusing on the safety of the individuals who are involved in the accident and of those who are working to resolve the situation. Do not discuss money or schedule impacts. As the event develops, you may provide information on injuries, damage, rescue or remediation efforts, and other situational details, but remember to keep the focus on safety. Do not rationalize that *"accidents happen,"* instead, say *"The cause is under investigation and our findings will be made available to you when complete."* Focus on the facts. Do not speculate on causes, damages, or hypothetical questions. Know how are going to end the interview. This can be a simple summary such as, *"Our primary concern right now is for the safety and well-being of the individuals involved. Everything else can wait. I will let you know when additional information is available. Thank you."*

10

Controlling the Project

"Everyone has a plan until they get punched in the face."

– Mike Tyson

The PM's job is to develop and deliver the project within the established constraints of budget, scope, and schedule. A straightforward objective. And then stuff happens.

There may be days you feel like Colonel George Custer making your own last stand, facing threats to your triple constraint from seemingly every side. Political indecision. Changing priorities. Evolving success criteria. Funding complications. Technical challenges. Extreme weather. Unanticipated stakeholder influence or involvement. Union strikes. Entropy. Inexperienced or unskilled team members. Staff turnover. Boneheaded decisions. Realizing unanticipated or inadequately accounted risks. Bureaucratic inertia. Some of these threats you can predict and control. Most you cannot.

These perhaps mostly external threats can then be further complicated by internal and contractual constraints of quality, resources, and risk that lay atop your triple constraint of budget, scope, and schedule. Effective PMs understand how to be successful amidst ambiguity. They have the innate ability to navigate uncertainty with the assurance of their internal compass guiding them through the unknown. This behavior and the associated results have nothing to do with chance or luck. Effective PMs actively monitor and control their projects.

10.1 Managing Expectations

"Meeting expectations is good. Exceeding expectations is better."

– Ron Kaufman

Managing expectations is one of the most important, and also one of the most overlooked, responsibilities of a PM. It is typically not taught in school, is difficult to measure, is incredibly time consuming, will likely not be considered on your performance evaluations, and yet is an essential component to any successful project.

Each stakeholder approaches the project from a different perspective, with different heartfelt and intellectual concerns, and with varying levels of passion and intensity. Each

Transportation Project Management, First Edition. Rob Tieman.
© 2023 John Wiley & Sons Ltd. Published 2023 by John Wiley & Sons Ltd.

stakeholder also approaches the project from a different intellectual and professional background, often assuming their professional and life experiences directly transfer to the transportation project workflow, which is often not the case. Even if all of these stakeholders unanimously agree on the goals of this project, there would likely still be a wide range of differing expectations. Elected officials may expect a ribbon cutting before the next election. Adjacent business owners may have expectations regarding construction hours, continuity of access, uninterrupted power and utilities, temporary signage, and construction noise. Nearby medical outpatient facilities may have expectations regarding notification in advance of blasting. Local schools may have expectations regarding continuity of bus access and accommodations for parent and day-care pick-up lines. And so forth. As PM, it is essential you identify, understand, and manage these expectations, which can vary widely and often conflict.

Experienced PMs know that one does not accidentally execute a successful public project. There are far too many bureaucratic challenges whose institutional inertia will quickly slow it enough that it can easily be derailed by competing interests, changing agendas, or apathy. Successful projects are not planted and then left to grow without care, they are cultivated. Projects require pruning, weeding, watering, and constant maintenance. When expectations are proactively and effectively managed, all stakeholders can feel good about the outcome of the project, even if the process or end results are different than they first expected. Conversely, if expectations are mishandled or disregarded, a project can come in on time and under budget and still be viewed by many as a failure.

10.1.1 How Are Goals and Expectations Different?

Goals are created from the project objectives and answer the *Who, What, When, Where*, and *Why* of the project. They should be well defined and consistent with the project's Purpose and Need. Typically, they are outward facing, and well documented in the project's Budget, Scope, and Schedule. Often, they include that which the project's performance metrics are measured against. It is common that project goals remain static throughout the life of the project.

By contrast, expectations are typically dynamic and evolve throughout the life of the project. While expectations can certainly be attached to the *Who*, *What*, *When*, and *Where* of a project, they are often concentrated in the *How*. The process, feel, and atmosphere of a citizen's information meeting can have a major impact on how the content of the message is received. Expectations are often formed by unexpressed preconceptions, and judged by perceptions and interpretations. They are emotional, are usually less transparent, and can be far more personal as they are often rooted in personal beliefs or hopes for a particular result, sometimes completely unrelated to the specifics of the project.

It is important every PM understand that expectations have two critical functions. First, expectations establish the Stakeholder baseline by which they will evaluate the project's success. Secondly, expectations often drive stakeholders' actions and decisions.

10.1.2 Four Approaches to Goals and Why They Matter

Most every project manager knows of SMART (Specific, Measurable, Attainable, Realistic, and Time-bound) goals. All goals can, and perhaps should, be SMART goals. However, the lines between goals and expectations can easily become blurred. There are four approaches to goals. Understanding the differences of how goals are established as compared to how they are evaluated will enable the PM to more effectively manage expectations.

10.1.2.1 Approach 1 – Quantitative Goals that Are Measured Quantitatively

Quantitative goals relate to measurable quantities that can be expressed in numeric terms or other clearly defined categories (e.g., a six-lane roadway of a specific typical section that can support 40,000 vehicles per day). Quantitative goals are easy to define and quickly become matters of engineering design. They typically drive the project, and are the specifics detailed in the Request for Proposal and Consultant's Scope of Services. These are the goals PMs focus upon with laser-like intensity. Quantitative goals are easy to monitor, and do not typically create a gap between the goal and related expectations when measured quantitatively.

10.1.2.2 Approach 2 – Quantitative Goals that Are Measured Qualitatively

Some stakeholders will measure quantitative goals against their own qualitative scale. Consider an intersection improvement that adds turn lanes and modifies the existing traffic signal to raise the AASHTO Level of Service from "F" to "B." Quantitatively this project may have met its design goals and be a success. However, a citizen who regularly drives straight through this intersection in off-peak hours and now experiences turning arrow delays to support the limited number of cars they see, who by the way never seemed to have a problem turning at that intersection before the improvements, may consider it a failure and complete waste of money. Similarly improving a park's recreational facilities by adding one new field and ninety additional parking spaces may meet the standards and satisfy the design goals but still be considered grossly inadequate by those struggling to find a parking space the Saturday morning of their child's twenty-team soccer tournament. When stakeholders use their own qualitative expectations to evaluate quantitative goals, they may not be swayed by reviewing the engineering reports, preferring to prioritize their own anecdotal experiences. Qualitatively measuring quantitative goals can easily exaggerate the gap between project goals and stakeholder expectations.

10.1.2.3 Approach 3 – Qualitative Goals that Are Measured Quantitatively

Some stakeholders will measure qualitative goals quantitatively. This can be used to assert success or failure of the project to forward their own agenda or perspectives. Consider opening a new bike path that was intended to revitalize a historic area. One group may assert it was a success because the opening weekend drew 187 visitors. Another group may say it was a failure because it didn't even draw 200 cyclists. Both positions may be quite convincing, but they are misguided. While their quantitative observations, projections, or studies may be accurate, it is often not appropriate to evaluate qualitative goals with quantitative analysis. It is analogous to mathematically trying to average averages – you get a number, but it is meaningless. More than the basis for unrealized expectations, this approach is often used as evidence to validate the stakeholder's expectations. Beware of those using this approach, both inside and outside of your organization.

10.1.2.4 Approach 4 – Qualitative Goals that Are Measured Qualitatively

Qualitative goals relate to the desired quality, essential nature, or character of the result. They are often felt rather than measured. Although harder to define, qualitative goals can be powerful and inspiring. Consider a streetscaping project that will rebuild a sense of community in the neighborhood, or a pedestrian facility to be a community amenity and encourage personal fitness. Because qualitative goals are inherently difficult to measure, it is not always clear if and when the goal has been achieved. Occasionally there is a straightforward consensus of success, but far more common is a range of perceived results. Qualitative goals often become secondary to quantitative goals because they are difficult to measure. As the old PM adage states, "*That which gets measured, gets done.*" While many PMs consider qualitative goals "*touchy-feely*" and less important than quantitative goals, exceptional PMs know you do this is at your own peril. Everyone would readily agree citizens should be treated fairly during acquisitions for a public project; however, different stakeholders approaching acquisitions from very different perspectives may hold vastly different expectations of what this process should look like. Similarly, adjacent residents may have very different opinions of what constitute reasonable construction hours. And so forth. The longer any expectation gap exists, the more entrenched it becomes. Most unrealized expectations begin with an admirable qualitative goal that is measured qualitatively. As PM, be especially aware of the risks associated with this approach.

10.1.3 The Expectation Gap

Lee Atwater famously said, "*Perception is reality.*" The Expectation Gap is the gap between reality and one's expectations. But before we discuss this gap, it is important to clarify which type of reality is being considered.

Consider the following situation. Your agency presents an offer to a property owner in a critical, complicated, and politically sensitive acquisition. One week later their attorney submits a counter offer that lists fourteen specific requests covering a wide variety of design issues. As PM, you understand the urgency and push hard for resolution. Research is done. Alternatives are analyzed. Meetings are held. And just four weeks later you have accomplished the remarkable task of securing official responses for all fourteen specific requests. Lesser PMs gauge reality based upon intended value. In this example, the PM's leadership and interpersonal skills may have forged consensus from multiple departments that rarely agree on anything. Thus, their intended value is providing the very best to that stakeholder. More sophisticated PMs gauge reality based on delivered value. In this example, the intentions and actions to be thorough and timely are irrelevant; the reality is the agency responded in four weeks. Stakeholders gauge reality based upon perceived value. What if after three weeks the citizen calls their attorney and asks for an update. The attorney replies that they had not yet received a response. So, when the carefully crafted response arrives a full month after they submitted the counteroffer, it is received with frustration. The stakeholder's perceived reality, the reality that matters, is that the agency is slow, unresponsive, and dismissive.

Intended value is usually higher than delivered value as most everyone has intentions to perform in ways that do not always materialize for a variety of reasons. Similarly, delivered value is often higher than perceived value. As you begin to consider Expectation Gap, understand you should be comparing the stakeholders' perceived reality to their expectations, as shown in Figure 10.1. To complicate matters, these are mostly concentrated on qualitative goals that are evaluated qualitatively.

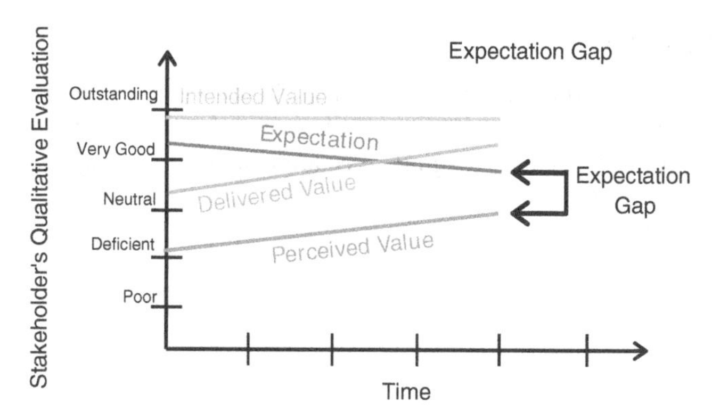

Figure 10.1 Expectation gap.

It can be beneficial to visualize this graph as you communicate with each stakeholder. What does their graph look like before and after your communication? Did the gap grow or shrink? Is the slope of the expectation line trending more favorably? The concepts in this graph can be a valuable tool in managing expectations that allows you to proactively identify troubling trends, and assists in prioritizing where to spend your valuable time.

The goal is for the perceived and delivered value to exceed expectations. This is often the case if the perceived value exceeds, or is equal to, delivered value. As Vin Scully said, "*Good is not good when better is expected.*" If there is a negative Expectation Gap, meaning expectations exceed perceived value, the PM can best close or eliminate this gap by:

1) Decreasing expectations
2) Increasing perceived value
3) Over time, maintain favorable slopes to close or eliminate this gap

In order to gauge a stakeholder's perceived value, you need to ask them. Often the act of initiating conversation and soliciting feedback will increase the perceived reality value in and of itself. Additionally, these interactions can be valuable self-reflection tools that illuminate our blind spots so we can be more effective moving forward. This is especially true if there are significant discrepancies between intended, delivered, and perceived realties versus expectations.

For most stakeholders, satisfaction is how their perceived reality measures against their expectations, not your intentions or how the project satisfied its purpose and need within contractual constraints. The Triple Constraint of budget, scope, and schedule, along with organizational performance metrics, are inconsequential to most stakeholders. These are mostly internal constraints and objectives. Exceptional PMs understand and leverage the differences.

Expectation Gaps may be project risks that should be logged on the Risk Register. This may be especially appropriate should critical stakeholders or leadership have significant expectation gaps, or if there seems to be a systemic expectation gap that is indicative of larger PM or organizational issues. Corrective actions can be documented in the project's risk response plan.

10.1.4 How to Manage Expectations

Managing expectations can be more art than science. Accomplished PMs do this exceptionally well. Conversely, most PMs never quite figure it out, and consequently are perceived to be mediocre, at best. Those who are skilled at managing expectation do the following nine things very well, as shown in Figure 10.2.

Figure 10.2 How to manage expectations.

1) Know your capabilities

Before you can effectively manage expectations, it is essential you know your own, your team's, and your organization's capabilities and limitations. You need to know what can and cannot be realistically produced. Nothing spoils existing and future relationships faster than over-committing and under-delivering. You need to thoroughly understand the PDP and workflow processes so you can correctly anticipate the intended and unintended consequences of your actions and decisions. You need to have a firm grasp on how the involved agencies interrelate so that you can coordinate them to the benefit of your project. You need to have command of the engineering principles involved so you can understand the foundational issues driving the technical challenges, which will enable you to effectively steer the ship toward resolution. It takes experience and wisdom to hear an idea or look at a set of plans and determine what won't work in the field. If you don't have that first-hand design and construction expertise, it is imperative you have someone that you trust on your team who does.

2) Be honest and forthright

Words mean things. Stakeholders can sense sincerity. Be yourself. Be genuine. Build trust and respect by demonstrating reliable credibility. Accordingly, never lie, ever. Nothing erodes trust faster than dishonesty or hypocrisy. "*I don't know*" is an acceptable answer. Exceptional PMs all have built a respected reputation as one who can be trusted. General George Patton perhaps said it best, "*Say what you mean and mean what you say.*"

Being honest, forthright, and realistic can be especially critical when expectations are still forming. Actively listen. Empathize. Convey to stakeholders that you acknowledge and appreciate their concerns. Consistent sincerity is noticed and valued. These actions build trust that can prove invaluable when conflict arises. It is very satisfying to receive calls from stakeholders saying they have heard troubling rumors, so they called you to find out the truth.

There are also times when the whole truth cannot be shared. This may be due to design decisions that are under debate, privacy issues, political considerations, potential private developments, pending legal actions, or a variety of other issues. Discretion is a professional necessity. In these situations, it is acceptable to divert the conversation, not to say anything, or to state you can't discuss that issue at this time. Whatever your response, never lie.

The pressure on PMs to develop and deliver projects can be immense. Mediocre PMs are easily manipulated to saying "*yes*" or making ill-advised concessions. This may especially common in public meetings or closed-door meetings where directives are given by leadership in a way that does not invite debate. However, in addition to being politically savvy, successful PMs strategically reshape the discussion into what is achievable and follow established change management procedures.

3) Capture expectations

Expectations can be deep and broad or singularly focuses and extremely specific. They can be formed by what you said, what you didn't say, or even how you said it. They can be formed by something someone else said or did, past histories, hopes, or desires. They may be rational or irrational, valid or unfounded, practical or impossible. Some stakeholders will directly tell you. More may allude to them. Others will remain silent and presume you should know.

Before you can effectively manage expectations, you first need to capture them. Empathy is critical. You need to hear what they are saying, and infer what they are not saying. If it still isn't clear, ask direct and probing questions. It is important that you fully understand their position because you need to not only identify the expectation, but also the driving force behind it. A citizen's unusually strong attachment to a particular tree makes sense when you discover it was planted in memory of a deceased loved one. A private school's illogical entrance request becomes reasonable when you understand the adjacent tract of land is to be bequeathed to them. You can more effectively resolve the issue at hand when you understand and consider the underlying motivations. When talking with stakeholders, it can be very efficacious to repeat the expectations and motivations back to the stakeholder to confirm understanding. For instance, "*I want to make sure I heard you correctly. I understand that your expectation is that there will be a median break on the proposed roadway at your business's side entrance because you are concerned about delivery trucks being able to make a U-turn and safely enter your site?*" This acknowledges understanding of their issue while inherently allowing flexibility in resolution. It is incredibly important to most stakeholders that they feel like they are being heard. It is imperative that you convey empathy for their situation and acknowledge their position. This is an essential step before you can genuinely reassure them you are committed to doing what you can to address their concerns. The final resolution on most issues will not satisfy all stakeholders. However, they are much more likely to accept an undesired outcome if they feel heard, understood, and part of the process.

Identifying and capturing expectations is an ongoing effort that needs to be sustained throughout the life of the project. Realities and opinions change. Expectations are a dynamic, constantly changing landscape. It is critical you continually keep you finger on the pulse of the expectations. On a large-scale public project, you cannot logistically gather the expectations of all involved stakeholders. If a project has ten adjacent property owners you can, and perhaps should, meet with each of them. If you have a hundred adjacent property owners, the PM will not have time to meet with each individually. You need to proactively identify the pivotal players and actively seek them out. Be it the corner business owner, the church's pastor, the HOA President, or the neighbor who knows everyone, find the stakeholders who influence others. If you don't know who they are, ask.

4) Prioritize expectations

You may find there are as many expectations as there are stakeholders. Eventually you need to prioritize them, assigning a relative importance and urgency to each. Some considerations include who holds the opinion, how many people this stakeholder may represent or influence, and the size of potential risk they pose to the scope, schedule, and budget of the project. These practical considerations must be appropriately weighted while striving to do the right thing the right way. Prioritizing expectations is a dynamic process throughout the life of the project. This is especially true for larger transportation projects where the schedules can stretch for years. More detailed information on Stakeholder Engagement is included in Chapter 9, *Communications*. Those expectations that generate risks should be appropriately documented and addressed, as detailed in Chapter 6, *Risk Management*.

5) Address conflict

Transportation projects often create conflict. Sometimes there appears to be winners and losers. If you are straightening a roadway curve for safety, it is moving closer to one side of the street but further away from the other side. If you are installing a new water line, some will benefit from the increased capacity while others will be inconvenienced by its construction.

Some have the ability to enter seemingly almost any situation and walk out with consensus, or at least a reluctant acceptance, of the decision. These individuals can have very different styles to bring two sides together, convince one side to adjust their thinking, or have one side accept the outcome. There are many outstanding resources on how to successfully navigate conflict to an acceptable and desirable resolution. But there is one commonality; when it comes to conflict, leaders confront it.

Whether you enjoy conflict, tolerate it, or dread it, conflict resolution is an essential PM skill. You cannot ignore it and hope it goes away. Like mold on forgotten leftovers in the back of the refrigerator, unless you take care of it, the mess will only grow. This is especially true with conflicting expectations. The longer a stakeholder holds an expectation, the more entrenched it becomes. If left unchecked, this anticipation of a perceived reality can transform into a powerful fact in their minds that becomes difficult to refute. If you become aware of stakeholders with conflicting expectations, promptly address them. Don't let them fester and grow into larger issues. The saying, "*an ounce of prevention is worth a pound of cure*" definitely applies here. Many conflicting expectations can be avoided with proper planning and proactive, consistent communications.

6) Communicate often

Any concerted effort to manage expectations includes proactive and consistent stakeholder communications. In general, frequency is much more important than length, or sometimes even content. Frequent and concise communication instills an assurance that you are on the job and actively looking out for their interests. These "touches" maintain a dialog throughout the project that builds into relationships that can become critical when and if priorities shift and concerns arise.

Within your communications, be honest, straightforward, and forthright. Most can sense sincerity. Don't bluff. Don't lie. Never assume. Be humble. Compliment others when speaking to groups. Praise in public and criticize in private.

Many PMs struggle with detailing technical and procedural knowledge to stakeholders in a way that fosters confidence while not appearing arrogant or condescending. It is vital that you can describe complex engineering situations in nontechnical terms. If this is not easy for you, practice. Practice describing a project issue without using any technical terms or acronyms. If you are not sure how to begin, pretend you are talking to a group of partially-interested fifth graders. To hold their attention and get your point across, you should be interesting, concise, and perhaps a little entertaining. When nervous, many naturally revert back to their technical comfort zone terminology. Resist this urge as it may make you appear insecure or aloof. Impressions matter.

One of the main purposes of an active communications plan is to avoid surprises. These surprises can be unpleasant questions from citizens or the media that blindside your elected leaders. They can be legal action by a property owner or activist group to slow or stop your project. They can be consultant delays in meeting a deliverable date. Many of these situations could have been avoided if the PM communicated effectively such that they had their finger on the pulse of the project.

The media is another tool that can be used to disseminate project updates to a large number of stakeholders. Proactively publicizing a traffic flow shift on an active construction site can minimize citizen frustrations. If your project just made a major milestone, let the media know the good news. If it is a slow news cycle, they may be searching for content and would be glad to profile how your important project is moving forward. Be creative and intentional.

Additional information can be found in Chapter 9, *Communications*. Section 9.2 discusses *Stakeholder Involvement* and *Stakeholder Management Plan*. Section 9.3 addresses *Media Relations*.

7) Influence expectations

It is very difficult to manage expectations if you cannot effectively influence expectations. If a stakeholder's Expectation Gap graph is unfavorable, there are only two ways to change the graph in your favor: (1) lower expectations, and/or (2) increase perceived reality.

The best sales professionals are outstanding listeners. They want to learn what motivates and matters to their customers. This will enable you to be more persuasive. What are their expectations? What are their assumptions? What are their desires and fears? What drives their expectations? What is it they want, and why? What will they accept? Like managing expectation tasks, influencing expectations is an on-going, dynamic process that benefits from persistent and concerted efforts.

To effectively influence others often requires acute emotional intelligence and a refined host of soft skills. Being charismatic, funny, and a good storyteller can all be very helpful character traits. While these can be developed, we are all blessed with different gifts. Don't focus on what you may be lacking. Concentrate on your strengths. Be yourself. Be genuine. Be logical. Listen. Empathize. Speak to others' emotions. Find common ground and emphasize shared values and goals.

Additional information on influencing others can be found in Chapter 11, *PM Soft Skills*. Section 11.5 details the *Seven Pillars of Influence Management*.

8) Monitor expectations

Stakeholder expectations will change over time. While Budget, Scope, and Schedule changes should be well documented, changing expectations often are not. You must monitor expectations. It is essential to keep your pulse on stakeholders' expectations so that you know when and how they are evolving. You cannot manage that which you do not know. So, ask questions. Listen to your stakeholders. These regular interactions also provide opportunities to identify incongruent expectations and influence them while at an infant stage before they become too entrenched.

9) Deliver

The most powerful impact in managing expectations is to deliver. Successful PMs shape expectations to mirror what they are able to produce. This approach fundamentally relies on being able to deliver. As such, it is essential you assemble a team with complementary skills, experience, knowledge, and technical expertise to get the job done. With public transportation projects, the ability to navigate through the sea of red tape is invaluable. Exceptional PMs have that unique combination of patience, persistence, tact, and hard-nosed determination to will something into fruition. Strive to develop and maintain a reputation for making realistic commitments, and delivering results based upon those expectations. This can be major project milestones, or promptly returning their phone call. There is no faster way to build trust and strengthen your credibility that doing what you say you will do.

10.1.5 Managing Up

Managing up can be a challenging and a high-risk proposition. But pending the project, there are times when a PM must manage up in order to keep the project moving forward.

You need to understand how your organization really works. Be aware the practical reality may significantly differ from the current org chart and published procedures. How are decisions made? Who makes the decisions? Who do those people rely on and trust? What are the big picture organizational and programmatic objectives leadership wants to advance? You need to have a mental map of the leadership and decision dynamics in order to effectively strategize how best to secure your desired results. You need to know who to influence, and when, to increase the likelihood your proposal is accepted or the right decision is made. You want to establish a layered foundation so decision discussions become a chorus of affirmative head nods. Bringing decision makers together prematurely, or before you really understand all the issues at play, can result in unexpected and unpredictable results that can directly impact your project's budget, scope, and schedule. As exceptional trial lawyers never ask a question to which they don't already know the answer, so PMs should not bring critical project issues to decision makers without some advance preparations.

In order to effectively manage up, those above you need to know and trust you. They need to be familiar with your work, your values, and your commitment. It is important to find ways to establish these connections and build social capital before you need to use it.

Realize that people are persuaded by different means. Some require a logical argument. Others look for passionate emotions. The most effective persuaders strategically combine facts and feelings, but in different measures pending their audience. Listen. Ask questions. Understand how to best tailor your message by leveraging and emphasizing the information and benefits.

Approaches that best resonate with feelings are *Reflecting* and *Asserting*. *Reflecting* is the ability to reflect back to the other person not only what they are saying, but what they are feeling. This relies on active listening and empathy. *Asserting* is the ability to express your own needs and desires in a compelling way that authentically leverages emotions. This relies on connecting the idea to deeper feelings. Both of these approaches can forge powerful connections to feeling-minded decision makers.

Approaches that best resonate with facts are *Questioning* and *Suggesting*. *Questioning* is the ability to ask neutral, open-ended, logical questions whose answers provide you insight into meaningful data or opinions held by others. *Suggesting* is the ability to propose a well-crafted case or business plan that is supported by facts and reason. Both of these approaches bring credibility and order to fact-minded decision makers.

Reflecting and *Questioning* are most applicable early in the influencing process as you are gathering information and preparing to advance your preferred course of action. *Asserting* and *Suggesting* are best used later in the influencing process, after you are satisfied the table is set for a favorable response. Since most decision makers consider combinations of facts and feelings, you may need to use all four approaches in varying degrees. The intention is not to manipulate anyone, but rather to ensure a fair case is made for your assertion in the way that most resonates with each individual.

10.1.6 Managing Down

Managing down relies on the same foundation of influence management. Listening and connecting to others is paramount. Pending your organizational structure, you may be empowered with positional authority where your entire team reports directly to you. More likely, different team members have different hierarchal ladders but everyone is coming together as resources for this project. It is very difficult to manage down if you do not know your team members. If it is a large team, focus first on the individuals that others respect and trust. This can quickly amplify your influence. While it may be easy to tell direct reports exactly what to do and by when, many PMs need to also rely on influence management for nondirect reports. Chapter 7, *Managing Resources*, particularly in Section 7.1, *Developing and Managing a Team*, and Section 7.2, *Personality Assessments*, provide additional information that is directly applicable to this effort.

As PM, you play a critical role in a project's information flow. As information, expectations, and feedback flow to and from leadership, stakeholders, and team members, you are uniquely positioned to act as a filter. Be an intentional and strategic buffer between leadership, stakeholders, and team members. Choose to filter out the unproductive static while conveying essential messages. Protect positive morale, be consistent with the truth, and look for ways to highlight others' contributions.

Where self-assurance can be advantageous in managing up, humility can be beneficial when managing down. Your team members and SMEs likely have expertise you do not. Be respectful of their unique skills and abilities. Be curious of their knowledge and processes. Find ways to make their job easier. Surprise them with acts that express appreciation for their important contributions. Invest the time and interest to get to know them as people, beyond their team role. Discuss the *who*, *what*, and *when*. Trust them enough to share the *why*, and figure out the *how*. Act as Tim McGraw sings in the song entitled, "*Be humble and kind.*"

10.2 Monitor/Control the Project

> "*If everything seems to be going well, you obviously don't know what is going on.*"
>
> – Edward Murphy

As PM, you will spend most of your time monitoring and controlling your project. Monitoring is the collection of performance data. How is the project progressing? What do the formal and informal metrics convey? Controlling is what you do with the performance data. This starts with comparing actual progress versus that which was planned, but then also includes analyzing the variances, identifying trends, and taking preventive or corrective action, if necessary. For example, if your monitoring shows the project schedule to be late, then controlling is analyzing the impact, making schedule adjustments, and pursuing change control processes if the delay is unrecoverable.

A monitoring and controlling mindset is critical to PM success. These are not part of your job; they are your job. The cyclical processes described in this section are to be continually practiced throughout the life of the project. This discipline and consistency will enable you to position your project for success, even in the most trying of circumstances.

10.2.1 The Three Tiers of Importance

All issues can be important. That doesn't mean they share the same priority. As PM, your most valuable and scarce resource is your time. It is difficult to successfully manage your project schedule if you cannot manage your own schedule. It is very common for many PMs to have more emails and voicemails and meetings than there are hours in a day. While all issues may warrant your attention, they do not all warrant the same amount of your time and thought. The three-tier system in Figure 10.3 will enable you to better discern where to spend most of your time and your best thoughts.

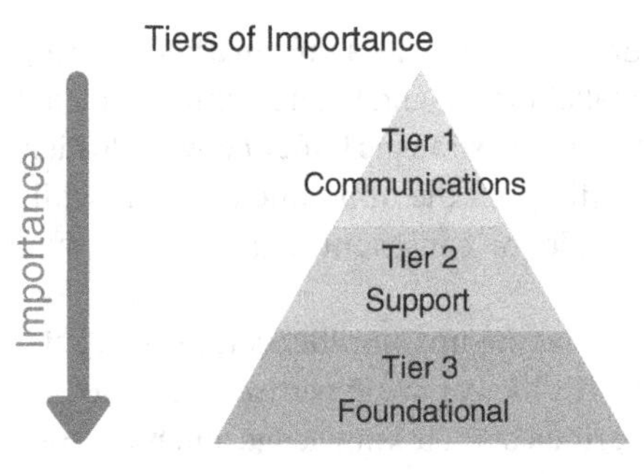

Figure 10.3 Tiers of importance.

10.2.1.1 Tier 1 – Communications

Tier 1 leverages project communications to focus on other's expectations. As PM, you need to have your ear to the ground so you can know and control the pulse of the project. Acute situational awareness will prove invaluable in positioning you to strategically manage change. The tasks themselves are rather simple to describe, but can be challenging to do well. To be effective, this cycle must be continually repeated throughout the life of the project. While important, Tier 1 issues may be the lowest priority.

Tier 1 control is repeating the cycle of engaging stakeholders, monitoring expectations, and managing expectations, as shown in Figure 10.4. This Tier 1 control cycle should include leadership, team members, internal stakeholders, and external stakeholders.

Figure 10.4 Managing expectations.

These actions should be performed in accordance with the Project Communications Plan and Stakeholder Engagement Plan. Specific details are discussed in Chapter 9, *Communications*, particularly in Section 9.1, *Communications Plan*, and Section 9.2, *Stakeholder Involvement and the Stakeholder Engagement Plan*. Understand that these plans should contain the minimum communications. Pending the situation or stakeholder, additional measures may be appropriate and beneficial.

10.2.1.2 Tier 2 – Support

Tier 2 focuses on risks, resources, and quality. These must be closely watched as they are all core secondary project issues that can directly impact more important Tier 3 Foundational project issues. If your Project Management Plan includes a Risk Management Plan, as discussed in Chapter 6, *Managing Risk*, Section 6.2, *Risk Management Plan*, a Resource Management Plan as discussed in Chapter 7, *Managing Resources*, Section 7.3, *Resource Planning*, and a Quality Management Plan is detailed in Chapter 8, *Managing Quality*, Section 8.2, *Monitoring and Controlling Quality*, then Tier 2 project monitoring should be relatively straightforward. If you do not have these management plans, you still will need to establish evolving ways to monitor and control these critical areas.

One can assert most all of project management somehow falls under a risk management umbrella. As such, it is absolutely essential that you actively monitor and control risks throughout the life of the project if you want to be successful. Similarly, resources and quality must be actively monitored and controlled throughout the project. This can best be done by repeating the cycle of evaluating risks, evaluating resources, and monitoring quality, as shown in Figure 10.5.

Figure 10.5 Managing risks.

10.2.1.3 Tier 3 – Foundational

Tier 3 are the foundational issues that define and guide your project. These are the triple constraint of budget, scope, and schedule. Due to the inherent interdependencies of these three critical constraints, any disruption to one impacts the other two. Any Tier 3 issue has the potential to partially or fully jeopardize the project if not properly managed through established and accepted change management

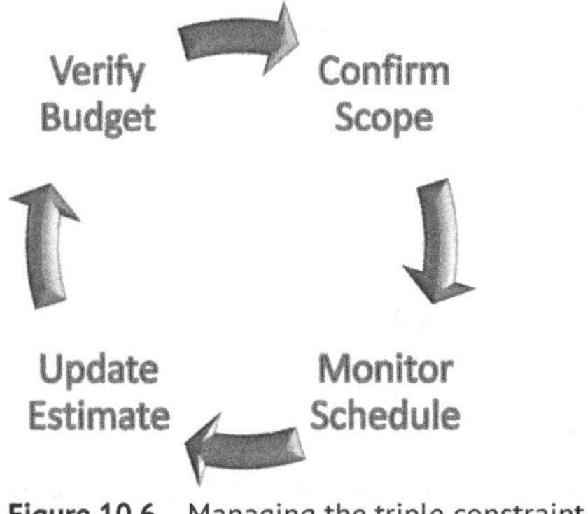

Figure 10.6 Managing the triple constraint.

procedures. As such, all Tier 3 issues should be a high PM priority.

In order to effectively monitor and control Tier 3, the PM should regularly repeat the integrated tasks below, as shown in Figure 10.6. Work with your team to confirm the scope. Monitor the Schedule. Use the current schedule to update the estimate. As the estimate changes, work with programming to verify the budget. Then restart the process. Remember one of the superpowers of the triple constraint is to enable your project to better satisfy the big picture program and portfolio goals of having the right amount of the right type of money in the right place at the right time to keep your project moving forward as detailed in your project schedule. You can only do this if you are actively monitoring and controlling Tier 3 issues of budget, scope, and schedule.

10.2.2 How to Discern Priorities

Tier 1 issues may impact Tier 2. Tier 2 issues may impact Tier 3. As PM, you need to actively monitor and control all three tiers. While situations may elevate the urgency of a Tier 1 issue, as PM you need to keep the proper perspective. Tier 1 issues are serious, but not as serious as Tier 2 issues. Similarly, a Tier 2 issue can be very serious, but it assumes a new urgency when it becomes Tier 3 concern. For example, a particularly involved stakeholder may be raising all kinds of public relations issues (Tier 1). Whatever the specifics, it deserves more of your attention if it becomes a risk that needs to be actively managed or requires restructuring resources (Tier 2). Similarly, the issue deserves your foremost attention if it elevates to the point where it is adversely impacting the project's triple constraint of budget, scope, or schedule (Tier 3).

10.2.3 When Crisis Occurs

Scott Adams said, "*One 'Oh Shit' can erase a thousand attaboys.*" If you are a PM and you haven't yet had one of these moments, just wait. The very nature of public transportation projects is inherently filled with stakeholders and risks that can be challenging to define, predict, and control. For many projects, the question is not if a crisis will occur, but rather when will it occur, what will it be, who will be involved, and how will you handle it.

When crisis occurs, it is imperative that you do not overact. You are the PM. The project is your responsibility. You need to be a steady and calming force in the midst of the storm. Even if you don't know where you are steering the ship, it is comforting to others to know someone is at the wheel. Intentionally exert an authoritative presence. When it hits the fan, take a deep breath and follow the five steps in Figure 10.7.

1) Think; don't react

When crisis occurs, chaos may ensue. Assumptions may be turned upside down. Paradigms may shift. Tempers may flare. Time can feel like it is speeding up as information, reactions,

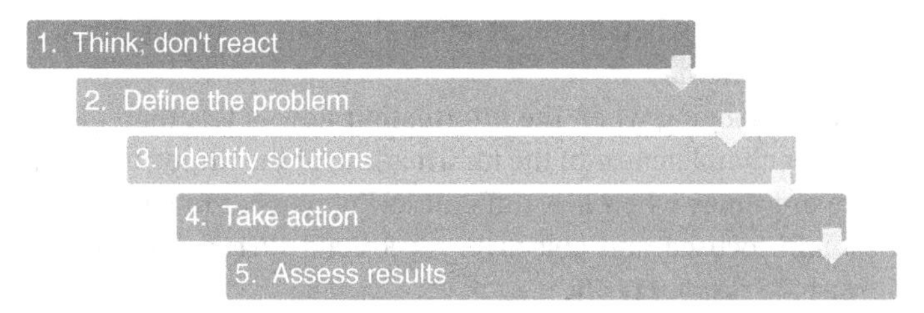

Figure 10.7 When crisis occurs.

and expectations are pouring on you faster than you can absorb it. It is in time like these that prudence is absolutely essential. You need to slow things down. Resist the urge to assess blame. There will be time for that later when it can be handled in a constructive manner. In the midst of the storm, you need to focus on the problem. This requires clarity of thought. Be fiercely protective of you limited time, prioritize critical information, and think.

Voltaire famously said, "*No problem can stand the assault of sustained thinking.*" But solutions do require thinking. In the moment, this can seem counterintuitive. In most sports, you ideally want to eventually think less and react to the situation in a way that relies on your strategic preparation and disciplined practice. This focused and instinctive response allows the athlete to quickly and nimbly adapt in a rapidly changing environment. Many of these premises are also true in business. As PM, there are plenty of times when you should rely on your experiences and the strategic game plan for you project. You should trust your instincts and be able to improvise with thoughtful and agile reactions. However, crises are not these situations. While you may very well rely on your own instincts to eventually make a decision, it is imperative you recognize the level of seriousness of the issue and identify the intended and unintended consequences of the action and reactions. This requires disciplined and intentional thought. When things seem to be falling apart around you, this is exactly when you need to stop, take a deep breath, call a time-out, and think.

2) Define the problem

You can't solve what you don't understand. What problem are you trying to solve? Answering this question may be challenging. For example, imagine you find out you didn't get your environmental permit because the plans changed and now show additional impacts that were not previously discussed or approved. Is the problem that you don't have your permit? Is the problem whatever design considerations caused the plan change to increase the environmental impacts? Is there a communications problem within your team? Is it a job performance problem in that the environmental agent didn't update the permit to show the design changes? Is the problem that this information will cause additional work for the consultant which will strain the already tight budget? Is the problem that any resolution will delay project advertisement, which will push up against funding deadlines on this politically charged project that was promised to be delivered on time? What is the problem? What if it is more than one of them?

Think back to proofs in geometry class. The first step is to define the givens. Do not allow yourself to be distracted. It is imperative you define which problems you are trying to solve. You can't do this effectively if you are reacting, and not thinking. You need to strategically determine what are the core issues. And when there are more than one, you must search for dependencies and then prioritize them.

Defining the problem can be challenging when the information flow is less than ideal. This may be due to the accuracy or completeness of the information, or the timing of when you become aware of it. There are times when it is difficult to know what to believe. In these instances, you must remain focused and objective. As Ronald Reagan said, paraphrasing the old Russian proverb, "*Trust but verify*."

In all cases, effective communication is critical. You need to reach out to those you trust, and to those who have the answers you need. As the situation develops, remember to keep communication as transparent as it is appropriate and productive. This means reaching out to your team, your peers, and your leadership. Strive to keep all those who need to be in the loop in the loop, be they below you, above you, or at your level in the organizational hierarchy.

3) Identify solutions

Once the problem is defined, then you can search for solutions. Resist the urge to put the cart before the horse and jump right to possible solutions before you have defined the problem you are trying to solve. Haste will often lead you down the wrong path.

There are three aspects to enacting solutions: (1) does the solution work, (2) what are the intended and unintended consequences, and (3) buy-in from the team and stakeholders on the new path forward. Searching for solutions is what engineers do well. This is what we are trained to do. Pending the size and complexity of the problem, PMs should resist the urge to do it all themselves. While a PM may be able to find the solution by themselves, and maybe they can even identify all the intended and unintended consequences of the solution by themselves, they most likely can't secure team and stakeholder buy-in by themselves. Project development is a team sport. Seek out different perspectives. Include key stakeholders and team members in the discussions. If there is time, this effort can foster collaboration and spurn innovation. Look beyond positional authority. When it hits the fan, seek those who you can trust to tell you the truth and provide sage advice and direction.

Just as there may be more than one problem, there may be more than one solution. In the previous project example, while the immediate problem is securing the permit, you may need to concurrently be adjusting budget realities and managing stakeholder schedule expectations. The strategic solution may include multiple aspects on parallel fronts. When this is the case, clear and consistent communication to the team, leadership, and stakeholders is essential.

While finding solutions, don't speed, but be speedy. Avoid the cliché of paralysis of analysis. Pending the situation, you will likely not know all the information you want to know before you need to move forward. Leadership is often a series of calculated and educated guesses. As such, spend your limited time wisely. In the previous project example, if it is a design issue that has caused the unanticipated encroachments, get the right people in the same room and figure it out. Target thinkers and experienced engineers. Consult the idealists, attorneys, and regulators to make sure you stay on track, but focus on the problem solvers. Spend your time with those who look for solutions, not those who look for problems.

Whatever your solutions, be sure to evaluate the implications to your project's triple constraint of budget, scope, and schedule. Whichever solution you pursue, you must practice appropriate change management in order to sustain project viability. Don't win the battle only to lose the war.

4) Take action

While it is important not to hastily rush to action, it is likewise unwise to doddle. *Fish or cut bait. Shit or get off the pot.* These clichés are true. There comes a clear time for action when you, as the PM, need to make a decision and move forward. Especially in times of crisis, the perception of authority is paramount. In the midst of a storm, the captain of the ship needs to be seen at the wheel.

How do you know when it is time to act? When facing tough decisions, General Colin Powell succinctly describes his 40/70 rule, "*Once the information is in the 40% to 70% range, go with your gut. Don't wait until you have enough facts to be 100% sure, because by then it is almost always too late.*" The 40/70 rule is shown in Figure 10.8.

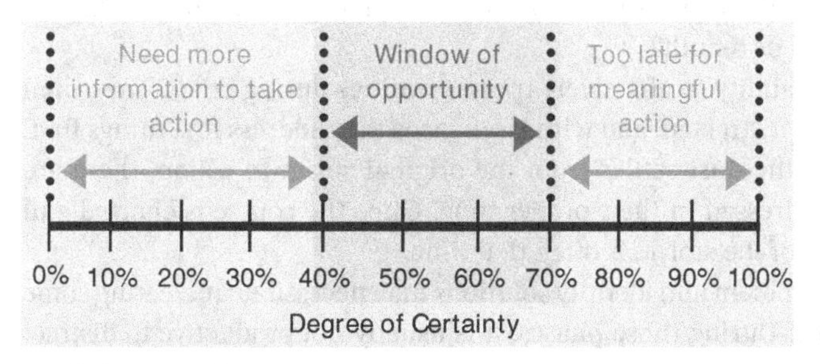

Figure 10.8 40/70 "Powell" rule.

As the Powell Rule details, sometimes more information is not helpful. Complex phenomenon and problems can often be narrowed down to a simple set of criteria to evaluate the underlying fundamental issues. At a point, extra layers of information may only confuse the issue. As Einstein said, "*If you can't explain it to a six-year-old, you probably don't understand it yourself.*"

Sometimes you won't know what to do. In these cases, you can seek wisdom from trusted advisors with whom you have established relationships of trust. The first impression of experts is worth more. Just as an experienced gray-haired engineer can look at a set of plans and quickly identify where there are looming construction issues, so can experienced leaders approach a complicated or contentious situation and offer valuable insights.

Once you decide what action to take, and when to take it, you must bring everyone into the loop who needs to be included. The old adage counsels that a passenger jet liner pilot can make dramatic and unexpected turns, but not without causing quite the raucous in the cabin. As pilot of your project, you need to provide warning to those who need it in the cabin, updates to controllers on the ground, and have your team prepared and ready to execute the changes so they can respond to any complications. Effective communication may not guarantee success, but lack of it will expedite failure.

In 1519, Spanish Conquistador Hernan Cortes landed on Mexico and promptly burned his ships. With their way home laying on the bottom of the ocean, his men were extremely motivated to succeed in their mission. While I am not suggesting you so drastically mark every decision as a point of no return, it is important that you and your team are properly motivated to move forward on the decided path. This can be especially challenging for those who may have been proponents of possible solutions that were not selected. In these instances, address the issue. Whether you meet with them individually or as a team, emphasize shared principles and the collective team goals. Projects are a team sport. Each member of the team has a vital role to play, and we must do our jobs well in order for the project as a whole to succeed. Even after these efforts, there may be some holdouts. It is then your responsibility to kindly remind them that not everyone agrees with every decision all the time, but when the flag goes up, we all need to get in line and salute.

5) Assess results

There is a time for everything.

After the problem is defined, solutions are identified, and actions are taken, then you should evaluate results. Are your solutions working? Are the intended and unintended consequences acceptable? Are there new risks that need to be managed? Are there relationships that need to be repaired or rebuilt?

Exceptional PMs have the ability to effectively triage priorities during crisis. They then do not allow themselves to be distracted, and with laser-focus they address first things first. That often means that after the dust settles from the original sprint to action, there are other issues that must be addressed in their proper time. Once the course is charted and you are steering the ship out of the storm, is often that time.

Your efforts to define the problem and identify solutions may necessitate recreating some course of events or decisions. During those phases, it is usually not productive to distract yourself with assessing blame or exacting retribution. The exceptions for this are if there is criminal activity, individual actions that warrant organizational disciplinary action, or if individuals' roles on the team need to be adjusted in order to effectively move forward.

Crises can be remarkable teaching moments. The stress, urgency, and stakes of the situation often sear these lessons learned into our professional fabric. Understand this unique opportunity, and strategically use these occasions to train team members and stakeholders. Be candid, kind, and constructive. Invest in them to strengthen your team for present and future success. This is what leaders do.

10.2.4 Juggling Multiple Projects

Most PMs are responsible for more than one project. Pending your organization and the size and complexity of each project, you may have five, ten, twenty, or more. How does that possibly work? There are only forty hours in a typical work week.

If everyone has the same number of hours in the day, why are some people so much more productive? This rhetorical question seems especially true when we ponder an extremely high achieving PM. How is it possible they can successfully manager that many projects? How can they seem to know current updates about each one? How can they keep track and follow-up on all of the outstanding issues? How do they seem to be able to break through the bureaucratic inertia to keep things moving? How do they seem to consistently develop and deliver projects on-time and on-budget?

This book is written from a single project perspective. That is very intentional. Each project should stand on its own accord. Transportation projects are typically funded by taxpayers and administered by public transportation organizations. As such, their development and delivery should remain above reproach. This inherently necessitates that each project be selected and funded on its own merit. Successfully managing multiple projects means successfully managing a collection of individual projects.

Effective time management is absolutely critical to success. Use the three-tier system of importance, described earlier in this section, to evaluate issues and discern priorities. Keep a proper perspective. Use the strategies discussed in Chapter 11, *PM Soft Skills*, Section 11.4, *Time Management*. Trust your team, and delegate. Be protective of your time. Make daily to-do lists. Schedule time each day to think and plan. Find balance. Use self-reflection to evaluate if you are being effective. Not busy, but effective. Strive toward self-awareness and self-improvement. Find a mentor. Learn from your mistakes. Read business and industry books. Strive to be exceptional at your craft.

Be aware of project interdependencies. Perhaps this is shared team members or the same design consultant. While each project and contract must stand on its own, these situations may present opportunities to optimize efficiencies. This does not mean muddying the waters by mixing resources, but it does mean being situationally aware to leverage resources where possible to realize increased efficiencies. Some examples may be strategically scheduling plan submissions to normalize workloads and expedite review times, or completing the hydraulic analysis of an upstream structure before finalizing the hydraulic analysis for the downstream structure on another project.

Perhaps most important, you need to find a way to see the whole picture. What is your entire workload and schedule of commitments? If you only concentrate on the fires of the day, you will never get anywhere. You need to keep a laser-like focus on the long-term goals, and managing interim goals to ensure you are positioned for success. Leverage your organization's project, program, and portfolio management tools. If they are lacking, find a way to overlay your project schedules so you can see future workload, upcoming milestones, and submission deadlines. Know, and be fiercely protective of, the triple constraint (budget, scope, and schedule) of each of your projects. A successful PM portfolio is built on individual projects that are developed and delivered on-time and on-budget.

10.3 Change Management

"The only constant in life is change."

– Heraclitus

While a project's Budget, Scope, and Schedule should be set at Scoping, a transportation project can resemble a living, breathing organism that changes with the political tides, economic and financial realities, adjacent community developments, upgraded industry standards and regulations, stakeholder expectations, public input, and evolving priorities. For example, if a project takes five years to develop and deliver, it is likely unreasonable to assume site conditions, community needs, and involved stakeholders would remain unchanged. A growing urban area can be completely transformed in a far shorter time with residences, retail, business, and industry that may completely change the anticipated needs, benefits, access points, multi-modal accommodations, and traffic flow of your project. Often the biggest challenge a PM faces is how to successfully and seamlessly incorporate these changes into your project.

It is the PM's job to manage the project within the triple constraint while balancing risk, resources, and quality. This can be extremely challenging to accomplish pending the proposed or dictated changes. To proactively manage these triple constraint risks of change, it is imperative the PM keep their finger on the pulse of stakeholder expectations, public input, and leaderships goals and priorities. Whether you see the change request coming a mile away or if it is thrust upon you at an unscheduled meeting Friday afternoon, your response should be the same.

One of your most important PM responsibilities is to faithfully and consistently practice effective change management procedures. Many organizations have established change management policies. If yours does, know it and follow it. If yours does not, I recommend you use something similar to general guidelines described below as it protects you, your organization, and the project.

Once the Budget, Scope, and Schedule are set, they are set, and it is your job to fiercely protect the triple constraint and work within it. All Budget, Scope, and Schedule change requests should be in writing. This documents who is requesting the change, when, and why. A change request form should then be completed that describes the request and the details the resulting Budget, Scope, and Schedule impacts. This should then be evaluated by the designated change control board/agent. This may be the project sponsor, a director, or other assigned individual empowered to make these decisions. While some organizations may allow the PM to serve as the control agent, this is not typical. Assuming the change request is approved, all change control documentation should be appropriately saved, necessary adjustments to the Budget, Scope, and Schedule should be made in the appropriate systems, the decision should be communicated to the team and all impacted stakeholders, and any associated project documentation (e.g., permits, risk register, etc.) should be updated. This should be followed for all requests, whether they adversely impact, or benefit, the triple constraint. Astute PMs realize that change request forms can be the most powerful and impactful tool in their arsenal. It can calm emotional agendas, reinforce clarity of purpose, promote collaborative progress, and bring order to chaos.

Changes are not necessarily bad. They are a normal and expected part of project management. However, a project can quickly get into trouble if changes are made without regard to, or approval of, the resultant impacts to the existing Budget, Scope, and Schedule. When an overeager or overaccommodating PM works to incorporate a requested change into their project without analyzing and documenting the resultant impacts to the triple constraint, they do so at their own risk and they introduce a significant project risk. Well-intentioned undocumented changes can quickly become a nightmare should there be staff turnover, the winds change course and request additional or conflicting changes, or if there are inconsistencies with federal, state, or funding boundaries, requirement, or established purpose and need. If the project ends up with Budget, Scope, or Schedule busts, undocumented changes are extremely difficult to explain and defend. In these circumstances, a PM can quickly find themselves alone at the end of very long and unstable limb. Even though it can be time consuming and some control agents may not want to attach their name to all requested changes, practicing good change management procedures is an essential PM responsibility.

10.3.1 An Engineer's Bias

Transportation organizations are notoriously terrible at saying "no." There may be a number of reasons for this reality. Sometimes there are simply too many cooks in the kitchen. Other times leadership or PMs are determined to avoid conflict and pursue appeasement. Other times there are simply not good ways to say "no" for a variety of reasons. It should be noted that there may be other factors as well.

Engineers love challenges. Many thrive on the opportunity to solve a difficult puzzle. As such, when ideas or issues arise in a project, it is not uncommon for an engineer or PM to think to themselves, "*I could do probably make that happen.*" Resist this urge.

Regardless of the reason for saying "yes" or "no" to project issues, it is incumbent on the PM to insist on practicing approved change management protocols.

10.3.2 Change Management Plan

Every project should have a Change Management Plan that details the acceptable change control process. Many organizations achieve this by deferring to their established change management plans and procedures. Change Management Plans do not need to be lengthy, but should cover the process by which the PM and team will evaluate, consider, decide, and implement change throughout the project. At a minimum, the Change Management Plan should address how the PM, team, or organization. Most Change Management plans generally follow the steps in Figure 10.9.

The PM is responsible to ensure changes are identified, evaluated, approved, and implemented using the organization's acceptable change control processes. One common foundational principle is consistency. This begins with a formalized change request that considers impacts of the proposed change to the existing budget, scope, and schedule. It is critical that the PM not act upon the change request until the request has been evaluated and approved by the designated change control authority. Another important role of the PM is to ensure all changes are properly implemented. This includes proactive communications to team members and stakeholders. This also includes updating all project documentation in the correct configuration, and with the proper versioning.

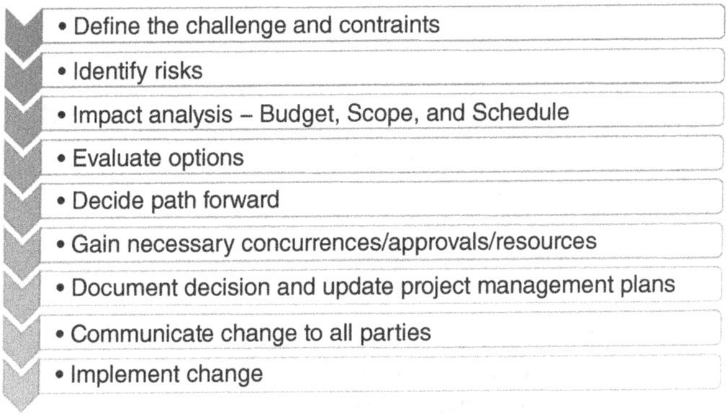

- Define the challenge and contraints
- Identify risks
- Impact analysis – Budget, Scope, and Schedule
- Evaluate options
- Decide path forward
- Gain necessary concurrences/approvals/resources
- Document decision and update project management plans
- Communicate change to all parties
- Implement change

Figure 10.9 Change management plan.

10.4 Balancing Innovation – Performance-Based Design

"Good design is good business."

– Thomas Watson, Jr.

The triple constraint dictates that you cannot change the budget, scope, or schedule without impacting the other two. This fundamental principle has underlying impacts on how a project solution is pursued.

The traditional approach to project development is quite formulaic. The transportation industry is overflowing with manuals, standards, and guidelines. Time tested and trusted, these standards can expedite design in a manner that enforces consistency and uniformity. The traditional approach is for the standards to dictate the scope. For example, this classification of roadway requires a specific pavement section and typical roadway section. Waivers and exceptions were discouraged. The solution was determined and then the needed money was programmed to fund it. Scope drove the budget and schedule.

Our transportation infrastructure and systems are aging. Our populations are growing. Citizen expectations are rising. Environmental regulations continue to expand. Costs are increasing. Revenues are not keeping up with demands. The bottom line is we have more transportation needs than available financial resources. As Aubrey Lane said, "*Limited transportation dollars have consequences.*" Today's reality is that the industry has transitioned to where the budget is often driving scope and schedule.

Something needed to change. The transportation industry has responded with a radical idea that requires a fundamental paradigm shift in transportation solutions. Performance-Based Design (PBD) focuses on pursuing context-sensitive solutions that satisfy the defined purpose and need. The traditional approach starts with the design standards, which dictate the scope. PBD starts with the purpose and need, considers project-specific conditions and constraints, and then develops a cost-effective scope. These different approaches are depicted in Figure 10.10.

This change in approach represents a seismic shift in thinking. Sound engineering principles haven't changed. PBD promotes finding new and innovative ways to apply them that may

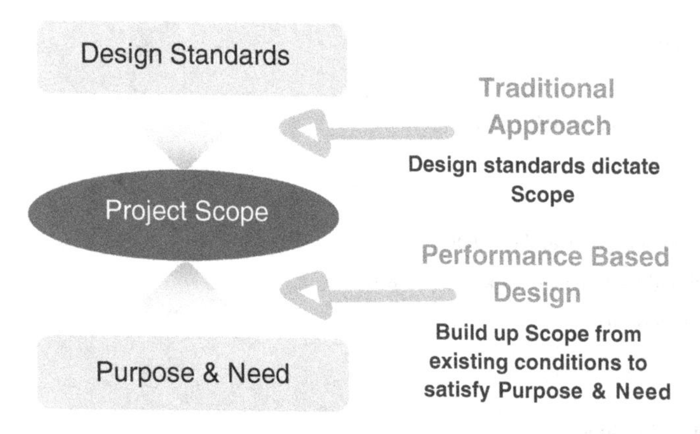

Figure 10.10 Performance-based design.

be outside of the manuals, while ensuring safety. It does not dismiss engineering standards, but rather emphasizes engineering judgment. PBD remains flexible in approach, unwavering in the objective, encourages innovation, and balances the solution with economic realities. The roadway geometrics and project solutions are based on needs, not blind application of standards.

The industry transition has been slow and steady. In 2006 FHWA released their "*Flexibility in Highway Design*" publication. Subsequent years saw many DOTs and organizations begin to embrace design exceptions, waivers, and deviations. Engineers began parsing the manuals for "*shall*" versus "*should*" conditions. In 2018, the AASHTO Green Book (*A Policy on Geometric Design of Highways and Streets*) shifted to Performance-Based Design. While this approach may not be applicable in every circumstance, the intention of this transition is to provide greater flexibility to the designers so organizations can optimize their Returns-On-Investments (ROI) of transportation dollars, improve safety and mobility, and ensure solutions are appropriate for the setting and satisfy the purpose and need.

The PBD mindset can be especially advantageous to the PM during change management circumstances. A growing number of criteria-based funding applications set the project budget earlier in the process. Some of these are set as early as funding application approval. At this early stage, preliminary plans may be available, but likely not beyond 30% design plans, if that. In these situations, the project risks may not be fully identified or evaluated. Project schedules are preliminary best guesses. The project is likely not yet scoped. If the budget leg of the triple constraint is fixed, then any subsequent changes to Scope or Schedule will require some creative change management solutions.

PBD thinking can be an invaluable tool to navigating these sticky wickets. The team will use the manuals and engineering standards, but seek context-sensitive solutions that are appropriate for the specific site conditions and satisfy the big picture purpose and need. Examples may include saving a pavement base instead of tearing it out and rebuilding, or integrating a diverging diamond interchange into the design, or shrinking bike path widths during critical sections of the improvement. While there likely is some crossover between this and Value Engineering, there are some differences. Value Engineering is typically a singular event involving professionals with little or no prior exposure to the project who review an existing design and seek to identify opportunities to reduce costs or expedite construction. PBD similarly seeks ways to reduce cost and expedite delivery, but also maintains a laser-like focus on the project's purpose and need, concentrating on the needs not the wants, while never compromising safety. PBD is also not a single event, but a mindset that can be integrated into the project from funding application through final design.

While PBD is a powerful design tool, it may not be applicable for every project. There are challenges in its evolving implementation. Engineering discipline and divisional silos need to be broken down. Collaboration needs to be embraced. Designers need to release their death grip on manuals and standards and truly think about each unique situation and solution. This approach takes courage. Engineers may not readily recommend a waiver or exception to standards as this bring additional organizational and professional risks. What if the idea doesn't work? The standard solution will work and is irrefutably defendable. It is said civil engineering is the one profession where our mistakes become monuments of incompetence for all to see. The degree to which an organization or individual pursues PBD is proportional to their commitment to the strategy and their risk appetite.

Innovation can also bring additional challenges. For example, where adding a roundabout into the design may improve safety and lower costs, it may ignite substantial public concern that necessitates significant stakeholder outreach efforts.

The PM is responsible to develop and deliver the project within the established budget, scope, and schedule. As such, the PM plays a pivotal role in balancing scope solution innovations that balance economic and schedule realities. They should remain focused on the scope, but flexible in approach. Especially during change management challenges, PBD can be a powerful tool the PM and designers can leverage to progress to success.

10.5 Goals and OKRs

> "*A goal should scare you a little, and excite you a lot.*"
>
> – Joe Vitale

The majority of engineers and PMs are inherently goal-oriented. Many of these are also disciplined, inwardly driven, and intensely competitive. This blend can create a lethal combination of dedication and productivity that can seemingly will a project forward. Other PMs have vastly different approaches, but experience similar results. While it may be possible to be a successful PM and not be goal-oriented, it is not common.

Most everyone is familiar with SMART goals. They should be Specific, meaning they state what you will do. They should be Measurable, meaning there is an objective way to evaluate progress and success. They should be Achievable, meaning they are possible to accomplish. They should be Relevant, meaning they are reasonably connected to your role and worth doing. And they should be Time-based, meaning they have a schedule. SMART goals are well-established tools that can be remarkably effective in forwarding short-term and long-term agendas. A PM should be experienced and accomplished at setting and achieving SMART goals.

Many forward companies have supplemented SMART goals with OKRs (Objectives and Key Results). The Objective should speak to the intrinsic value of the work itself, and be inspirational in nature. Each Objective should be paired with three to five Key Results. These are the real-world, metric-driven realities that balance quantity and quality. They should be specific enough to be accurately measured. An OKR often takes the format of (objective) "*as measured by*" (key results). Employees should have no more than three to five OKRs at one time, which should be consistent with established short-term and long-term goals. This allows them to come to work and easily know and focus on their top priorities.

Great leaders have uncluttered thinking that allows them to clearly drill down on what needs to happen to move forward. OKRs can be operational and aspirational. Operational OKRs are those that need to be achieved within an established schedule and budget. Examples would include many of the project development tasks. Aspirational OKRs are the audacious goals that ignite our passions. Examples may include innovative engineering solutions. Particularly when formed with team member input, individual and team OKRs can be a powerful force to focus, align, track, and stretch your team.

Both goals and OKRs should drive us to be innovative, efficient, and productive. At their best, goals and OKRs should push us beyond our comfort zone, leading us to achievements that live between our abilities and our dreams. Timid goals that intentionally underestimate potential to ensure success will not improve you or your team. Similarly, establishing goals or OKRs with low value objectives that really don't matter to anyone, or key results that are nearly impossible to actually measure is an academic exercise of futility. While these sandbagging tactics may seem successful if your goal is to Cover Your own Ass (CYA), they will never drive or meaningfully improve your operations.

One key to success with SMART goals or OKRs is often the ability to say no to things that distract or detract. This can include our own pride or expectations. As Voltaire said, "*Don't allow the perfect to be the enemy of the good.*" Know when good enough is good enough.

Much like SMART goals, the power of OKR is in the mindset that drives success with these tools. OKRs inherently value merit over seniority. What you know is less important than what you can do with what you know. What do you produce? It is simple and straightforward. What is the objective, and what are the specific key results to know if you achieved your objective? This "can-do," "get-it-done" approach is especially important in transportation project management.

Transportation PMs are often faced with goals set for us that are not SMART goals. What then? While these circumstances can be frustrating and challenging, they are not impossible. Perhaps there is no silver bullet, but combining a number of tools in this book can position you for success. Communication, stakeholder engagement, managing expectations, managing the project's triple constraint, practicing change management, and leading the team are all tools that will enable you to set and achieve individual, team, and project SMART goals or OKRs.

10.6 Project Management Methodologies

> "*Management is, above all, a practice where art, science, and craft meet.*"
>
> – Henry Mintzberg

10.6.1 Changing Conditions

Many transportation organizations continue to use waterfall policies and practices that have served them well for many years and decades. This relies on a well-defined Work Breakdown Structure of carefully arranged tasks within a Critical Path Methodology. Whether it is rooted in history, precedence, pride, or just trying to keep their heads above water, many transportation organizations are hesitant, or outright resistant, to revisiting their well-established project development processes. In some cases, this may be an intentional decision. In many other cases, this may more so be the lack of clarity or confidence that considered changes would produce more good than harm it may cause. Not all public agencies are nimble and aggressive in pursuing increased efficiencies. While most desire to improve, they carefully weigh uncertainties against full workloads, established workflows, and valued tradition. "*We've always done it that way*" resonates in a culture that relies on institutional knowledge. However, the reality is the transportation world is changing before our eyes.

Traditionally, the scope was fixed in transportation projects. If we were going to build a bridge carrying a four-lane, divided roadway over the river, we would build a bridge that carried a four-lane, divided roadway over the river. The Scope would be defined by this objective and conformance with applicable design and safety standards. If the project took a little longer, or cost a little more to construct, that was accepted as long as we built a bridge that carried a four-lane, divided roadway over the river that complied with all applicable standards. The scope was fixed, and the budget and schedule components of the triple constraint were the more flexible constraints.

This approach is rapidly changing. As the transportation moves toward more objective, application-based project selection, often the budget or schedule becomes the fixed components, with the Scope becoming the flexible constraint, as shown in Figure 10.11.

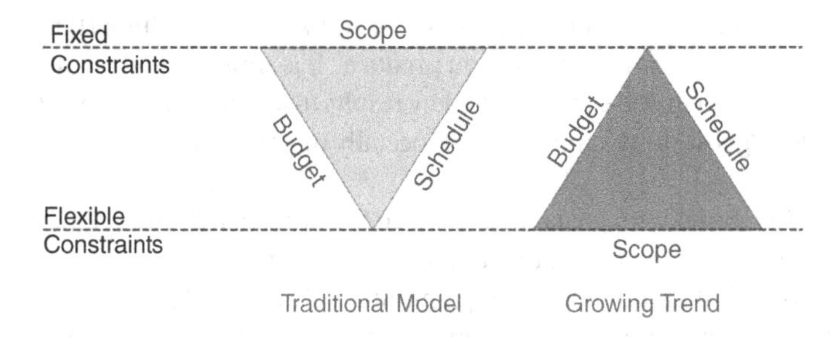

Figure 10.11 Fixed and flexible constraints.

This trend will continue to grow. Limited resources have consequences. Engineers are encouraged now more than ever to pursue innovative designs, using the project's Purpose and Need as their guide. Design waivers and exceptions are no longer viewed as something to be avoided at all costs, but rather an approach to be leveraged to promote context-sensitive solutions. This power of flexibility is the basis of Performance-Based Design, as described and advanced in AASHTO's Green Book. Design is no longer driven by blind applications of standards in all circumstances. Design is increasingly encouraged to accommodate a project's unique circumstances.

Upon this landscape is an ever-growing list of success criteria that transportation organizations and PMs are compelled to address. Organizations, and their resulting performance metrics, often focus on a project being on-time and on-budget. These are tremendously important priorities, but they are not the only success criteria. Some expanding criteria lends itself to being objectively quantified, such as safety, congestion mitigation, and economic development. Others are more subjective in nature, such as accessibility and wealth creation. Still others have local or political priorities, such as maintaining the character of a corridor or opening the new roadway to traffic before the start of a new school year. There are technical and social aspects to these new realities.

Technically, transportation systems and assets have continued to become increasingly complex. Transportation projects are so much more than concrete and steel. Integrated traffic signals and intelligent traffic systems have transformed our roadways. The amount of automated and physical data collected on projects and systems is staggering. Drones can capture seemingly unlimited data points during a remote bridge inspection. Integrated Geographic Information Systems (GIS) can display layers of seemingly every asset from extensive data sources. We are designing roads in 3D models that can create remarkable life-like simulations of facilities not yet constructed. Success criteria on many transportation projects now extends far beyond opening up new pavement to traffic.

Most transportation organizations have many more needs and wants than resources. Infrastructures are aging. In lean environments, preventive maintenance is often the first effort to be pushed back. This short-term solution often brings longer-term problems as deteriorating assets become more expensive to repair, or may reach the point of being functionally obsolete and require complete replacement. Unpredictable market conditions, such as inflation, supply chain issues, or other economic constraints, only exacerbate the situation. All of this further emphasizes the need to be flexible in individual project solutions.

Socially, there is a growing appreciation of transportation systems. This realization manifests itself in many ways. Take the holistic understanding that transportation assets are more than roads; they are a fundamental and shared community asset. This simple assertion can bring with it many cascading success criteria. Subjective concepts like community building, expanding access, wealth creation, environmental stewardship, and economic justice are now commonly emphasized transportation project success criteria. These ideas are often bolstered by guiding legislation and political agendas. These subjective goals may clash with objective design criteria, and often surface during stakeholder engagement. Virtual stakeholder engagement has expanded the avenues by which citizens can become involved in the process. While transportation organizations and projects benefit from increased transparency and stakeholder involvement, it does necessitate an increased ability to shift project development in a more Agile way.

10.6.2 Waterfall Project Management

Waterfall project management is what most non-IT folks think of as traditional project management practices. It is a clearly defined, linear process of carefully sequenced tasks. The Critical Path (CPM) methodology fits nicely within a waterfall mindset. The project flows down a prescribed workflow, like water flowing down a well-worn path.

Waterfall is terrific approach for highly regulated and complex processes. It more easily accommodates staff turnover, and encourages documentation. There are clear gates through which one must pass before continuing, be it for funding or development. Objectives are clearly defined, and documentation is embedded into established workflows such that the organization maintains process and project control. This predictability and clear division of labor is well suited for most transportation projects, which is why most transportation projects, programs, and portfolios are administered using Waterfall.

The primary disadvantage of Waterfall is its inherent inflexibility. Once a project has passed a milestone, it becomes troublesome to go back and revisit those decisions or directions. It also has difficulty incorporating ongoing stakeholder feedback into the project design throughout the life of the project. This inflexibility is fundamentally challenged when requirements change, or success criteria evolves.

Even with these disadvantages, the nature of transportation funding and workflows necessitate that many of the fundamental tenets of Waterfall project management are here to stay. However, much of Waterfall does seem to contradict the changing conditions in the transportation world.

10.6.3 Agile Project Management

Agile project management takes a completely different approach, with vastly different values and objectives. At its core, Agile is a management mindset for project development. The four values that capture this approach are:

1) Individuals over processes
2) Working solutions over comprehensive documentation
3) Stakeholder collaboration over contract negotiations
4) Responding to change over following a pre-set plan

These four values are further clarified by twelve principles that are defined in the Agile Manifesto. This approach was born in the software industry to satisfy a need to be more adaptive and responsive to industry trends and market conditions. While mainly concentrated in the computer and software sectors, aspects of inherent flexibility are transferable to most industries.

There are many different approaches by which Agile can be implemented. The most common is SCRUM, which typically relies on a collaborative team that focuses to achieve short-term objectives within a series of two-week sprint that are consistent with larger objectives. Every sprint intends to produce a tangible or working product. While there is a master plan at the outset, priorities can change during each sprint with the constraint that resources are fixed. The nimble organizational structure enables promptly concentrating on evolving priorities. This is one of Agile's primary strengths, but also one of its biggest disadvantages.

10.6.4 A Hybrid Approach

Modern-day Transportation project management is neither purely Waterfall nor purely Agile. In Waterfall, the scope is fixed and the budget and schedule are estimated constraints. In Agile, the budget and schedule are fixed, and the scope features are estimated constraints. Most transportation projects now lie somewhere in between these two extremes within the Hybrid zone, as shown in Figure 10.12.

Generally speaking, Waterfall is often most appropriate where there are clear requirements and processes. Meanwhile, Agile is often most appropriate when requirements are changing and

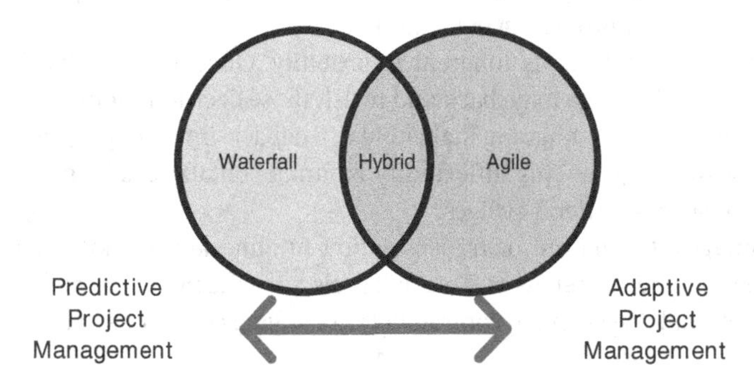

Figure 10.12 Project management methodologies.

processes are evolving, as shown in Figure 10.13. The nebulous Hybrid lives in-between and tries to leverage the strengths and constraints of each to improve efficiencies.

Figure 10.14 compares these two management approaches by highlighting some of their key differences.

When considering the spectrum of transportation projects, from large and complex megaprojects to local streetscapes and everything in-between, it is obvious that one size does not fit all. Pending the size,

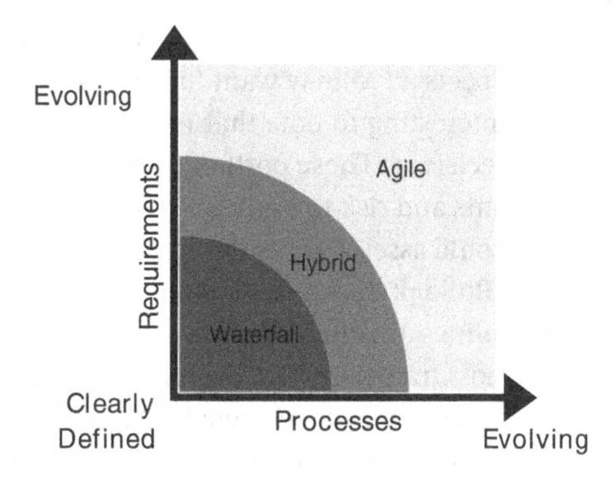

Figure 10.13 Project approach comparison.

	Waterfall (Traditional)	Agile
Approach	Predictive	Adaptive
Focus	Process-centric	People-centric
Prioritization	Process over product	Product over process
Team Mindset	Disciplined	Innovative
Management Style	Controlling	Facilitating
Management Philosophy	Command and Control	Leadership and Collaboration
Goal	Predictability and optimization	Innovation and adaptability
Organizational Values...	Governance	Work Done
Methodology Building Blocks	Work Breakdown Structure	Feature Breakdown
Project Requirements	Clearly defined very early	Evolving during project
Project Scope	Fixed	Variable
Project Scope Risk	Low	High
Project Schedule	Variable	Fixed
Project Schedule Risk	High	Low
Project Budget	Variable	Fixed
Project Budget Risk	High	Low
Team Composition	Specialists	Cross-functional
Workflows	Linear and Sequential	Iterative
Stakeholder Involvement	Low	High
Attitude toward change	Slow and resistant to change	Flexible and embraces change
Ability to respond to change	Time-consuming and difficult	Quick and straightforward
Guiding question	"How" things get done	"Why" things get done
Control Approach	Documented and Quantitative	Internalized and Qualitative
Size	Larger teams and projects	Smaller teams and projects
Issue Escalation and Decisions	Time-consuming	Prompt
Documentation Requirements	High	Low

Figure 10.14 Waterfall compared to Agile.

specifics, and complexity of any transportation project, there may be certain Waterfall or Agile aspects a PM may want to leverage and embrace.

It is interesting to note that many of the comparisons in the Figure 10.14 are organizational decisions. These portfolio-wide decisions are often based on the company's cultural precedents and risk tolerances.

One could assert that Agile strengths are already being leveraged in a hybrid approach in Design-Build project development and delivery. In these projects, schedules can be crashed as disciplines are often pressed to work concurrently, and in a more iterative fashion.

For most transportation owner-operators, on Design-Bid-Build projects, there are some inescapable truths that entrench waterfall methodologies. The project development process is paramount as it is carefully crafted to satisfy all applicable federal, state, and local requirements and design standards within a wide range of subject areas. Project durations can be lengthy. Team turnover can be high. Documentation is paramount to satisfy the various federal and state reviewing and auditing entities. The flow of federal and state funds is extremely prescribed. Yet, stakeholder engagement can and does influence the project scope halfway through design. Performance-Based Design does encourage innovation and flexibility in solution approaching throughout the design. Changing market conditions does require flexibility and nimbleness in order balance resources and priorities to successfully develop and deliver projects, programs, and portfolios. So how do we leverage some of the benefits and practices of Agile into the necessary Waterfall structure, processes, and roles?

10.6.5 Altitude Matters

As an organization struggles with how to leverage and embrace aspects of Agile project management, it is important to recognize that approaches and implementations may be different at different altitudes within an organization. This means there may intentionally be differences at the project, program, and portfolio levels.

Especially for larger and more mature transportation organizations, Waterfall may be entrenched with extensive procedures, practices, and PDP. There are good reasons why this is so, and should remain so at the portfolio level. The portfolio big picture is to pursue your organization's mission by successfully executing program objectives. Inherently, this requires accurate waterfall schedules to ensure the right amount of the right kind of money is at the right place at the right time to keep all projects moving forward. The culture at the portfolio level may be inherently at odds with Agile. Meanwhile, at the other end of the spectrum, many organizations encourage innovation and enable flexibility at the project team level. They want projects to be nimble and adaptive to stakeholder input and changing conditions. As such, PMs often seek to leverage and incorporate Agile strengths. In this scenario, the program level may be caught in the middle trying to balance portfolio Waterfall requirements with project Agile characteristics, as shown in Figure 10.15. One example is a waterfall schedule of meaningful task spans of significance (e.g. – the spans between Scoping meeting (30% plans), Public Hearing/Design Approval (50% plans), Field Inspection (60% plans), Pre-Advertisement Conference (90% plans), and Advertisement (100% plans), while granting project teams more flexibility between these milestones. Culture, leadership, acknowledgment of realities, and vertical concurrence of success criteria are crucial elements to successfully implementing Hybrid solutions. While it may seem unwise to embrace different project management approaches at the project, program, and portfolio level, this stratified approach can be quite effective provided each level is consistent in objective, flexible in approach, and collaborative and supportive in practice.

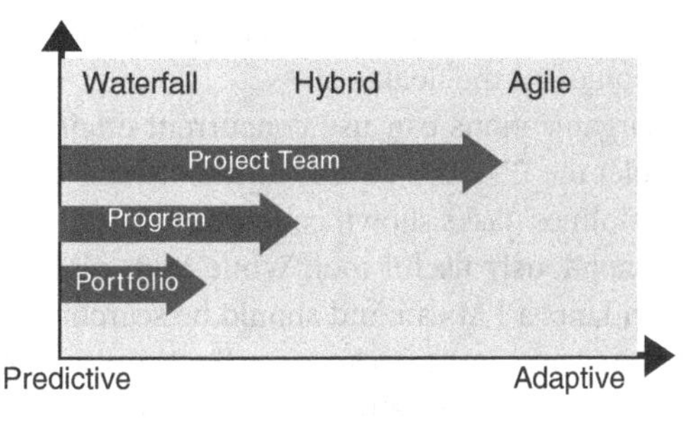

Figure 10.15 Varying PM methodologies at different levels.

Understanding these realities, one significant challenge can be identifying and implementing practical ways to secure vertical concurrence of goals while enabling flexibility of approach. One needs to first acknowledge contextual realities, givens, and constraints in order to create a culture where scope, schedule, and budget changes can be candidly discussed and addressed. Especially within public agencies, it can be challenging to create a safe space for work product, which encourages positioning and game-playing and is enormously counterproductive and inefficient. Pursuing transparency, responsible public stewardship, and active stakeholder engagement can unintentionally inhibit candid conversations of project change. It can be difficult to break down organizational and process silos to embrace an iterative mindset that relies on nimbleness of cross-functional teams. While each transportation organization that moves toward Hybrid may settle on slightly different solutions, there will be consistent similarities. They will all almost certainly modify their change management thresholds and practices, adjust their performance metrics, and initiate a bold cultural change.

10.6.6 Concurrent Engineering

Most transportation organizations have an established Project Development Process (PDP). These are almost all inherently Waterfall, and for good reason. Many of the tasks in a transportation PDP are prompted or driven by a federal, state, and local policies, procedures, and standards.

Concurrent engineering focuses on conducting tasks in a simultaneous, rather than consecutive, manner. The ultimate purpose of this is to increase efficiencies, optimize resources, and expedite project development. Concurrent engineering does not inherently change the existing PDP. It contains the same tasks, but looks at the PDP from a slightly different perspective. One may assert concurrent engineering is just practicing smart schedule management principles. And they would be right. Concurrent engineering aggressively leverages Critical Path scheduling to identify concurrent tasks that will minimize the critical path.

The reality is that this does not always naturally happen. Many organizations, PMs, and team members work to the schedules. They don't always act proactively, or in a way that is best for the project. If a hydraulics study takes 2 months, they may choose to wait until they receive all the survey data. Another option is to start the hydraulics study when a critical mass of survey data is received such that progress can be made in an unimpeded way until the additional data is received. Another common practice is for the hydraulics study to be delivered within the two-month deadline, even if it is completed in one month. Task owners don't own the float. PMs don't own the float. In project centric environments, the organization owns the float.

Organizations can use concurrent engineering to create fantastic ways to graphically depict the PDP is a slightly different way. This is most often done by discipline, with each disciplines' tasks shown concurrently with the others. Especially for the PM, this can be a tremendously useful tool. While each discipline is making progress, swimming in their own lane, a PM can and should be searching for ways to expedite the schedule by identifying opportunities to leverage floats in a way that encourages concurrent progress to optimize the critical path. This focus is empowered when paired with other good project management practices, such as proactive risk management and preemptive stakeholder engagement.

10.6.7 Changing Performance Metrics

One of the biggest questions a transportation organization faces when moving toward hybrid project management is what to do about performance metrics. This is important because many organizations' performance metrics do more than reflect realities, they often drive progress, shape culture, and influence behaviors. It can be challenging to strike an acceptable balance between meaningful project development process milestones, funding requirements, focused organizational objectives, and agile methodologies. One reason many organizations struggle is the underlying workflows. Where waterfall is predictable, agile is inherently iterative, see Figure 10.16.

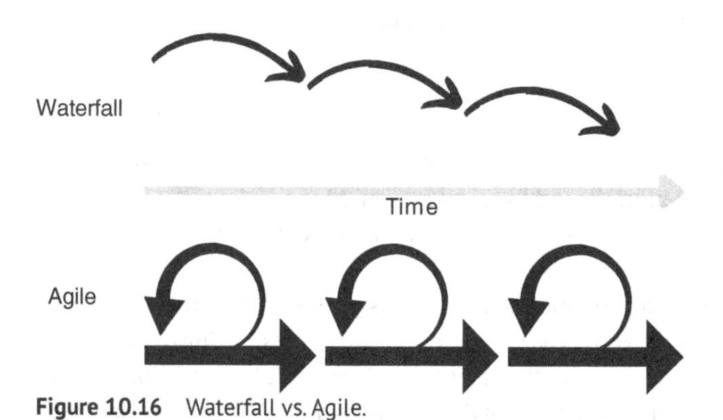

Figure 10.16 Waterfall vs. Agile.

This creates a clear paradigm conflict for performance metrics. By nature, many performance metrics compare what actually happened to that which was expected. While there can be significant contextual nuances with each of these concepts and compared fields, we will refer to that which is expected as the baseline. Be it a schedule, budget, or some other entity, the baseline is that by which actuals, or more current information, are compared to determine acceptable progress or success, however that may be defined or endorsed by the organization.

There are many ways in which organizations assign and prioritize baselines. Some organizations capture budget and schedule baselines at Scoping. In these instances, these three triple constraint components are formalized in a scoping document that passes through a prescribed approval process. Scoping approval then sets the project's scope, schedule, and budget baselines. Other organizations may use other triggers to capture baselines. Many organizations set performance metric schedule and budget baselines based upon the planned Fiscal Year for advertisement. In these instances, previous baselines may be captured, but are not prioritized as those for the current Fiscal Year. Still other organizations use different triggers to capture scope, schedule, and budget baselines at different points of project development. For instance, scope may be captured at time of project selection from an application-based prioritization, the prioritized schedule baseline may be captured at Public Hearing, while the prioritized budget baseline may be when the project is in the planned current Fiscal Year for advertisement.

There are a few fundamental decision points. When will baseline(s) be captured? What are the performance metric tolerances? And under what circumstances will the organization's change management policy enable adjustments of the baseline(s)?

The iterative nature of Agile encourages multiple points to "reset" baselines. Scoping is obvious. Design Approval, which is after Public Hearing, is another reasonable choice. At this point, the public has been engaged, and by its very nature, Design Approval provides concurrence with the design and approval to move forward. Other potentially logical choices may be the major 30%, 60%, and 90% design reviews. At each of these milestones, the different disciplines come together and the project is momentarily "static" during the plan reviews. Organizations should strive to intentionally choose points that are logical and reasonable. One key factor is when and how data is collected. You may decide a baseline(s) should be captured at Design Approval, but that only works if your systems and workflows support capturing meaningful baseline(s) at that point in development.

Another choice organizations must make is whether to capture scope, schedule, and budget baselines at the same time, or individually capture them at different points of development. As long as it is an intentional decision with sound logic that supports your organization's mission and objectives, there may not be any absolute, or more correct, option. Ideally, you would only reset baseline(s) at a point of clarity where risks have been mitigated, design has significantly progressed, or key decisions were secured.

Another way organizations can implement a hybrid approach in performance metrics is to expand their change management procedures. This often resonates with more mature waterfall organizations that value formal approvals and robust documentation. By expanding the criteria by which scope, schedule, or budget baselines can be revisited, an organization can retain process control while embracing flexibility. This approach can encourage innovation, while discouraging a performance metric "get-out-of-jail-free" card by automatically granting a baseline reset that accommodates poor project management or progress.

Another option is to integrate contextual realities with agile approaches by leveraging tolerance logic in existing waterfall metrics. Consider the budget baseline. Let's assume an organization acknowledges the reality and adopts a cone of uncertainty. If their cone of uncertainty states a preliminary cost estimate is +/− 30%, then performance metrics should reflect this understanding. Similarly, if the cone of uncertainty specifies an expected accuracy of +/− 15% at Public Hearing, then performance logic should reflect this understanding. The logic behind automated performance metrics can be modified to accommodate changing estimates within these organizationally approved tolerances.

10.6.8 Conclusion

Transportation project development is predominately Waterfall, and will remain so. However, industry changes beckon for a more Agile approach. It is now common for project changes in budget, scope, or schedule to be dictated to the project, throughout the project. When this occurs, the processes need to be nimble and adaptive enough to respond. More mature organizations recognize the inherent inconsistencies and opportunities within these changing industry dynamics, and are choosing to pursue a Hybrid approach.

The basis of hybrid project management is to acknowledge contextual realities, and then leverage the strengths of Waterfall and Agile to improve efficiencies and streamline development. While the challenges transportation owner-operators face may be similar, each organization will likely follow their own path as they strive to responsively and responsibly move forward. It will be fascinating to see how the industry evolves and embraces the best of Waterfall and Agile.

10.7 Performance Metrics

"Not everything that counts can be counted, and not everything that can be counted counts."

– Albert Einstein

10.7.1 The End Goal

The idea of performance metrics is imminently reasonable and seemingly rather straightforward. But then it quickly gets complicated.

Is your project successful? This is a relatively simple question that can be very difficult to answer. Is your project healthy? An even simpler question that can be even more challenging to answer. How do you know? What criteria is important? What do you measure? How do you measure it? Are you measuring the right things, at the right times, for the right reasons? And to what end?

Performance Metrics should collect data that succinctly and accurately answers specific questions. Start at the end and work backward. What is the question you want answered? Just because data has always been, or can be, collected and measured doesn't make it valuable. In order to create meaningful metrics, you need to ensure you are collecting the right data and using in the right way. There are potential dangers if you go the other way. If you start from the data you have or can collect, one can easily stretch the use of that data to infer answers to questions it cannot support.

Over time, the use of a metric may evolve. This may create a situation where the practical purpose of a metric has shifted to a result that extends beyond what the original data can support. Beware these intended and unintended consequences. Consider the desire to track educational competencies in public schools. Standardized tests are created. Rewards are established to encourage higher individual and school performances. Positive results and trend lines are publicized to celebrate successes. All well-intentioned efforts. But at what point do some schools stop teaching the entirety of the subject to concentrate on focused test preparation? At what point does the metric stop reflecting reality and start driving schools' decisions and actions? Is this a good thing or a bad thing? Some may assert a bit of both. Regardless, the metric of standardized test scores has evolved to empower the standardized test writers to drive the curriculum and modify teaching practices and priorities in many schools. Similarly, does your organization's on-time metrics reflect reality or drive decisions and influence behaviors? What master are you serving – good business or the metric? Are you making good decisions, or is the metric driving your actions? Is the data being used to answer questions that stretch beyond the data's meaning and reasonable influence? Beware the tail wagging the dog.

It is easy to lose the forest for the trees. If a project makes its schedule but blows its budget, was it a success? If the PE is completed under budget and the project is advertised on-time, but the quality of the plans is such that construction becomes a drawn-out, change-order infested quagmire, was the project a success? Would your answer change if you were the PM who brought the project to Ad on-time and under-budget before handing it over to Construction? There is rarely a single performance metric that tells the whole story.

10.7.2 The Road to Hell Is Paved with Good Intentions

The old cliché says, "*That which gets measured gets done.*" Peter Drucker later modified this to, "*What gets measured gets managed.*" One important question to answer regarding any performance metric is, *to what end*? If done right they can provide a picture of your progress, and insight on how to enhance your efficiencies. They can bring clarity and focus on what you do well and where you have growth opportunities. They can encourage transparency and promote accountability. They can document successes and generate healthy competition. They can balance reality and drive productivity and innovation. However, always remember that performance measures are not your product. Performance measures are not your deliverables. The process draws you in. In the world of performance metrics, it is easy to become overly focused on the production of information to the point you drown in the data. If not careful, the servant can quickly become the master. Remember also that metrics can subtly or overtly shift the focus, change the culture, or alter professional behavior in an organization, program, or individual projects. Be observant and aware of a metric's intended and unintended consequences.

10.7.3 Organizational Maturity

There are different data science, or data analytics, maturity models, but most of them share the same underlying fundamental tenets. For our purposes, the specifics of the different models are less important than the general overtones. As your organization's data maturity level rises, so should the tangible value and benefits you realize. However, each level also brings inherent constraints. This is important to understand as most transportation organizations are solidly positioned on the left side of Figure 10.17.

Descriptive Analytics strive to answer the question of "What Happened?" Typically, this reactive reporting is static or passive in nature. Organizations at this maturity level are systematically or unintentionally decentralized. Underlying the data analytics is a fundamental

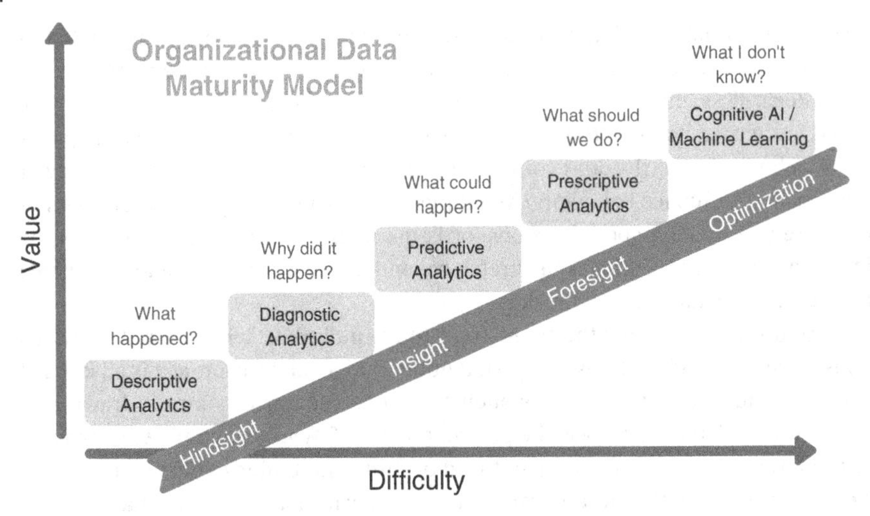

Figure 10.17 Performance metrics organizational maturity.

reliance on manually created and manipulated reports, often in simple or highly customized spreadsheets. Consequently, individuals at every level create their own spreadsheets to track their own operational issues. This inconsistent environment inevitably leads to endless conversations as to the origin and accuracy of one's numbers, at the expense of focusing on what story the data can convey.

Diagnostic Analytics strive to answer the question of "Why it Happened?" This dynamic reporting is a giant leap forward for an organization as it fundamentally shifts to using data analytics to guide decisions and drive progress. Data owners are defined. Data sharing is encouraged. Business Intelligence (BI) tools are leveraged so everyone has the same data. Dashboards are created and embraced. While originally conceived to convey a simplified view (think of an old car dashboard with a handful of dials clearly conveying specific information), many become a behemoth of a data dump to answer a growing list of questions, enabling users to dive as deep as they desire. This means more users need to become analysts to discern what is really going on. BI and reporting expertise become concentrated in smaller, siloed pockets as users struggle to reconcile the accuracy of the data as filtered through complicated business logic against the story the Dashboard is telling. Any apparent disconnects between the Dashboard results and the perceived reality can create skepticism of the tools and advance the idea of multiple sources of truth. Remember the end goal is not a flashy report or integrated Dashboard. Just because the data is accessible, doesn't mean it is worth reporting. What question are you asking? And are you really answering it?

Predictive Analytics strive to answer the question of "What could happen?" This level of maturity surfaces when organizations transition to focus on understanding what information different employee roles need to be successful. Here the data is telling a story, bringing IT and the Business together to serve the employees by answering their most relevant questions. This requires aggressively combining different systemic data sources across the organization. However, as opposed to working to get as much data in everyone's hands as possible, this strips down the data to that which is essential for different roles. Strategically filtering the data streamlines performance and allows more focused decision making. This data evolution requires a consistent view of data throughout the organization, and ultimately enables more proactive insights and measurable results.

Prescriptive Analytics strive to answer the question of "What should we do?" Here intelligence emerges from the data that can rapidly add value to your organization. Employees should feel supported by proactive information the data provides that can even be tailored down to the individual. This maturity level relies on collaborative, cross-functional cooperation, which is built upon a reliable trust of the data.

Cognitive AI/Machine Learning strive to answer the question of "What I don't know?" At this maturity level, the organization is using AI and Machine Learning in real and meaningful ways. This transformational environment should allow the organization to reach new levels of optimization by focusing employees on high-value efforts and automating low-value tasks. Approval hierarchies and organizational structures become flatter. Governance becomes less restrictive and more cross-functional. Employees are empowered to nimbly and effectively use the data-driven culture and resources. Know that this maturity level remains aspirational for most organizations and industries.

10.7.4 Apples to Apples

One basic tenet of performance metrics is comparison. To measure progress, evaluate performance, and predict future trends, you must be able to compare one data marker to another. The characteristics, precision, and accuracy of the data is paramount. We all know you shouldn't compare apples to oranges. Yet this happens all the time in metrics. Obtaining a true comparison of apples to apples can be difficult, especially over time. Sometimes you have to settle for comparing different varieties of apples, but you can't compare apples to oranges and draw meaning or insightful conclusions. Yet this is what many organizations regularly do. This situation is often exacerbated as those creating, assembling, and reporting the metrics often do not fully understand the power, meaning, and limitations of the data they are processing.

Every engineer learns of significant figures, and how they detail the precision of the number. A young engineer may design the curb elevation detail to a thousandth of an inch. While this may be accurate in the computer model, the meaningful design elevation is only as accurate as the contractor can build it. And the true elevation is only as accurate as the surveyors can document the as-builts. Similarly, someone may determine the exact location of an underground utility based upon the fact GIS clearly shows the bend to be triangulated 4.87′ from feature one, 13.24′ from feature two, and 9.58′ from feature three. Would you be as confident of the location if you knew the GIS utility layer had an accuracy of plus/minus three feet and the aerial photographs were slightly stretched to match the GIS-referenced map tiles? GIS and other applications are powerful and incredibly useful tools. But there is a reason you still send out ground survey crews and dig test holes. Detailed design requires a higher level of accuracy than most GIS models can provide. Experienced engineers know and understand significant figures during design, using the correct data precision for the task at hand. Yet this concept is often disregarded in metrics. You can compare a metric of 8.6 to one of 9.843, and readily calculate the delta is 1.243. In this example, the 0.043 of the difference has no meaning, while providing an incorrect sense of accuracy. Yet many organizations make decisions and take actions based on this 0.043.

10.7.5 Innumeracy

John Allen Paulos' book, *Innumeracy, Mathematical Illiteracy and Its Consequences*, details how even well-educated people often struggle to understand basic mathematical principles, the meaning of numbers, and their associated probabilities. This pervasive innumeracy regularly results in misinformed decisions at all levels. Most metrics are not well understood. In large part, this is due to innumeracy in those who create and interpret the metrics.

Magnitude matters. Engineers often use units and assume the data consumers understand. In reality, most engineers can likely not truly grasp the magnitude difference between an environmental contaminant reading in ppm (parts per million) vs. ppb (parts per billion). The general public most assuredly would be less enlightened. The same is true for other units. Are 12,000 vehicles per day reasonable or excessive for your project's ADT (Average Daily Traffic) volume? Is 18,000 cfs (cubic feet per second) peak stream flow reasonable or excessive for the proposed road crossing? Is a $1.5 million change order reasonable or excessive on a $20 million project? Most struggle to fully grasp the magnitude of numbers, particularly in context. This means most can be swayed to accept an asserted agenda that cites most any number.

Many lack foundational arithmetic, algebraic, and geometric skills. This can translate to basic miscalculations in those that create, interpret, and consume data and performance metrics. This is particularly tricky when paired with moral equivalency. For instance, an evaluation of the anticipated benefits of light rail or biking facilities may appear convincing when compared to automobiles; however, most do not normalize for cost, usage, and volume. You cannot compare siloed data in an equivalent way without properly scaling the comparison.

Percentages are especially susceptible to misinterpretations. *Traffic congestion is down 15%. Projects are now being delivered 20% faster. Project development is now 75% on target.* One may assume these three examples are conveying good news. Are they? Benjamin Disraeli said, "*There are three kinds of lies: lies, damned lies, and statistics.*" On their own, these three examples convey no context as to what is really being evaluated, and to what it is being compared. You would have no idea if this is good news or not? To conclude otherwise you must disregard logic or fill in the many gaps with your assumptions or predispositions.

There are two common innumeracy traps that can make percentages especially misleading. First, the denominator matters. If a couple has their child come home from college, their volume of dirty dishes may increase by 50%, but when the child goes back to college the couple only experiences a 33% reduction in dirty dishes. Assuming equal contributions, the volume of dirty dishes generated by the child didn't change, but the percentages don't appear to add up. Second, you can't average averages, but that doesn't stop many from doing it. If look closely, you will likely be stunned how many metrics and reports rely on averaging averages, a clear violation of mathematical principles.

10.7.6 Objective vs. Subjective Measures

Difficulties often arise when organizations attempt to convey subjective matters in objective truths. In these situations, it can be easy for performance metric creators and consumers to overvalue the resulting numeric information. This is often exacerbated when metric creators attempt to oversimplify the context and meaning of the data. Some situations cannot be accurately simplified. *Road rage is up 25%.* Can you accurately define, determine, and quantify the collective subjective emotional state to consistently define and delineate road rage in an accurate and comparative manner?

As a general rule, objective data can be measured objectively while subjective data should be measured subjectively. Expressing traffic volumes in a detailed statistical analysis makes sense as traffic data is inherently numbers based. Accurately capturing the tenor, urgency, and level of passion of citizens at a Public Hearing is a different kind of challenge. Consider a public hearing in which 100 people attend. 60 submit comment sheets. Based upon the subjective check boxes (strongly disagree, disagree, neutral, agree, strongly agree), it appears 30 are generally in favor of the project. Of the 60 comments, 120 different specific issues were mentioned. 30 people spoke. 10 endorsed the project; 20 spoke against some aspect of the project. All speakers inferred they were speaking on behalf of others. Did the public hearing go well? Is the public in favor of the project? Many times, the provided answer frames the data to support the preference of whoever is responding. It can be challenging for an objective statistical analysis of the data to accurately tell the whole story.

Attempting to objectively summarize subjective issues often results in the temptation to attach weight to metrics that they were not intended to carry or cannot accurately support. Math, definitions, assumptions, and reality matter.

10.7.7 Outputs vs. Outcomes

One underlying tenet of performance metrics is to clearly delineate outputs versus outcomes. Lack of clarity can cause systemic confusion, frustration, and cynicism.

Outputs are the tangible product or progress. These are deliverables and the things that happen. They are quantitative, measurable, and well-defined activities that requires consumption of resources. When did the Field Inspection meeting occur? Was the project advertised on time? Outputs are primarily handled at the project management level.

Outcomes are the results that reflect the impact or value of the product. These are the consequences of the outputs, and should convey the impacts and effectiveness to achieve the organizational mission and goals. They can be qualitative in nature. Did the funding source satisfy its objectives? Are your customers happy? Outcomes are primarily handled at the Program Management level.

Outputs precede outcomes. Outputs are the products, services, and activities of the organization that lead to achieving the outcomes. For instance, outputs of the number of miles paved and number of bridges repaired may lead to the outcomes of number of accidents and reduced travel time for the traveling public.

Difficulties arise when outputs are confused with outcomes that can permeate an organization and its culture. An organization can quickly, and unknowingly, value and pursue the wrong things. A good place to start is asking if your organization has made a conscious distinction between outputs and outcomes. There is value in deliberately debating and deciding at what level should outputs and outcomes be evaluated. Reason dictates there is benefit in appropriately filtering hierarchically nested information. For instance, does it make sense for executive leadership to be concentrating on outputs such as individual project schedule tasks and details on deliverables? Similarly, does it make sense for PMs to be focusing on the outcomes of how efficiently the organization is executing the multiyear improvement plan? Carefully crafted outputs should consistently aggregate upward to outcomes that can satisfy your organization's missions and overarching goals.

10.7.8 Organizational Motivations for Performance Metrics

One aspect of performance metrics that is often overlooked is the organization's motivation to create and value the effort. Some are motivated by a desire to provide greater accountability, transparency, and visibility to the public. Some are motivated to build or restore credibility and demonstrate proper management and responsible leadership. Some are motivated to educate the public and improve the perception of the organization's performance. Some are motivated to enhance prestige or expertise among other agencies or peers. Some are motivated to streamline operations and optimize resources and their organization's positive impact. Some are motivated by a desire to evaluate employees. Some are motivated by pride. Some by fear. Some by strategic design. Of course, all of these may be internally driven or externally dictated. Regardless, an organization's motivations will likely shape how the metric is created, how it is valued among staff, and how resulting insights are utilized by leadership. It is important to note that an organization's motivation may change over time. Mature organizations will be self-aware of these changes, and consider modifying their metrics to remain consistent with their shifting motivations.

An organization's motivations may also dictate how a metric is framed. Is it a mirror that reflects your current state? Is it a carrot that inspires and drives your organization toward excellence? Or is it a stick that your employees consider punitive. Intense focus on metric-based achievement can bring unintended consequences to an organization, its culture, and the behavior of its employees.

10.7.9 Metric Creators vs. Consumers

As organizations mature in their data science advances, it is common for the metric creators to be concentrated in areas where this expertise can be best utilized across the organization. Complicated and impressive looking reports and dashboards are created by the data manipulation experts to be consumed by the end user. While this certainly can make good business sense, in some cases those creating the metric do not fully understand the nature of the data they are leveraging. Reports should answer questions, not be created because the data can readily be packaged in a clever and visually appealing way. Data scientists often do not understand the nuances of the data upon which they are reporting. There are some basic questions that are often not adequately asked or answered.

- Is the data accurate? For example, you may have very accurate GIS data search results but then summarily dismiss the inherent accuracy of the source GIS data.
- Is the data being entered in a timely manner? For example, are schedule task finish dates being entered into the schedule the day the task is completed, or does the promptness of the data entry vary?
- Does the workflow inherently constrain the accuracy or timeliness of the data? For example, how long is reasonable to wait in your organization before reconciling all invoice and payment information before one summarizes current project financial metrics?
- Are you using the data field in the same way as the end user would define the field? For example, is the project budget reflecting allocated amounts, obligated amounts, the latest estimate with or without contingencies and inflation, or something completely different?

10.7.10 Metric Logic

The practical realities of constructing consistent and comprehensive metric logic can quickly complicate simple questions. The logic behind a metric is an aspect that those who consume the data rarely consider. Yet the assumptions and logic rules that build the metric greatly influence the observed trends, ensuing insights, resulting conclusions, and subsequent actions.

Before you even begin, what data should be included? A simple question with a complicated answer. Think of a project. Or start with an even simpler level...when is a task on-time or late? This is a simple question with a complicated and qualification-filled answer. The simple answer is, was it completed before or after it was supposed to be completed? But compared to what? For this example, let's assume there is universal agreement on what constitutes the completion of the task, whose responsibility it is to complete the task, and whose responsibility it is to report its completion, and everyone fulfills their role in a timely manner. How do you determine when the task was supposed to be completed? When and how do you establish the baseline against which progress is compared? Is this consistent across all projects and programs? Do your change management policies and procedures allow for modifications that support or erode the integrity of the data and resulting metrics and summaries? And for these summaries, when is progress counted, when it is completed, or when it was planned to be completed? The answer to these sample questions can greatly impact project and program decisions in order to prioritize short-term results and long-term sustainability. Consistency matters. For instance, switching definitions midstream can dramatically improve short-term results while jeopardizing long-term stability, while tainting the historic data for accurate trend analysis.

Performance metrics use data to tell a story. Even when the right data is selected, the resulting metric can still tell a misleading story to the data consumers. Talk with the everyday practitioners to ensure data is promptly and accurately captured. When developing a metric or dashboard, it is critical to include the right Subject Matter Experts (SMEs) to ensure you are measuring the right data, in the correct context, and applying logic in a way that retains the integrity of the data and tells a meaningful and accurate story.

10.7.11 Beware Simpson's Paradox

Simpson's Paradox is a statistical phenomenon in which consistent trends appear in discreet groups, but disappear or reverse when the data groups are combined. This is different than apparent statistical inconsistencies within the same data set. An example of this may be when evaluating the durations of project schedule tasks. Pending the specifics of the single sample set, you may find a task's average duration is longer than the 75th percentile. The Simpson's Paradox involves multiple data sets or variables. Consider two high school baseball players, John and Rick. John has a better batting average than Rick every year, yet Rick has a higher career average over these same four years, as shown in Figure 10.18.

At first glance, the above chart makes no sense. It may seem counterintuitive at best, or even impossible, but it isn't when you look at the math behind the percentages, as shown in Figure 10.19.

Player	Year 1	Year 2	Year 3	Year 4	Career
John	0.250	0.273	0.375	0.421	0.317
Rick	0.235	0.242	0.344	0.403	0.339

Figure 10.18 Simpson's paradox – table 1.

Player	Year 1	Year 2	Year 3	Year 4	Career
John	0.250 7/28	0.273 12/44	0.375 12/32	0.421 8/19	0.317 39/123
Rick	0.235 4/17	0.242 8/33	0.344 21/61	0.403 29/72	0.339 62/183

Figure 10.19 Simpson's paradox – table 2.

Understanding that this reality does occasionally occur should prompt us to confront the fundamental question of which data should we value, the aggregated or the partitioned? The answer depends upon the story behind the data. This cuts to the heart of effective performance metrics, that being the insights from the resulting metric should not extend beyond the inherent constraints of the data itself. This is another reason why those creating the performance metric should work closely with the data generators and data owners to ensure the weight and meaning of the data is accurately reflected and leveraged.

Pavlides and Perlman demonstrated that in a random $2 \times 2 \times 2$ table with uniform distribution, Simpson's paradox will occur once every sixty times. Meanwhile Kock demonstrated in path models (i.e., models generated by path analysis) with two predictors and one criterion variable, Simpson's Paradox would occur once every eight times. While Simpson's Paradox is not common, it does occur. As such, experienced performance metric creators and consumers resist the urge to quickly infer causation, and double check the assumptions and logic before waving the statistically significant banner.

10.7.12 What Works

Many transportation organizations are data rich and information poor. Among these, a common business objective is to better leverage and utilize the available data. Ideally it wouldn't stop there. Organizations should strive to advance up the data science maturity ladder. Accurate data should become meaningful information that leads to thoughtful insights that allows strategic foresight and enables the progression to new levels of organizational optimization. The experience should become knowledge that translates into wisdom.

Transportation projects can be far more challenging to track and measure than many other engineering disciplines. One of the foundational tenets of traditional Industrial Engineering is to define a process with discreet tasks that can be meticulously measured. This data can then be mathematically evaluated and optimized to ultimately decrease cost while increasing the speed, efficiency, and quality of the process and product. This can be a challenge when making widgets in a factory. It becomes significantly more challenging when evaluating a transportation project that has multiple funding sources, extends for years, and has a significant number of stakeholders, many of which have conflicting goals and success criteria. This is complicated as many of these factors in a transportation project are outside of your direct control. Don't let yourself get overwhelmed. Focus on what you know is true.

The PM is responsible to develop and deliver the project within the triple constraint of budget, scope, and schedule. Accordingly, transportation project performance metrics should start here. The PM must also consider the resulting impacts to risk, resources, and quality. Your organization may choose to extend performance metrics to these areas as well. Then you may wish to expand to others.

Choosing the right metrics is as much an art as it is a science. Resist the urge to pick metrics that are indicative of progress. Instead choose metrics that drive progress. For instance, does Federal Right-of-Way Authorization worth tracking? From a funding perspective...absolutely. To ensure the authorization is secured before proceeding with associated Right-of-Way costs...absolutely. To track the development of the project...not so much. Since the federal authorization creates a schedule start-no-earlier-than constraint for some downstream tasks, it is important; but it does not drive progress since securing the authorization may occur long before other tasks. While it may create a schedule constraint, it rarely drives the critical path.

The specific metrics your organization will choose are highly dependent upon your available data and priorities. The higher the stakes, the more important it is to track progress. It can be challenging to weigh competing metrics. For instance, a project that is delivered on-time and on-budget may be a short-term success but long-term failure if the quality is substandard. At the program and portfolio level, the focus should shift from individual project progress to efficiently executing the organization's multiyear improvement plan. The big picture is to develop the projects so you can balance the program so the right money is the right place at the right time to develop and deliver projects that improve safety and positively impact our communities and commerce. Governments bear an ethical obligation to be responsible stewards of public funds. The same schedule, budget, and scope metrics at the project level can often be scaled up to a program or portfolio level. Various Business Intelligence objects allow detailed diagnostic analysis that can easily transition from program-wide views down to project specifics.

10.7.13 Earned Value Management (EVM)

Earned Value Management (EVM)is the Holy Grail of performance metrics when it comes to capturing the overall health of a project. That is because it cleverly combines time and money in a single metric. EVM can be a challenge to fully implement, but the results can be well worth it. Tailored EVM is common on unique Design-Build projects, bur far more difficult to enact across programs or portfolios. It usually requires a robust systemic data infrastructure and more mature organizational processes and procedures.

Figure 10.20 shows a typical EVM graph. Figure 10.21 provides explanations of what is shown. Understanding these relationships unlocks the full power of this tool.

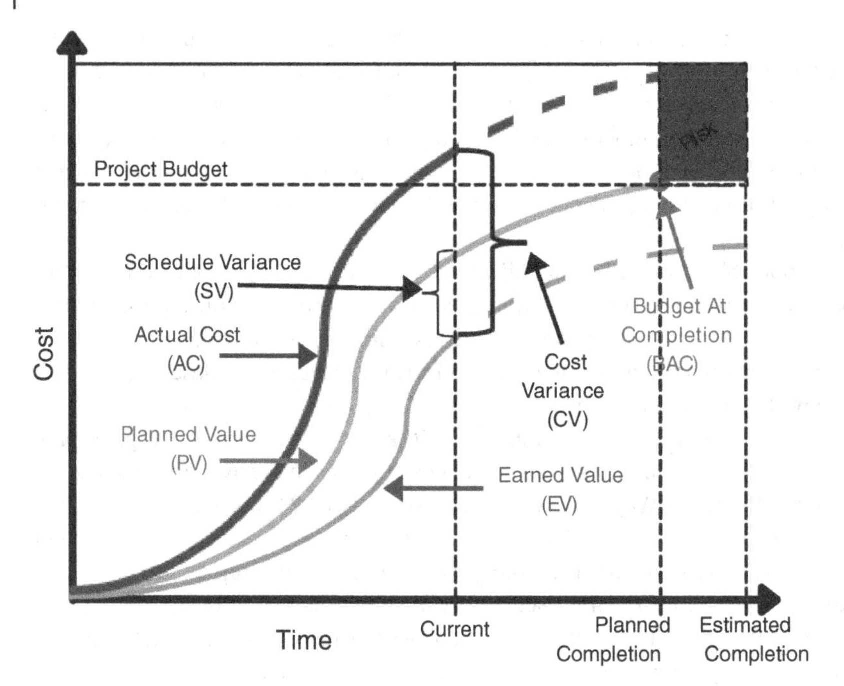

Figure 10.20 EVM overview.

Name	Abbreviation	Description
Planned Value	PV	Value of planned work scheduled to be completed
Earned Value	EV	Planned value of all work that has been completed
Actual Costs	AC	Actual cost of all work completed
Budget At Completion	BAC	Total value of the planned work (e.g., project budget)

Figure 10.21 EVM definitions.

EVM is built upon four fundamental ideas that are measured in dollars on the "Cost" y-axis across "Time" x-axis. These four measures can then be manipulated to provide meaningful analysis and valuable forecasts.

PV is determined by summing the value of the planned work that is to be completed at any point on the baseline schedule. This can be calculated by the organization as a template based on past experience and specific project criteria, or by summing the budget assigned to each work breakdown structure component, or scheduled task, and cross-referencing it to the baseline schedule. The PV serves as the EVM baseline, and enables schedule progress analysis.

EV is the measure of work that has actually been completed. The EV is the corresponding sum of completed tasks. This is graphed on EVM, and then compared to the PV to determine project health by enabling schedule and budget progress analysis.

AC are the actual costs incurred by the project at any time. This is graphed on EVM and used to determine the financial health of the project by enabling budget progress analysis.

BAC is the project budget. This is a specific point on the EVM graph that is a specific budget amount at specific time on the schedule, that being at Award or at project completion pending the span of the EVM analysis.

These four measures can be used to the following standard variances that can be especially insightful to PMs. Cost Variance (CV) is described in Figure 10.22. Schedule Variance (SV) is described in Figure 10.23. They are graphically represented in Figure 10.20.

Cost Variance (CV) compares the Earned Value (EV)to the Actual Costs (AC). This represents how the project budget is performing relative to the work performed.

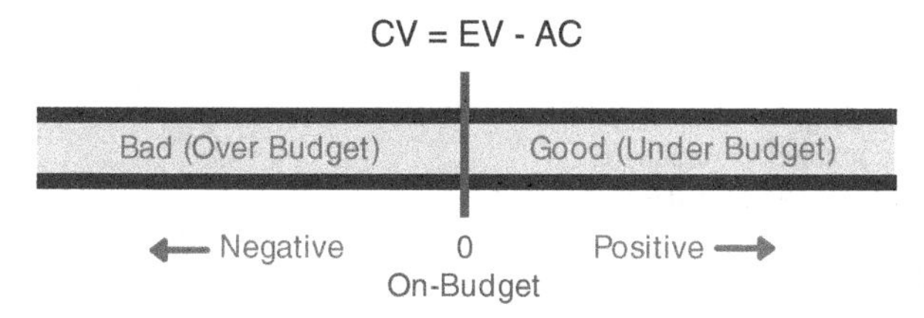

Figure 10.22 Cost Variance (CV).

Schedule Variance (SV) compares the Earned Value (EV) to the planned Value (PV). This represents the value of work by which the project is ahead or behind of the planned schedule.

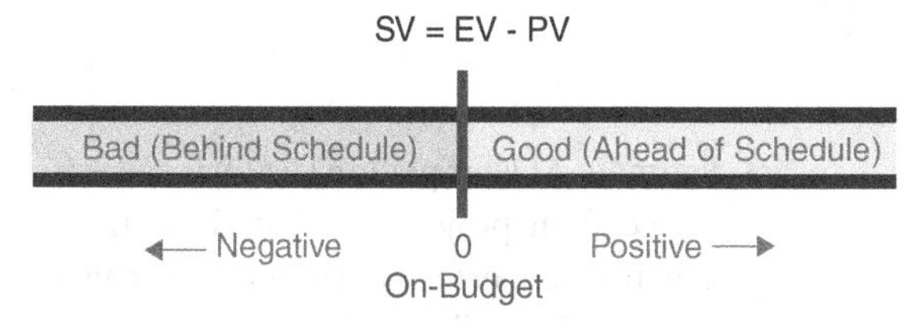

Figure 10.23 Schedule Variance (SV).

These fundamental measures can also be used to calculate standard indexes that provide additional insight into the health of the project. Cost Performance Index (CPI) is described in Figure 10.24. Schedule Performance Index (SPI) is described in Figure 10.25. Figure 10.26 shows how these two indexes can provide project insights.

EVM also provides established ways to forecast the project's progression that the PM can use to proactively identify situations that may require corrective action. The following are the EVM standard forward-looking measurements.

- Variance At Completion (VAC) is the forecast of what the CV will be specifically at the point of project completion. This represents the expected amount the final cost will be over or under the original budget. The formula is VAC = BAC − EAC. This essentially is the old budget minus the projected budget.

Cost Performance Index (CPI) is a budget analysis ratio that compares the amo unt of money scheduled to be spent (PV) against that which was actually spent (AC).

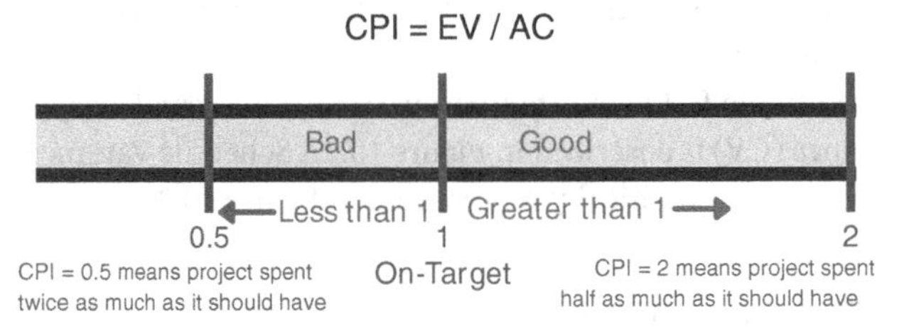

Figure 10.24 Cost Performance Index (CPI).

Schedule Performance Index (SPI) is a schedule analysis ratio that compares the value of accomplished tasks (EV) against the value scheduled to be realized at this time (PV) .

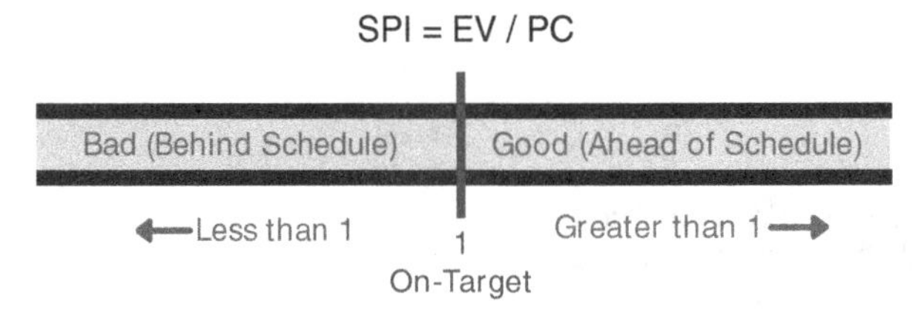

Figure 10.25 Schedule Performance Index (SPI).

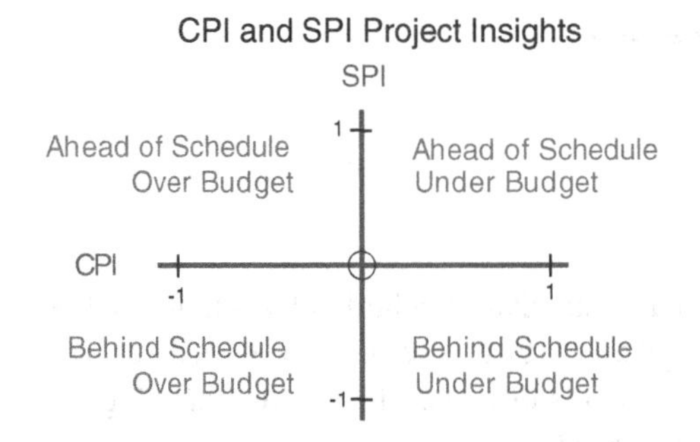

Figure 10.26 CPI and SPI project insights.

- Estimate At Completion (EAC) is the estimated final project cost based upon the past project performance. It can be calculated by four different methods to evaluate different future scenarios.
  1) Scenario one is when there is an outlier cost event that is not anticipated to be repeated. EAC is determined by adding the work remaining to the Actual Cost. This assumes the future work will be done at the budgeted rate. Use this formula: $EAC = AC + (BC - EV)$
  2) Scenario two is when you expect future performance to be similar to past performance. Use the formula: $EAC = AC + [(BAC - EV)/CPI]$
  3) Scenario three is when you assume the costs and schedule performance will continue, and the schedule has a direct impact on costs. This is meant for slow performing projects, and is not applicable for projects ahead of schedule. Use this formula: $EAC = AC + [(BAC - EV)/(CPI \times SPI)]$

4) Scenario four is when events are such that you need to start over to determine a new accurate estimate. You take the Actual Cost and add a newly calculated ETC. Use this formula: EAC = AC + ETC

- Estimate To Complete (ETC) is the expected remaining cost to complete the project. This is not the expected total project cost (EAC), but rather what is yet to spend to complete the project. ETC can be calculated one of two ways.
  1) Based on past performance, use the formula: ETC = (BAC – EV)/CPI
  2) ETC = a new revised estimate that is prepare for all remaining tasks
- To Complete Performance Index (TCPI) represents the CPI efficiency that is needed to complete the project by the scheduled completion. This is an indication of whether the project's baseline schedule is recoverable. TCPI is calculated one of the following two ways. In both cases, you divide the remaining work by the remaining cash.
  1) Scenario one is when you assume the original BAC is still achievable. Use this formula: TCPI = (BAC – EV)/(BAC – AC)
  2) Scenario two is when you assume the original BAC is not achievable, and the EAC is now the new budget target. Use this formula: TCPI = (BAC – EV)/(EAC – AC)

  A TCPI of 1.3 means the project must be 30% more efficient to finish on budget.

10.7.14 Lessons Learned

An organization with a performance metric-oriented mindset will experience the intentional or unintentional cultural paradigm shift as to why data is collected. As it advances, the organization will move away from collecting data for diagnostic analysis that is more analogous to an audit mindset, to collecting data for predictive analysis that is more a conscious plan to improve performance. Regardless of where your organization is on the data science maturity spectrum, below are a number of time-tested lessons learned for performance metrics:

- Performance metric difficulties are often rooted in a lack of clarity of an organization's goals.
- Those creating the metric should involve the data owners and practitioners to better ensure the metric reflects the meaning of the data and accommodates the inherent constraints of the data and workflows.
- Metrics that are designed to reflect results can end up changing the culture and driving behavior in unexpected ways.
- Many metrics are used in ways that exceed their inherent constraints, resulting in the wrong answer to the right question or the right answer to the wrong question.
- Organizations often create and value new performance metrics because they can, never stopping to ask if they should.
- Less is more. Too many metrics can burden staff, dilute collective focus, and promote apathy. Concentrate on the core outputs and outcomes that really matter.
- The cost, effort, and difficulty of collecting and reporting data is regularly underestimated.
- A simple, intuitive, and compelling graphic visualization can be powerful factor in an organization's willingness to embrace and value a metric.
- The success of a performance metric could, and perhaps should, be the degree to which it influences decisions and business operations, especially resource allocation.
- Performance-based planning is a never-ending process amidst evolving political guidance, public expectations, and advancing technologies.

11

PM Soft Skills

"Soft skills have more to do with who we are than what we know."

– Marcel M. Robles

Soft skills drive business success. A study from Boston College, Harvard, and the University of Michigan determined that soft skill training delivers a 250% Return-On-Investment based on productivity and retention. Given the importance of these topics, perhaps Soft Skills would be better labeled Human Skills.

11.1 The Importance of Soft Skills

"It's what you learn after you know it all that counts."

– John Wooden

Engineering is the application of mathematic and scientific principles to improve and control our lives and the world around us. It is a noble profession with grand aspirations. Engineering is a strange combination of technical expertise, creativity, and stubborn determination. Most non-engineers have a stereotype image of engineers. Whether the profession attracts or amplifies the stereotypical personality is an interesting question. Regardless, there is something different about most engineers compared to other professions. A large part of this is because engineering is about more than solving problems; it is a mindset in how we define problems and pursue solutions.

Let's start by looking at the college engineering education. While there are exceptions, most engineering students scored very high on their Math SAT/ACT, and perhaps less so on the English sections. Most took advanced science in high school, and did quite well. Most have an inherent interest and aptitude in math, combined with a natural curiosity to know *Why*, and understand *How*, things work. They get to college and work at a major that typically requires more credit hours to graduate, and has more demanding coursework than many other majors. A wise engineer once told me that an engineer's college education was structured in phases to intentionally shape how we think, each tweaking our thinking just enough to prepare us for the next phase. They teach us math because it is the language of engineers, and so that we can think in different ways than we could before. Once we

Transportation Project Management, First Edition. Rob Tieman.
© 2023 John Wiley & Sons Ltd. Published 2023 by John Wiley & Sons Ltd.

understand the math, they teach us physics, which uses math to describe the world we observe around us. How to define the problem. Once our mind has shifted enough, they teach us Theoretical and Applied Mechanics (TAM), or other applied engineering sciences so that we can learn the principles of statics, dynamics, thermodynamics, materials, and others. How to manipulate the world. How to approach solutioning. Once we understand those fundamental theories, we shift into our focused engineering disciplines where we can apply all we have learned in the form of engineering. If we were to jump right into the Engineering courses, while we may be able to solve the problems, we wouldn't understand how to think about the problems so that we can apply it correctly to other situations. Engineering education is focused on training engineers how to think. How to define problems. How to assign risk. How to describe assumptions. How to think about solution strategies. How to optimize resources. How to calculate and build solutions.

These are aspects that excite engineers. These are why most engineers became engineers. We grew up building overly complex Lego creations and took things apart to figure out how they worked. We asked *Why*...a lot. We played strategy games and enjoyed competing to get the right answer. While there are certainly exceptions, there are undeniable commonalities among most engineers. And I don't know of any engineer who went through the gauntlet of their engineering education with the aspiration of pushing papers. We want to solve problems. We want to build things. It is how we are hard-wired.

This passion and technical expertise are fundamental to being a good PM. But they do not ensure you will be a great PM. In order for you to be a good PM, you must have both technical proficiency and the knowledge and understanding of the Project Development Process. These are foundational for you to have the opportunity to be a great PM.

But in order to be a great PM, you must excel at skills that stereotypical engineers traditionally lack. While engineering educations are heavy on math, science, TAM, and engineering, they may not include formal training in, or even recognize the importance of, interpersonal soft skills. As you advance in your careers, emotional intelligence and interposal skills can become even more valuable than your technical base. This is perhaps nowhere more evident than as a PM, where most of your job is collaborating and communicating with your team, leadership, and stakeholders.

Below is an unprioritized list of essential PM soft skills that are required for you to reach your full potential and be a great PM:

- Leadership – the ability to transform vision into reality by motivating, mobilizing, and guiding a group of people to act toward achieving a common goal
- Team Building – the ability to bring individuals together into a cohesive unit that is working interdependently and cooperatively to accomplish a common goal
- Coaching – the ability to promote individual development in a supportive and challenging manner that equips and empowers by providing personalized training and guidance
- Conflict Management – the ability to identify and constructively resolve conflict promptly, sensibly, fairly, and efficiently
- Trust Building – establishing credibility and building trust to develop positive, productive, and efficient working relationships with team members, leadership, and stakeholders
- Negotiation – the ability to effectively work with others to compromise and resolve differences in a way that reaches a mutual agreement or finalizes a deal

- Emotional Intelligence – the ability to monitor and discern others' and one's own emotions, and then use this information to guide your thinking and behavior
- Political and Cultural Awareness – understanding political realities and cultural differences such that your future actions consider and leverage these constraints, traditions, and opportunities
- Decision Making – the ability to evaluate options and wisely choose solutions that forward project, program, and organization objectives
- Influencing – the power to be persuasive and have a meaningful impact on someone's decisions or actions, or the outcome of a process
- Communication – the ability to actively listen and effectively convey ideas and information to others in written and verbal formats
- Motivation – the ability to instill and sustain in yourself and others the desire to act in service of a goal

It is important to recognize no one is as skilled and proficient as they would like in all these areas. However, this does not diminish the importance of each essential soft skill. As discussed later in this chapter, our individual professional development should be intentionally focused on these critical skills.

11.2 Characteristics of a Successful PM

"*What you do speaks so loudly that I cannot hear what you say.*"

– Ralph Waldo Emerson

The PM is a critical and visible position for project development and program delivery. Your team, leadership, and stakeholders are watching. Consistency and dependability are paramount. Exceptional PMs realize there is more to this role than satisfying the project criteria. While developing and delivery of your project is fundamental, how you do things can be just as important as what you do. If you deliver your project on time and on budget but lose your team and burn every relationship bridge in the process, you will likely struggle with long-term, sustained success. It is a career-crippling mistake to assume you need to sacrifice one to achieve the other.

Throughout a project, the PM may fill a variety of roles. From task master to cheerleader. From delegator to party planner. From accountant to engineer. From planner to firefighter. From strategist to micromanager. From negotiator to mediator. From spokesman to scribe. From coach to mentee. From instigator to peacekeeper. From decision maker to seeing which way the wind is blowing. In one meeting you may feel like the General issuing orders to the troops, in the next you may feel more like a Private taking orders they don't yet fully understand. It is an exciting and challenging role.

A successful PM is able to seamlessly shift between the three critically important roles of Leader, Manager, and Entrepreneur. A leader ensures the organization is doing the right things. A manager ensures the organization is doing things the right way. An entrepreneur ensures the organization is innovative and able to do old things in new and more efficient ways. Pending the audience or situation, a PM should provide the role that is needed to best forward the project at any given time. Figure 11.1 lists core competencies for each of these roles, in no particular order.

Core Competencies		
Leader	**Manager**	**Entrepreneur**
Integrity/Honesty	Dependable/Predictable	Creative Thinker
Accountable	Involves Others/Develops Staff	Persistence/Determination
Decisive/Courageous	Communication Skills	Assertive
Political Savvy	Builds Trust/Respect	Problem Solver
Inspires/Empowers Others	Prioritizes	Innovative
Focused Vision/Deliberate	Sets Realistic Goals	Commitment
Communication Skills	Follow Through	Initiative/Opportunity Seeking
Shows Respect for Others	Resolves Conflict	Follow Through
Strategic Thinker	Manages Time/Workload	Resourceful
Relationship Builder	Technically Proficient	Self-Aware
Resilient	Helps Others Succeed	Charismatic/Persuasive
Effective Negotiator/Influencer	Anticipates Future Needs	Self-Confident
Adaptable	Organized	Communication Skills
Maintains Balance	Emotional Intelligence	Situational Awareness
Effective Change Agent	Goal Oriented	Risk Tolerant

Figure 11.1 Core competencies.

While there are stark differences between these three roles, there are also commonalities. Common threads of integrity, honesty, interpersonal skills, communication skills, follow through, and an inner public service motivation run through each of them. Use these as a solid foundation to successfully transition between them.

Long-term PM success and sanity requires being able to efficiently transition from one role to another. This begins with accepting the reality. You may not like it, but success demands it. Filling these roles spans a wide range of characteristics, skills, and abilities. You won't have them all, and you don't necessarily need to in order to experience success. Spend the time and energy to be self-aware. Various personality assessments are discussed in Chapter 7 of this book. Take them. Know your own strengths, weaknesses, and motivations, as well as those for key team members. Leverage your strengths. Manage your weaknesses.

It is important to realize a PM has very different success criteria based upon the viewpoint. A project's budget, scope, and schedule are absolutely critical, but there is more. Leadership considers a PM's ability to lead a team, and organizational loyalty. Past performance is valued, along with technical expertise, process knowledge, and a good business and political senses. Charisma, stakeholder acumen, and availability also matter.

As does a PM's ability to think on their feet and present polished verbal and written communications. A project's team members may have very different hopes for the PM. Typically, they want a technically competent leader who is forthright, follows-through, and is respectful of their subject matter expertise. They value a constructive, as opposed to destructive, communicator who praises in public, criticizes in private, and defends the staff. They want to know they are being heard and their concerns are being considered, and that a timely decision will be made.

Much like your projects, you are a work in progress. You should be a different PM five years from now than you are now. If you aren't, that's a problem. Never stop learning. Never stop growing. Be intentional in your professional development. Concentrate on refining your strengths and improving your weaknesses. Learn something new every day. Allow yourself the freedom to extend beyond your comfort zone so that you struggle or fail. Learn from your mistakes. Find a trusted mentor. Share your experiences with others who can benefit from your wisdom. Foster positive and productive working relationships that are built on trust, respect, and productivity. Appreciate the different styles and skill sets that generate results. Be thankful for your natural skills and abilities. Embrace every day with joy and determination.

11.3 Meetings

"People who enjoy meeting should not be in charge of anything."

– Thomas Sowell

Experienced PMs know not to confuse meetings with progress. While effective meetings can be one of your greatest tools to advance a project, ineffective meetings can be an enormous time sink that casts your project and team into a quagmire of unfocused, demoralized disarray. So, what's the difference?

As a PM progresses in their career and completes more complex projects, two truths become self-evident. First, your most valuable commodity is time. Second, your most valuable assets are your knowledge, experience, and relationships. Exceptional PMs use meetings to strategically leverage their time to learn and disseminate critical knowledge that forwards project and program objectives.

Every meeting should have a purpose. As PM you are the captain of the ship, and it is not acceptable to wander about rudderless or float wherever the tides and winds take you. A PM uses the project's Budget-Scope-Schedule as the North Star to proactively chart to success, and then strategically reacts to real-time challenges to stay the course. Meetings play a critical role in both proactive navigation and responding to stay the course. This is especially true as you work with your team to deliver the project. As Warren Buffet said, "*You will never see eye-to-eye if you never meet face-to-face.*"

As you formulate the meeting's purpose, it can be helpful if you also consider the meeting's intent, the meeting format, the expected participants, and the desired results. Bear in mind that every meeting should have both human connection and work product components. Intentionally considering and strategically selecting these aspects of meetings will increase team morale while driving your productivity.

11.3.1 Types of Meetings

There are generally five types of meetings a PM plans, leads, or attends. Each serves a specific function, and has its own unique strengths and limitations. Differentiating these will enable you to better approach the conversation and accomplish more in less time. While most meetings may contain aspects of more than one meeting type, at their core, each meeting should fall into one of these categories.

1) Status Update Meetings

Regularly scheduled status update meetings are an essential part of most every PM's work life. They often take two forms. One is to keep the PM up to speed on the team's progress. The other is for the PM to keep those above them informed. As the PM, it is imperative you establish both team meetings and executive briefings. Team status updates are often held on a daily, weekly, or monthly schedule. Executive briefings are often held on a monthly or quarterly basis. In both instances, strive to keep the meetings short and focused. A standing agenda can be helpful. It often consists of what has been accomplished in the last time interval, what is to be accomplished in the upcoming time interval, upcoming major milestones, review of project risks, and outstanding decisions.

2) Information Sharing Meetings

There are meetings when the purpose is to share information. Be it to the public, to your team members, to coworkers, to executive leadership, to industry, or others. In these instances, be intentional about the chosen visual aids and level of detail you present. Speak to the audience. The goal is not to demonstrate how much you know. Remain focused on the information they need to know, and then present it at an appropriate and meaningful level of detail. These meetings also provide tremendous opportunities to listen.

3) Decision-Making Meetings

Decision-Making meetings are the forks in the road, and should be laser-focused on either making a decision or securing a needed commitment toward progress. A commitment to move forward may take a variety of forms, including agreement on the process by which a decision will be made. These are incredibly important meetings as aspects of the project, or the entire project, may slow or come to a screeching halt until a key decision is made or commitment is secured. Situations such as these pose enormous schedule risk. An experienced PM can almost hear the proverbial time sucking sound, so they should be promptly scheduled and prioritized accordingly. Focused Decision-Making meetings can be extremely effective in mitigating this slowdown and jumpstarting action before bureaucratic inertia takes hold. The biggest factor in a successful Decision-Making meeting is ensuring the decision makers are actually in the room. This may be harder than it sounds. As a result, the timing, format, and setting of the meeting should accommodate their schedules.

4) Problem-Solving Meetings

Unexpected challenges are a part of most every project. These can range from inconvenient to crisis. There are times when the PM must gather the right Subject Matter Experts (SMEs) in a room and figure it out. Remember, this is why many of us became engineers, to solve problems. These Problem-Solving meetings may be rushed and often take a work session format. It is not uncommon for them to be overshadowed with stress of budget, scope, or schedule constraints or anticipated busts. In these instances, it is imperative you remain calm, focused, and actively maintain communications to effectively manage expectations.

5) Team-Building Meetings

Transportation project development and delivery is a team sport. Over the life of a project, the PM may play the role of coach, team captain, quarterback, equipment manager, and water boy. Through it all, an effective PM values that a cohesive, competent, and cooperating team will produce results. Team-Building meetings can be an invaluable tool to create and maintain strong project teams. These meetings can assume a variety of forms, ranging from one-on-one lunches, to retrospectives, to retreats, to team-building exercises, to happy hours, to dinners or events to celebrate milestones or specific project successes. Team-Building meetings should fortify your team, strengthen relationships, facilitate understanding, build trust, and encourage collaboration. As with most things of value in life, they must be earned. Collaborative success, consensus building, and realizing shared goals are powerful ways to strengthen relationships and increase trust. A cohesive and unified team is a remarkably powerful force to establishing and maintaining a sustained, successful Transportation development and delivery program.

A PM rarely gets to choose his own team. Team-building activities can be especially important when the team is dysfunctional or newly formed. If the dysfunction is concentrated to select individuals, it is imperative you reach out to them. You should strive to identify the core issue(s), eliminate the dysfunction, remove the isolation, and bring them back into the fold. This will likely also include reaching out the rest of the team to encourage a welcoming environment. Proactive team building is especially critical when a new team is being formed. One effective tactic can be to have everyone share some personal information (e.g., where they grew up, how many siblings, how many children, pets, etc.), their life priorities outside of work (e.g., family, travel, working out, etc.), work values (e.g., honesty, timeliness, professionalism, respect, etc.), and project goals (e.g., on-time, on-budget, satisfy scope, etc.). This can be a powerful way to quickly lay the foundations of trust as the team realizes their common life priorities, work values, and project goals.

11.3.2 Meeting Formats/Styles

Different formats and styles better serve different meeting types. Location, setting, resources, corporate culture, tradition, and characteristics of anticipated attendees can all directly or indirectly influence the formal or informal nature of meetings. Another major factor is often participant perception of social anxiety or professional risk. Most meetings fall somewhere between the formal Board meetings that strictly follows Robert's Rules of Order and the informal stand-up meeting with a team member on the way back to their desk from the coffee machine. When considering the proper format and style of a meeting, you may wish to consider the strength of governing rules and rituals, as well as the tolerance for surprise.

In general, strive to limit the number of attendees. If someone isn't a decision maker or a contributing Subject Matter Expert (SME), do they really need to be there? Many experts advocate that meetings are generally most productive if attendance is limited to five or less. While this may be a good rule-of-thumb for certain meeting types, others require significantly larger attendance in order to accomplish the objectives. Be intentional with the invite list so that it can enhance, and not detract, from the purpose of the meeting.

It is worth noting that virtual meetings are becoming far more common with decentralized teams. In these instances, it is even more critical that you actively manage the conversation. Without the full benefit of nonverbal communications combined with the potential delay of talking over each other, it is beneficial for the meeting leader to clearly recap key discussion points, action items, and responsible parties.

11.3.3 Characteristics of Successful Meetings

The best way to start any meeting is by stating the objectives. This forthright approach focuses the attendees on goals, deliverables, and decisions that should be achieved. It also can curb tangential discussion, and provides an easy mechanism by which the meeting leader can refocus errant discussions toward the purpose of the gathering.

The content should drive most meetings. Beware overusing carefully crafted presentations. While 3D models and glossy finished visual aids may be appropriate at a Public Hearing, a generic-looking agenda attached to the emailed invitation may be all that is necessary for a remote daily status update meeting. A smart PM strategically chooses the visual and digital technology that is appropriate for the meeting and intended audience. As a general rule, visual aids are a tool to be used when it can add value to the discussion. There are certainly settings where the message must be carefully crafted and skillfully presented in a polished and professional way, but these aren't most meetings. For most project meetings, the content, not the presentation, should drive the discussion. Do not rely on PowerPoint; focus on the issue at hand. While PowerPoint can be an incredibly valuable presentation aid, overreliance can quickly disengage the participants as they stare at a screen and decreasingly pay attention. I have found a group of engineers huddled over a paper roll-plot stretched out on a conference room table can do more to stimulate thinking, and generate innovative solutions than any PowerPoint.

It is said, "*If you didn't write it down, it never happened.*" PMs must prioritize meeting documentation. Every scheduled meeting should have an agenda. If at all possible, that agenda should be distributed to all meeting participants far enough ahead of time so that all may be adequately prepared. This demonstrates you value their time, and better enables productive discussion.

The agenda should list the meeting time and place, the purpose of the meeting, anticipated attendees, discussion topics, and desired outcomes. Meeting notes document the discussion and decisions should be created and distributed to all attendees. In larger meetings, the PM should assign a scribe to create official meeting minutes. It is not reasonable to assume a PM can effectively run a meeting and take meeting notes. Neither will be done well. The PM should use common sense or organizational standards to tailor the level of formality and completeness for the agenda and meeting notes. But both should be done for all meetings.

Corporate cultures can vary widely. My own opinion is that meetings should start and end on time. This demonstrates responsibility, conveys you value the time of the meeting attendees, and encourages a culture of delivering on your commitments. Another way to demonstrate you value others' time is to limit meetings to the duration that is necessary.

Generally speaking, most meetings lasting more than an hour experience diminishing return in productivity.

If applicable, always end meetings with recapping any action items for the groups. Action Items are to-do list items that are only complete if they include:

1) What needs to be done?
2) Who is going to do it?
3) By when?

11.3.4 Characteristics of Successful Meeting Leaders

Every successful meeting has a leader. Pending the type, format, and intention of the meeting, this may necessitate the leader to assume four very different roles: Facilitator, Negotiator, Task Master, and Team Builder/Cheerleader. Each of these roles can be critical to moving a meeting or project forward. Leading a meeting is vastly different than being an attendee, and it is rare for someone to naturally feel comfortable in all of these roles. But a PM will assume each of these roles, sometimes within the same meeting.

When leading a meeting, there are a few general guidelines that will help you be effective. Stay focused and engaged. Don't daydream or multitask by checking email, being on your phone, or doing other work. The meeting deserves your attention, or you shouldn't be there. Remain aware of what is being discussed, be it the issue at hand or positioning for other projects or initiatives. An effective PM recognizes these power plays and refocuses the discussion to the project at hand. They also recognize when to let the conversation run its course, and when tangents are headed down a rabbit hole and need to be refocused. Effective meeting leaders also curtail too much participation from any individuals, while encouraging participation by all. Praise in public. Criticize in private. Use soft words and hard arguments. Seek first to understand, then to be understood. Know and embrace your own style of leadership. Leverage your skills and abilities, and trust your gut. Remain professional and in control. As Thomas Jefferson said, "*Nothing gives one person so much advantage over another as to remain cool and unruffled under all circumstances.*" And as Desmond Tutu said, "*Don't raise your voice, improve your argument.*"

11.4 Time Management

> "*Time is our scarcest resource and unless it is managed nothing else can be managed.*"
> – Peter Druker

Time management is an intensely personal issue. Project schedules should be well-known and transparent. Team schedules should be collaborative with clear milestones and shared expectations. How you keep and manage your own work schedule may be a completely different dynamic. There are no one-size-fits-all silver bullets. Everyone's own personality, skills, strengths, weaknesses, and preferences impact how we each manage our own time. You need to find the right fit of what works for you. But it is imperative you find it. Your, and your projects', success depends on you being efficient and productive with your time.

Below are some business and project management concepts and tools that may help you refine your own style to be even more efficient and productive.

11.4.1 The Triple Constraint

A PM's primary responsibility is to develop and deliver their project within the established budget, scope, and schedule. You cannot do this effectively if you do not know the budget, scope, and schedule. These must be actively managed on all of your projects. If you don't know your projects' triple constraints, you will waste time responding to issues as they arise.

Dale Carnegie said, "*Knowledge isn't power until it's applied.*" You cannot capture and leverage the power of your project's budget, scope, and schedule unless you apply it. And you can't apply it unless you know it. Spend time in your projects' budgets, scopes, and schedules. That is your domain. That is your responsibility. Position yourself so you can quickly identify what can be accommodated within the triple constraint and when change management is appropriate. Position yourself to readily and accurately evaluate risks. Position yourself to more effectively identify and exploit opportunities. Position yourself to more strategically manage your team. Invest your time in your projects' triple constraints so you can position yourself and your project for success.

11.4.2 The Eisenhower Matrix

Dwight D. Eisenhower was an amazingly productive man. In World War II, this five-star General served as Supreme Commander of the Allied Forces in Europe. As America's 34th President he spearheaded the construction of the Interstate Highway System, created NASA, ended the Korean War, signed into law the first major Civil Rights legislation since the end of the Civil War, and oversaw Alaska and Hawaii becoming the 49th and 50th states. He famously said, "*What is important is seldom urgent, and what is urgent is seldom important.*" This idea has been expanded to a time management mainstay that now bears his name, the Eisenhower Matrix, as shown in Figure 11.2.

When evaluating your schedule and issues of the day, this matrix can be helpful. Important tasks are those that contribute to your long-term goals and add value to your life, mission, or objectives. Urgent tasks are time sensitive and demand your immediate attention. These are qualitative measures, but each task should fall into only one of the four quadrants.

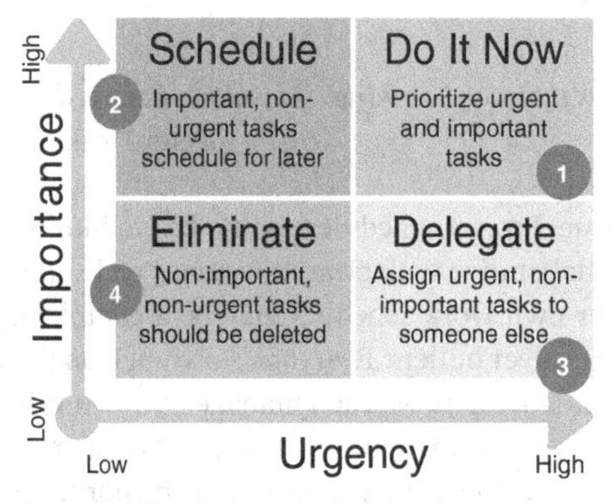

Figure 11.2 Eisenhower matrix.

- Quadrant 1 represents those tasks deemed important and urgent. These add value, are consistent with your long-term goals, and are time-sensitive. These tasks should be your first priority.
- Quadrant 2 represents those tasks that are important, but not urgent. You should schedule time on your calendar to ensure progress continues on these tasks that add value and are consistent with your long-term goals.

- Quadrant 3 represents tasks that are time sensitive but they do not advance your long-term goals. Where possible, these tasks should be delegated to others so you can focus on more important tasks.
- Quadrant 4 represents tasks that are neither urgent nor important. Avoid doing these tasks as they are likely not worth your time.

When determining which tasks to keep, and which tasks to delegate, this general rule may be helpful. Delegate repetitive tasks that take time, keep tasks that require seasoned judgment.

11.4.3 The Pareto Principle (80/20 Rule)

Joseph M. Juran proposed the 80/20 rule in 1896, and named it after the Italian economist Vilfredo Pareto. This business management axiom asserts that in many systems 20% of the input produces 80% of the output. This means 20% of the work will produce 80% of the results, as shown in Figure 11.3. While this 80/20 ratio may not always apply in transportation project management, it is clear that some tasks are more impactful than others. You can use this approach to help determine which tasks are important and should be prioritized. Focus on that 20%.

11.4.4 Common Sense and Lessons Learned

Be intentional with your time. Strategically manage your calendar. Start each day focusing on your short-term and long-term goals. Minimize interruptions. Time is most often wasted in minutes, not hours. Don't overbook your schedule. Schedule blocks of time when you can be uninterrupted to tackle larger tasks. Schedule times to return emails and voicemails. Plan to succeed.

Create and maintain an effective To-Do list. At the end of each workday, make your list for the following day. You may benefit from different categories or designations, but have one calendar and one To-Do list. While calendars are often dictated by the organization, find the To-Do list system, tools, or app that works for you.

Be strategic with meetings. Leverage status update meetings to also focus on issue resolution, risks, and opportunities. Adhere to meeting agendas and scheduled timeframes. Concentrate on those with whom you are meeting, granting them the respect of your full attention. This will more quickly build and strengthen relationships, and enable more productive discussions and decisions. As PM, you need to understand the issues and implications. You can't effectively lead or contribute to a discussion to evaluate risks if you are not fully engaged to access the intended and unintended consequences.

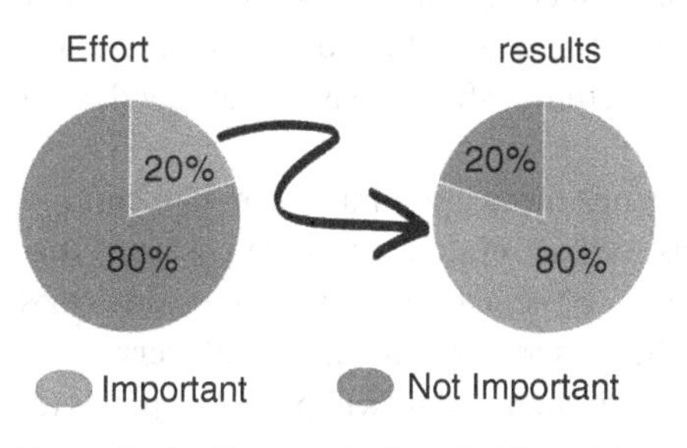

Figure 11.3 Pareto principle (80/20 rule).

Trust and rely on your team. Delegate appropriate tasks. Don't micromanage. Avoid doing others' work. Consider interpersonal and organizational dynamics. Understand and leverage task dependencies. If helpful, decompose larger tasks into more manageable interim steps. It doesn't have to be formalized, but crafting a Work Breakdown Structure (WBS) can provide clarity of direction. If possible, touch each issue only once and then move on to the next one.

Find and refine your own personal time management style, approach, and tools. Strive to ensure your actions are adding value to your projects, your team, and your organization. Busy can be the enemy of achievement. Ensure you are being productive, not just active.

Maintain a healthy work-life balance. Take care of yourself, physically, emotionally, and spiritually. Stay true to your morals, ethics, objectives, and expectations. Spend time with your family. Eat healthy. Exercise regularly. Get plenty of sleep. There is no softer pillow than a clear conscious. A life in harmony can be exceptional.

11.5 Influence Management

"*Diplomacy is the art of letting someone else have your way.*"

– American Proverb

Project Management Institute's (PMI) *Project Management Book of Knowledge* (*PMBOK*) estimates a PM spends 75%–90% of their time communicating. This is staggering, and perhaps intimidating to the stereotypical introverted engineer. However, there is no avoiding this reality. Successful PMs must practice consistent and effective communications with a variety of stakeholders. Much of this communication is to shape others' perceptions and realities. An essential skill set of an exceptional PM's is Influence Management.

Be it in a conference room, at a public hearing, sitting in a living room for an HOA meeting, standing in a citizen's front yard, leaning over an engineering workstation, or discussing a plan spread out on the tailgate of a truck, PMs commonly must practice influence management in order to advance project objectives.

There are an abundance of available resources on how to influence others. Many of them provide a magic bullet formula. While these can provide valuable information on tools and tactics, influence management is an acquired skill that often requires time to observe others who do it well, and patience to find and refine your own style. We each have our own unique personality, communication skills, risk tolerance, and comfort with conflict. It is essential that you find your own style, using as your guide the simple test of what works while being consistent with your values.

The product of influence management is often consensus or compliance. It is important to distinguish that consensus does not necessitate agreement. Informed consent is real. You may not need someone to agree with the idea. They don't even have to like the idea. It may be just as good if they can live with the idea. Often the goal is to avoid a thumbs-down response, which means a thumbs-sideways or shoulder shrug may be as good as a thumbs-up.

Transportation project management success depends more upon influence than control. This includes your abilities to persuade others, as well as refute alternative positions that would adversely impact the project's budget, scope, or schedule. This chapter describes some basic principles of effective debate, along with less sophisticated, but still very effective tactics to influence others and refute opposing arguments. This will provide a powerful set of persuasion weapons from which you can choose to arm your influencing arsenal, as well as the knowledge to recognize what tactic is being used against you, and how best to counter it.

11.5.1 There Is Nothing New under the Sun

It is said that good leaders guide the willing and persuade the stubborn. This is not a new insight. Aristotle divided the strategies of persuasion into three separate spheres: Ethos, Pathos, and Logos. Each of these focuses on different parts of the human experience. Aristotle knew there was not a single tactic to persuade everyone, so he would shift between these three approaches depending upon which one best resonated with the audience. Experienced PMs often work in aspects of each of these areas when talking with stakeholder, team members, or leadership.

Ethos uses character, credibility, and ethics to persuade. The purpose of ethos is to get another to trust you by establishing your personal credentials. Ethos appeals to your own believability, qualifications, and character. This is often accomplished by emphasizing your authority, expertise, or experience. You want the other to believe you are trustworthy, knowledgeable, a good person, and that you genuinely care. Ethos is all about the public persona you present. As such, reliable techniques include personal branding, citing credible or authoritative sources, success stories, titles, reputations, and personal anecdotes that illustrate why you are uniquely positioned to help. Confidence in delivery is also critical to ethos. Maintain eye contact, use positive and engaging body language, vary your vocal delivery, practice empathy, all while being polished and poised.

Pathos uses emotions, values, and passion to persuade. The purpose of pathos is to get another to feel by inspiring an emotional response. Pathos appeals to another's heart and emotions. Often this involves stories, inspirational quotes, and vivid and compelling language. To varying degrees, humans are emotional beings. Tapping into this innate reservoir can be a powerful way to motivate and sway others. Pathos is all about the emotions you stir in your audience. As such, reliable techniques include positive and negative stories that evoke strong emotions, your own thoughts and feelings, heart-tugging anecdotes, sweeping idyllic visions and dreams, appealing to all five senses, reliving dialog in your stories, and painting vivid mental pictures with your words. Coherence in delivery is critical to pathos. Your voice and body language should complement the story, while supporting and accentuating the emotions you are stirring in others.

Logos uses logical reasoning, truth, and evidence to persuade. The purpose of logos is to get another to think by arguing based on reason and facts. Logos appeals to another's mind and intellect. Often this involves a structured presentation flow, references to studies or statistics, and other reasoning methods such as comparisons, analogies, and metaphors. You want others to decide your aims, objectives, and conclusions are not only reasonable, but are the correct ones to pursue. Logos is all about facts, logic, and the resulting conclusions of what is presented. As such, reliable techniques include facts, figures, numbers, data, schematics, statistics, scientific research, processes, and collaborating and supporting evidence. You are leading the audience to a specific conclusion based on the evidence. Inserting some logos can be a very effective way to enhance ethos.

11.5.2 Seven Pillars of Influence Management

There are many ways leaders influence their teams and others. Some find success through fear and intimidation. Others rely on the transactional reciprocity of favors. Still others are skilled manipulators where you want to do thing for them, but somehow you don't feel good about it. Others have perfected the carrot of rewards and sticks of retribution.

The most effective form of influence management is when others have a choice and willingly act on your behalf or in your best interest. As football coach Mike Tomlin said, "*we prefer volunteers, not hostages.*" A PM has tremendous power when others accept your ideas and act as you ask because they want to be supportive of your efforts and are vested in your success. Often when others feel this loyalty to a colleague, they will do even more than what was asked. This approach to influence management is based upon the following seven pillars, all built upon a foundation of humility, as shown in Figure 11.4.

11.5.3 Foundation of Humility

The goal of Influence Management is to influence others, not manipulate them. Most people are perceptive enough to eventually discern the difference. Effective leaders are clear and consistent in their motivations. True influence doesn't come from an impassioned speech or grand gesture; it is earned by your day-to-day interactions, attitudes, and actions. Integrity is not a switch you turn on and off given the circumstance. As C.S. Lewis said, "*Integrity is doing the right thing, even when no one is watching.*" Consistency matters. When thinking about being strategic in influence management, starting from a place of humility best ensures authenticity that will lead to success with your team and networks.

True humility means respecting others and valuing their contributions. Allow others to have the spotlight. Listen to their thoughts with your full and undivided attention. Use active listening skills. Encourage constructive discussion. Say thank you...often! A leader who continually practices true gratitude and humility with his team will inspire others.

In the book *The Servant Leader*, the authors Ken Blanchard and Phil Hodges describe how when our head, heart, hands, and habits are in alignment it produces trust, loyalty, and productivity. Underlying all of this is a proper perspective of humility. As C.S. Lewis said, "*Humility is not thinking less of yourself, it is thinking of yourself less.*"

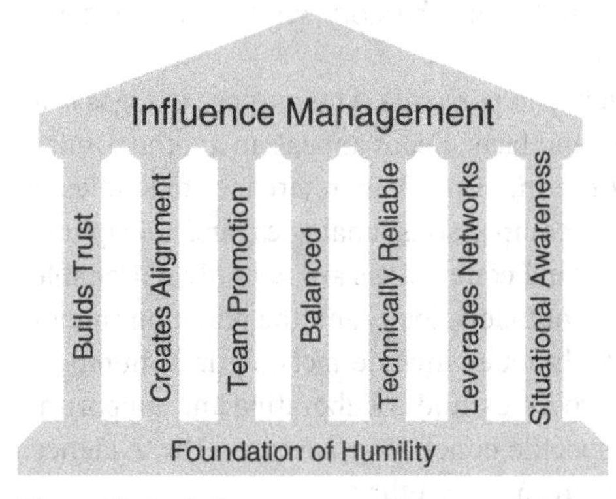

Figure 11.4 Influence management.

11.5.3.1 Pillar 1 – Builds Trust

A PM can have extraordinary talents, be a charismatic leader and compelling speaker, but if others don't trust you, they won't follow you. Trust cannot be dictated or assigned; it is earned. Trust is a bridge that connects you to others in cooperative ways that can span turbulent waters. This is why trust comes first. Especially when navigating difficult times or challenging situations, trust is essential. Without it, your influence is inherently constrained.

You may be able to force compliance, but the power of true commitment only comes with trust.

Trust is not a skill or ability; it is a feeling of confidence and assurance that you instill in others. Figure 11.5 shows how trust is composed of three main components that all intersect to engender trust in others: integrity, dependability, and benevolence.

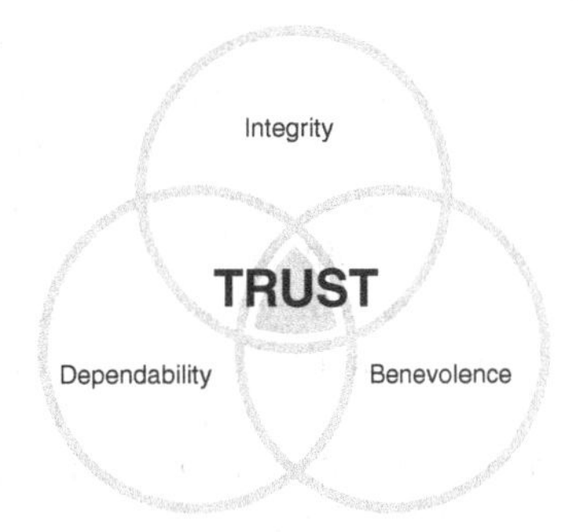

Figure 11.5 Trust components.

Integrity is consistently doing the right things at the right times for the right reasons. It is a moral compass that guides your decisions and actions. Dependability is predictable and consistent reliability. Benevolence is a well-meaning kindness that genuinely cares for others and wants to selflessly assist in pursuits for their best interests.

As PM you have enormous power to create a team atmosphere that promotes or discourages trust. Listen. Validate and value all contributions. Be courteous, respectful, and professional. Be generous, fair, and loyal. Be tenacious, but supportive and polite. Be focused, but adaptable and resilient. Another key aspect of trust is the willing relinquishment of control. Trust involves letting go of micromanaging people and processes, and being supportive and open to the freedom of their creative results.

11.5.3.2 Pillar 2 – Creates Alignment

Effective influence managers find ways to create alignment between individuals, divisions, and organizations. In projects and organizations, everyone may be generally moving toward their objectives, but often they are drifting in different directions. There is tremendous power in finding and implementing win-win solutions. Creating alignment can transform a contentious situation, improve morale, and generate positive momentum. Everyone wants to be a part of a positive solution. Momentum is a powerful tool that attracts others. Pursue it and strategically leverage it.

There are significant differences between cooperation and collaboration. Cooperation is interacting in a helpful and supportive manner. Collaboration is working together to achieve common goals. Cooperation is working together and providing assistance from your independent roles. Collaboration recognizes your inherent interdependencies and shared objectives. Cooperation relies on mutual respect and transparency. Collaboration relies on mutual trust and vulnerability. Cooperation retains individual accountability. Collaboration embraces group accountability. Cooperation is a low risk, low reward proposition. Collaboration can be a high-risk, high-reward proposition.

Creating alignment involves finding ways to cooperate and collaborate with others. Creating alignment is a mindset that is constantly searching for mutual benefit in all human interactions. It is the conscious decision to see challenges in your projects as cooperative opportunities, not competitive games to be won. This approach transforms your actions and others' perceptions. Persuasion and negotiations are not manipulative, but constructive. Challenging individuals become peers, not targets. Short-term transactional interactions transform into the basis for building longer term productive working relationships.

Steven Covey said, "*Win-win is a belief in the third alternative. It's not your way or my way; it's a better way, a higher way.*" This approach takes time and work, and is easy to sacrifice if we let ourselves be overly competitive or singularly focused. But this is an investment that will pay dividends. Finding win-win solutions is a solid foundation for a long-term relationship. This is fundamentally different than reciprocity, which is transactional and relies on both parties having something the other needs or wants. This approach focuses on the relationships and shared goals. PMs are uniquely positioned to facilitate these solutions on projects as they can be working with multiple disciplines to resolve an issue, all needing to stay within the larger contextual constraints of project budget, scope, and schedule.

In order to find win-win solutions during conflict, you need to listen. Separate people from the problem. Empathetically listen to separate out their perceptions, emotions, and motivations. Focus on interests not positions, and define the needs and wants. This will help shape the framework for potential solutions with mutual gain. It is important you emphasize this is a mutual problem we need to collectively solve. Use "*we*," not "*I*." Be flexible and maintain an open mind. Don't be overly attached to your own preferred solution. Recognize you may not own the entirety of the problem. Have the courage to identify and address personal feelings and emotions. Reach agreement on the objectives a solution should satisfy and then seek the detailed solution that meets the agreed criteria. Even if the final solution is no one's first choice, if it satisfies the agreed criteria needs, then the group can feel successful and the experience can strengthen these relationships. Synergistic solutions are like the rising tide that lifts all boats.

11.5.3.3 Pillar 3 – Team Promotion

Quarterback is the most important and impactful position in all of team sport. Think about the hall of fame quarterbacks we admire as great leaders and what they say in interviews after the game. When their team wins, they deflect the attention and accolades. They are quick to credit their coaches for an effective game plan. They praise the offensive line for an outstanding effort. They single out receivers or running backs that had a great game. They even praise the defense for putting them in a position to win. When their team loses, they talk about themselves not playing as well as they needed to win, never calling out anyone or blaming others. These public displays of leadership build loyalty, and are as effective in the work world as they are in the sports world.

Self-promotion is often viewed as selfish and shallow, quickly spreading discord among the team. It is rarely received well. By contrast, team promotion is so much more than a thinly veiled attempt to advance one's own career. It can produce pride, ignite motivation, create opportunities, and encourage collaboration.

Team promotion fundamentally relies on your ability to strategically and actively promote your progress and success. Find ways to cut through the chaotic daily noise to authentically and credibly promote your projects and achievements. You may be doing amazing things, but unless you tell your story, others may never know. You must intentionally gather and find audiences. This may require you to reach beyond your comfort zone. Find or create ways to tell your story, be it in person, to groups, at events, or in print. These acts may strike you as bragging, but the motivations should reach far beyond petty, self-serving objectives.

As PM you are in a unique position to praise others and lead the celebration of project successes. You set the tone. Make it fun. Make it productive. Make it supportive. Create a space where team members know their successes will be celebrated and appreciated. It is demoralizing to work hard on a task only to have it ignored, or watch another take credit for it and claim it as their own. Conversely, it is invigorating to have your work recognized. Gratitude is a powerful action that is both free and priceless. Create a visible platform where your team feels safe and confident in driving innovation and productivity. Be a champion for your team and team members. This simple act can ignite motivation and build fierce loyalty.

11.5.3.4 Pillar 4 – Balanced

Project and work demands can be stressful. Unexpected challenges can incite the fight or flight emotions in the most reasonable of people. Effective influence managers remain balanced in turbulent times. They find a way to maintain composure. The more emotions erupt in the room, the calmer they become. They speak respectfully, thoughtfully, and with a measured perspective. They approach conflict with humble confidence and courage. Experience grants the wisdom to know when to lead from the front and when to temporarily step back.

A predictable, steady demeanor engenders trust. Influence management is long-term game. Consistency matters. Deliver what you promise. Do the right things in the right ways. Having, living, and exuding the perspective that the highs should be celebrated and the lows will pass, builds an enviable reputation. There are inherent interpersonal and professional risks in trusting others. A balanced demeanor and track record of dependability makes you approachable, predictable, and reliable. It is hard to trust a loose cannon. Influence management is built on long-term relationships that encourage clarity, leverage strengths, and emphasize mutually beneficial interactions. Be the consistent anchor of a balanced temperament, healthy perspective, and reasoned thought that others seek out and rely on to navigate difficult situations.

11.5.3.5 Pillar 5 – Technically Reliable

In engineering, there are fewer professional stains that are harder to cover up than the reputation of a fool. The fake-it-till-you-make-it approach may work in certain fields, but more experienced engineers have a pretty accurate Bullshit-meter. As PM you don't need to have all the technical answers, but you do need to be able to understand the issues and reasonably discuss possible solutions. Don't pretend you know more than you do. Don't make assumptions. Don't throw around standards, acronyms, or units if you don't understand them. If your team members conclude you don't know or can't understand the technical aspects of a situation, you've lost them.

The reality is the PM will likely not, and probably should not, know as much about any one technical aspect of their project as the technical lead or Subject Matter Expert (SME). It matters how you handle these situations. When the SME presents you with a problem, monopolize the listening. Validate their concerns. Ask questions. Figure out why this is important and who else needs to be involved in the solution. What are the impacts to the project's budget, scope, and schedule. Stay in your lane. Rely on the expertise of your SME, but ensure the overall objectives of the project are being honored and pursued.

The PM doesn't have to know everything in order to build trust and influence engineers, but they do need to understand the big picture. They need to be able to comprehend the intended and unintended consequences of possible solutions. They need to appreciate the complexities and intertwined nature of issues that arise. They need to involve the right disciplines at the right time in the solution. These situations are tremendous opportunities for the PM to work more closely with team members and gain great stories to share with others.

11.5.3.6 Pillar 6 – Leverages Networks

Networks are a force multiplier that can amplify your influence and impact. Influential leaders intentionally cultivate their personal and professional networks. Organizations, industries, and careers are dynamic. Likewise, our networks are ever-changing and require attention to strengthen and grow.

Some people know a lot of people. This mile-wide, six-inches deep approach can be effective for distributing information, but the relationships remain shallow. For impactful influence management during challenging times, quality trumps quantity. However, your networks will naturally have circles of contacts that are at different levels of maturity. This is normal, and healthy. Networking builds professional and social capital. It is an investment that will continue to pay increasing dividends as your career advances. As such, concentrate on making connections at your current level. Then as you and your peers advance, you will have more mature relationships upon which you can depend when you and they are in higher positions.

Effective networking is easy for some, and less so for others. It is important you know your social strengths and weaknesses. Not everyone is comfortably in their element at a corporate happy hour or conference networking reception. While many introverted engineers struggle with these larger events, they can still have extremely broad and effective networks. Influence management relies on relationships built on trust. Larger receptions or industry events are perfect opportunities to make initial connections and reconnect with existing contacts. Strengthening relationships is a more personal experience. It could be having dinner together in a smaller group, working together on a task force or cross-divisional committee, or talking on the sidelines as your kids play together in a sports league. These relationships are made one-on-one, and are often based on sincerity and authenticity. These are qualities that most effective PMs do very well. If you are not gifted in this area, find those who are portals to others. Find individuals who know other key individuals and can arrange introductions. Volunteer for roles outside of your comfort zone. Become involved in industry organizations. Have lunch with your coworkers and those in other divisions. Attend conferences, seminars, and holiday events. View networking opportunities as opportunities and not obligations. Remember names, their spouse's names, and their children's activities. Discover their priorities, passions, and values. Take notes after the event if it helps. Create a spreadsheet that keeps track of when you met who and their role. Find a system that works for you. Make a plan and then execute it.

Your professional networks are your responsibility. You are solely responsible for building and maintaining them. And you are solely responsible for knowing how and when to strategically access and leverage your network assets. You are also solely responsible for the positive benefits and results you realize from your network. Building and maintaining an effective professional network is a long-term commitment. It takes time and patience. You may not know how one connection will benefit either of you, but as our career advances our professional network webs become increasingly intertwined. You are laying the foundational framework for future success. The dividends from your investments may not be realized for some time, but they can also position you to help others and do amazing things that you could never accomplish on your own.

11.5.3.7 Pillar 7 – Situational Awareness

The Roman philosopher Seneca said, "*Luck is what happens when preparation meets opportunity.*" A PM skilled in organization intelligence and situational awareness can make accomplishments look easy. Maybe they just have good timing, good luck, or good karma? More likely they were prepared and then identified and seized the opportunity.

Organizational intelligence and situational awareness are understanding how to get things done. It is grasping organizations' realities, strengths, and limitations. All organizations have a formal org chart and informal channels. Astute PMs understand positional and hierarchical authority, but are also well connected to the organizational influencers. They understand both how things are supposed to get done, and how they really work.

Situational awareness is both a mindset and a skill set. You should view challenges through your situational awareness glasses. Is there more to the story or this issue at hand? Why are they asserting that position now? Why is this manager always negative in the afternoons but more receptive to ideas in the morning? Situational awareness begins with being curious and observant. Think before you speak. Pay attention to nonverbal clues and communications. Listen and seek to understand. Don't win the battle only to lose the war. It is the ability to read a situation, know the cultural and political pulse, and discern the real issues at play. It is introducing an initiative at the right time to the right people in the right context. It is advancing solutions that solve the immediate problem and the one looming on the horizon. It is proactively taking care of your team members. This mindset and knowledge will sharpen your intuition that will allow you to more strategically and effectively advance your projects and initiatives.

PMs with great situational awareness attract attention. They leverage these insights to build and maintain that all important project momentum. There may be nothing more damaging to a public transportation project's budget, scope, and schedule than bureaucratic inertia. However, even in challenging situations, PMs with situational awareness find ways to make critical contributions that add value to the projects, processes, and organization. This reality and reputation can unlock tremendous opportunities in influence management.

11.6 Advancing and Refuting Arguments

"*Persuasion is not a science, but an art.*"

– William Bernbach

Have you noticed that being right is often not enough to persuade another? Maybe they refuse the truth? Or maybe they refuse to acknowledge the truth from you? Or maybe there is more going on? Analytic reasoning is a starting point for effective persuasion, but it doesn't end there.

11.6.1 Propositions

Before you can effectively assert or refute arguments, it is beneficial to understand the three general types of propositions, or assertions.

Propositions of fact assert a truth or falsity of some matter. The intention is to convince you to accept a premise as factually true or false. "*Traffic is worse in our subdivision now that it was five years ago.*" When advocating or refuting a proposition of fact, it is critical to understand definitions. What defines "*worse*" – peak traffic flow, total traffic volume, more through traffic, more truck traffic, etc.? Are these definitions industry standards, personal experience, historic precedence, by comparison with other areas, by a recognized expert, or some other foundation? Pay attention. He who controls the definitions controls the argument.

Propositions of value are aimed to make an evaluative judgment as to whether a matter is morally good or bad. These propositions rely on a calculated or perceived value, and often compare two alternatives and then ask which is right and which is wrong. "*How can you not put a sidewalk along this roadway? You are intentionally forcing our children to ride their bikes on this busy street which isn't responsible or safe.*" When advocating or refuting propositions of value, it is important to apply criteria to establish worth, and study the tools and tactics presented later in this chapter.

Propositions of policy advocate a specific course of action. These often take the form of a policy endorsement or commitment to a specific desired future that could be achieved if everyone travels down their desired path. "*I think it essential that prior to construction your organization meets with our home owner's association to discuss potential impacts so we can reach mutually agreeable hours of construction.*"

Discussions can naturally expand from a single issue to an unusual collection of issues. If possible, focus and resolve issues one at a time. When that is not possible conversations risk quickly devolving and becoming confusing and aimless. In these instances, discover how the issues interrelate so you can strategically group and prioritize them in a way that strengthens you position. Some common groupings include:

- Chronological – This approach organizes in time sequence.
- Spatial – This approach spatially organizes the issues, often in a linear order of workflow.
- Topical – This approach organizes by categories that lead from one to the next, think Venn diagrams.
- Psychological – This approach focuses on the strongest issues first, with other issues inherently being relegated to be less important.
- Logical – This strategic approach builds a structured argument that is carefully crafted so subsequent issues emphasize previous issues and conclusions, thus strengthening the overarching position.

11.6.2 Deductive vs. Inductive Reasoning

It is important to recognize and understand the difference between deductive and inductive reasoning, as shown in Figure 11.6.

Deductive reasoning starts with a general premise and asserts it must apply to all specific situations. This thinking follows the formulaic format of major premise, logic, and conclusion. For example, *roundabouts are safer than traffic lights...this intersection has a high accident rate... this intersection should be converted to a roundabout*. Deductive reasoning is excellent for asserting propositions of value, past fact, and future possibilities. This method can be especially effective to analytical, math-minded thinkers. The logic can be enticing, hard to refute, and easy to defend. It is also a terrific choice for speeches or when addressing larger groups.

The validity of a deductive reasoning relies on the credibility of the general premise(s) and the accuracy of the logic to leap from the

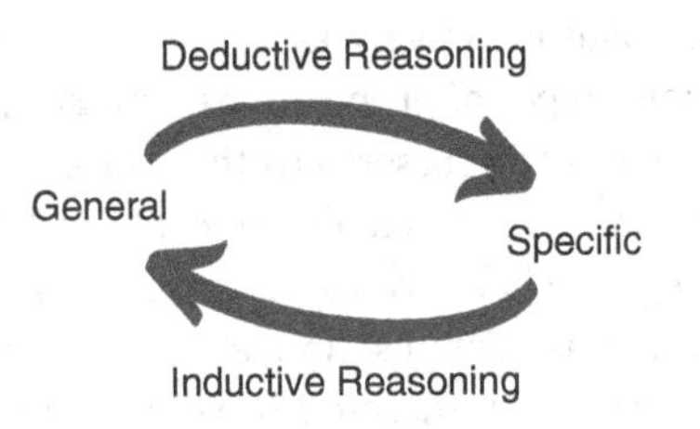

Figure 11.6 Deductive vs. Inductive reasoning.

general to the particular. Deductive arguments have two areas of weakness to guard against, or attack. If one agrees with the premise and cannot find fault with the logic, then they must agree with the conclusion. Conversely, the conclusions are only as strong as the premise and logic. Often the premise involves definitions or industry standards, which makes the logic especially critical to the validity of the conclusion. Beware using an inductive conclusion as the premise in your deductive argument. Also beware deductive arguments based on emotional premise or upon personal feelings or experiences. You must first address the implied emotion before you can reasonably disagree.

Inductive reasoning starts with specific observations and then infers general conclusions for all circumstances. For instance, an adjacent property owner may say, "*Your organization completely botched right-of-way negotiations with me. I'm sure you did with everyone else too. How do we even know the proposed traffic signal design will work?*" The strength and weakness of this approach is whether the particular examples can logically scale to be evidence of the overarching assertions.

This approach can be an especially persuasive if one pairs historical data with anecdotal experiences. The noted preacher Henry Ward Beecher said, "*An illustration is a window in an argument that lets in light.*" A well-crafted personal anecdote can vividly awake emotions and create a remarkable loyalty to a specific position. This is a powerful tool of persuasion, particularly when combined with the authoritative credibility of historic statistics.

The primary weakness of inductive reasoning is when the conclusion goes beyond what the example supports, thus making the conclusion conjecture. When refuting inductive positions, consider the following. The example must be factual. It can't be hypothetical. Is it based on emotions? Is the example applicable to the issue at hand? Is the example a typical representation of reality, or is it an outlier? As the saying goes, hard cases make bad law. Are there exceptions? Remember, it only takes one exception to disprove the logic leap from the specific to the general. Are the historical data and statistics accurate, timely, and applicable? Is the generalized conclusion logical, or is it slanted to advance an agenda? Is the asserted conclusion relevant and worthwhile?

11.6.3 Influencing through Evidence

Presenting evidence can be an extremely effective way to influence others. Below are three common types of evidence, and some thoughts on how to validate or refute what is presented:

- Factual Examples – Is it a reliable source? Is the information current and applicable to the situation? Are the definitions and assumptions accurate and applicable?
- Statistics – Are the units defined? Do the units provide an accurate index of what we want to know? Are the units comparable, for example you can't accurately compare rates? Do the statistics support another reasonable conclusion? Are the statistical assumptions valid? Is the sample set realistic and appropriately representative? Are there other studies that contradict the asserted conclusions?
- Authoritative Opinions – Is the endorsement relevant? Is the authority a credible expert? Is the endorsement supported by evidence, or is it an opinion? Are they free from prejudice and exaggeration? Are their opinions consistent? Are there conflicting opinions from other authorities?

11.6.4 Cause and Effect

Cause-and-Effect and Chain-Reaction arguments can be extremely powerful tactics, particularly when asserting that present policies should be changed, or a course of action altered. It can show a new solution will not only remove existing shortcomings, but realize additional benefits.

Whether advancing or refuting this line of thinking, there are some key points to consider. The logic flow between the cause and effect is paramount. Does the cited cause have the power or influence to produce the desired effect? Are there other causes that drive the effect? Is the cause significant? Is the cause original, or is it just contributing? Are there counteracting causes? Will new counteracting causes be created with the proposed solution? Beware assuming a cause is valid just because of timing or coincidence. If the cause and effect is being cited from a similar situation elsewhere, ensure the underlying assumptions and framework are transferrable.

11.6.5 Refuting Other's Positions

Just as important as being able to advance your own position is the ability to refute another. Some methods can be applied at any time. Question definitions. Are they using words with multiple definitions? Challenge assumptions. Challenge analysis. Minimize the issue. Contest the evidence. Be contrarian and demonstrate the opposite conclusions can, and should, be drawn from their argument. Use precedents, traditions, and accepted practices to frame the assertion as a fringe idea. Leverage the others' lack of knowledge and assert a truth they cannot refute. Ask a fallacious question, with loaded assumptions where either answer impugns. And never forget your most powerful weapon, common sense.

More effective methods require strong active listening skills to discern vulnerabilities. Expose reasoning or logic flaws. Attack circular arguments, where two unproven or questionable assertions are used to support each other. Expose inconsistencies. Shift the conversation to a more general truth which incorporates the point at issue. Shift premises. Identify irrelevant points that do not support or apply to the issue at hand. Hijack their logic to support your position. Demonstrate their assertion would result in a finite number of solutions, all of which are untenable. List the possible solutions along with the advantages and disadvantages of each. This can be very effective to focus the conversation and gain consensus in a group. Reduce their argument to an absurdity by applying it to a specific case to make it look ridiculous. This can be extremely powerful once mastered. Ben Franklin famously used this tactic to refute that one must own property to vote by asserting, "*You say I should own property in order to vote; suppose I own a jackass, I then own property and thus I can vote. But suppose I sell the jackass; I no longer own property, and therefore I cannot vote. Thus, the vote represents not me, but the jackass.*"

11.6.6 Negative Thinkers

Negative thinkers can provide unique challenges to a PM whose objectives include finding solutions and securing results. Four common types of negative thinkers are egotists, perfectionists, procrastinators, and distractors. Egotists sometimes tend to be argumentative, and may easily insult others. The motivation of this behavior is often to elevate their own position by lowering that of others. Perfectionists often see the world in absolute, black-and-white, terms. This approach often encourages them to go to extremes to pursue perfection while readily dismissing the reality that most decisions are compromises that may both positively and negatively impact the projects' budget, scope, and schedule. Procrastinators may lack follow-through or simply enjoy the adrenaline of pressing toward stress-filled deadlines. This can also be a tactic to create conflict or assert one's own importance by forcing issues to become critical path tasks or be addressed by a larger group. Distractors are those who pursue tangential ideas at the expense of the focused objectives.

When dealing with negative thinkers, it can be beneficial to try to ascertain the motivation behind the negativity, and then utilize previously mentioned tactics to combat them. It should be noted that this is best approached as a process, and not hoping for a single eye-opening realization. One's pessimism may be rooted in deeper psychological and personal webs than just the project at hand. Conversion from pessimism takes time and trust.

11.6.7 Persuasion Tactics

Leverage your expertise and credibility. We trust doctors because they are experts in medical knowledge. In the world of transportation project management, as PM, that is you. You have the battle scars and war stories of success against overwhelming odds. You know the SMEs and all the key team players. When viewed collectively, you should have a massive level of credibility. Use this as an assurance to others to strengthen your position.

Use the bandwagon approach to marginalize the opposition. It isolates the outlier. Nine out of ten dentists recommend something. Most people naturally want to be part of the majority. Draw a clear distinction between the majority opinion and the minority fringe outliers. This plays on the insecure who seek acceptance and validation through others.

Most people will follow a leader. For many, the destination is less important than safe passage. A powerful and decisive presence of control may naturally encourage agreement or compliance.

The Slippery Slope tactic presupposes an event may likely push the entire project or process down a slippery slope, resulting in a chain reaction of negative consequences that are out of your control, and cannot be stopped. The power of this is that while the immediate action may be tenable, the end result is unacceptable. This can be an effective way to bring a discussion back on point.

The False Dilemma frames the problem with a clear choice of fixed solutions, usually two; when in actuality there may be many more possible resolutions. This can be an effective way to lead a conversation toward your intended conclusion.

Glittering generalities avoid specifics while making broad generalities that appear unassailable. These often convey emotion, and can be extremely effective in setting acceptable, and unacceptable, parameters for the discussion.

People want to feel informed, smart, and understood. There may be no faster way to lose a crowd in a public involvement forum than to use technical jargon and acronyms, or to speak in a condescending manner conveying they surely can't understand. Make them feel smart. Others won't trust you if they don't understand you. Use simple illustrations and analogies. Talk in their vernacular to address their issues.

People rarely make their choices based on facts and logic. Most make emotional decisions and then search for the logic to justify their choice. Psychological and emotional appeals can be very powerful. When employing this approach, it is critical you capture and hold others' attention. You can appeal to their basic wants, their sentimental heartstrings, or their blocked desire, i.e. – the town needs an enema. You can appeal to beloved symbols or sentiments, using stories and emotionally loaded vocabulary. You can also appeal to their wants, fears, angers, or compassion to properly frame the context of the conversation.

You may choose to rely on the persuasiveness of the speaker. Positional authority and a respected reputation can be leveraged. Being genuine, sincere, and kind can go a long way to establishing credibility. An appropriate sense of humor can be disarming and engaging. A strong and pleasing voice with an interesting pitch, rate, volume, and cadence helps. Active listening body language, consistent eye contact, good posture, reasonable hand and body movement all add to your message. These can be powerful tools when combined with a sincere and enthusiastic desire to obtain a response.

Your style of speech also matters. Choose effective, image-bearing words. Use names and personal pronouns. Use simple words, avoid acronyms, and choose action verbs. Use simple, balanced sentence structures. Avoid sentences where the meaning is not clear until the end. When listing things, try to limit it to three. Strategically work in rhetorical devices like the direct question, figures of speech, applicable quotes, comparisons, rhetorical questions, and human-interest stories. Never underestimate the power of repetition. Tell them what you are going to say, say it, and then tell them what you told them.

11.7 Professional Development

"The best investment you can make is in yourself."

– Warren Buffet

You are responsible for your own career. You are responsible for your own professional development. Employers and industry organizations can provide tremendous resources and guidance, but you are charting your own course. If you are not advancing or developing as you would like, look in the mirror.

If you are an engineer, earning the licensure of being a registered Professional Engineer (PE) will likely open professional doors for you. The contrary is perhaps even more true; you may find not having your PE will eventually limit your career advancement. If you are a PM, earning a Project Management Institute (PMI) certification may present you with additional opportunities. For instance, becoming a certified Project Management Professional (PMP) adds an industry-recognized credential that increases your marketability. It is also likely the process of becoming registered or certified will increase your knowledge, skills, and abilities that will enable you to be more productive and efficient in your work. Licensures and Certifications are important steps in your career, but they are not the end goal. They form a solid foundation upon which you can build your career.

Most licenses and certifications require some ongoing Professional Development Hours (PDH) or Continuing Education Units (CEU) to remain in good standing with the issuing body. Choose to make these much more than just obligations you seek to most expeditiously satisfy. Eagerly pursue knowledge for the sake of knowledge. Make a plan for your own personal development, and then be intentional in achieving it. You are solely responsible for your own personal development.

The first step in effective self-improvement and professional development is to be self-aware. Many tend to minimize their weaknesses with excuses or blaming others, while overestimating their strengths. Others tend to not appreciate their strengths and focus on their weaknesses. And most everyone has blind spots that they just don't realize. Like all planning, you can't get to where you want to go if you don't know where you are, or what the path is to get there. Take the personality tests described in Chapter 7 of this book. Participate in 360-degree feedback evaluations. Then purposefully work to refine your strengths and enhance your weaknesses. Decide what you want, and then go get it.

There are also many ways to increase your knowledge and improve your skills outside of organized classes and instruction. Be well read. Read trade journals. Be aware of industry trends. Be curious. Ask *why*, often. Schedule time each day just to think. This may be your most important appointment. Read business and industry books. Listen to lectures. Never stop learning. Attend conferences. Find a mentor. Have lunch with a wise man or woman. As Henry Wadsworth Longfellow said, "*A single conversation across a table with a wise man is better than ten years mere study of books.*" Explore your tangential ponderings. Volunteer at your church or with community organizations in ways that push you out of your comfort zone and stretch your skills. What you gain in one area of your life is likely transferable to others. Keep the self-improvement momentum moving forward. Improve your verbal and written language skills.

It is a professional reality for many that the higher we professionally progress the less we accomplish by ourselves, and the more we accomplish through others. We may spend less time on technical work, replacing it with increased management, and eventually leadership, responsibilities. As such, it is imperative that we continue to develop our softs skills. This includes learning how to collaborate, how to build and direct teams, and how to delegate.

Remember that with the exception of licenses or some certifications, most professional development is cumulative in nature. There often are no immediate or foreseeable returns on your investment of time and energy. And much of this effort will be time after work that you can easily choose to spend in other ways. That is your choice, but then readily accept the consequences of your actions. Don't complain you are not advancing if you aren't working to improve yourself.

Think about your own motivations and discover your own internal "*WHY*" you do things. Why do you work? Why do you work where you work? Why did you choose your profession? Why did you apply to be in your current role? We only have so many years on earth, and you have chosen to spend your workdays where you are, doing what you are doing. Why? Don't waste this time or watch the clock slowly tick to the end of day. Know why you are where you are, doing what you are doing; then embrace it. Grow where you are planted. Make a difference. Feel good about your positive contribution. Define your passions and then find fulfillment in cultivating your passions. As you pursue your career goals and professional development, remember work-life balance is critical to your joy, contentment, and long-term personal and professional satisfaction in life.

12

Real-World Challenges

"Life is what we make it, always has been, always will be."

– Grandma Moses

12.1 Ethics

"Relativity applies to physics, not ethics."

– Albert Einstein

Nurses, doctors, and pharmacists consistently rank as the most trusted and ethical professions, followed closely by civil engineers. I suspect healthcare professions top these lists because we want to believe those who are directly caring for our health are altruistically acting in our best interests. But then civil engineers? Why is this the public perception? Is it warranted? Do we live up to these expectations?

Ethics are the values and moral principles that guide our decisions and govern our behavior. They are the parameters within which we operate. Our character should reliably reveal our personal ethics. Malcolm Forbes said, "*You can easily judge the character of others by how they treat those who they think can do nothing for them.*" My favorite definition of character is that it is who you are when no one but God is watching. Consistency is paramount. As Vince Lombardi said, "*Watch your actions, they become your habits. Watch your habits, they become your character.*" Hypocrisy should be abhorred, both privately and professionally. Lombardi also said, "*Practice does not make perfect. Only perfect practice makes perfect.*"

Our personal ethics are the foundation upon which our professional ethics should be built. While there are certainly professional situations and decisions where right and wrong are clearly evident, others live within the vast shades of gray. It is here where one's character and ethics become invaluable, serving as the North Star to safely navigate the seas of ambiguity.

Public transportation projects have a direct and substantial impact on the community's quality of life. In the development and delivery of transportation projects, we serve as

Transportation Project Management, First Edition. Rob Tieman.
© 2023 John Wiley & Sons Ltd. Published 2023 by John Wiley & Sons Ltd.

public agents and responsible stewards of public funds. As such, PMs and engineers should hold themselves to the highest of standards, fiercely guarding and protecting the public trust. This is especially true if they are licensed Professional Engineers (PE). In addition to the many business and legal reasons, it is the right thing to do. Or as Mark Twain said, "*Always do the right thing – this will gratify some and astonish the rest.*"

A civil engineer's ethical obligation is first to the public, then to their employer and client, and finally to other professionals. The National Society of Professional Engineers (NSPE) has identified six fundamental canons of their Code of Ethics that impact each obligation.

1) Engineers should recognize that the general public's quality of life is directly dependent upon their designs, decisions, and engineering judgments. As such it is imperative engineers hold paramount public safety, health, and welfare.
2) Engineers should practice professional competence by preparing, performing, sealing, and overseeing work only in areas for which they are qualified.
3) Engineers should maintain objectivity and truth in all professional services including design, reports, statements, and testimony. All relevant and pertinent information should be considered so that a sound technical decision can be made based upon technical expertise, relevant experience, and honest conviction.
4) Engineers should be a faithful agent or trustee for their employer and client, and avoid all potential conflicts of interest.
5) Engineers' words and actions should be professional and remain above reproach by preserving confidentiality and even the appearance of impropriety. Engineers should build their professional reputation on the demonstrated merit of their expertise and provided services.
6) Engineers should act in an honest, forthright, and responsible manner that enhances and advances our noble profession

In ethical matters, do the right thing, all the time. Public transportation projects are by definition, public projects. Assume all of your actions and decisions may be questioned or investigated in the local news. The ends do not justify the means. PMs and PEs must fiercely guard their credibility. As Abraham Lincoln said, "*Be sure you put your feet in the right place, then stand firm.*"

12.2 Project Continuity

"*Strategy must have continuity. It can't be constantly reinvented.*"

– Michael Porter

Public Transportation Projects can take a while to develop and deliver. Boston's Big Dig began in 1982 and was completed in 2007. There are plenty of other high-profile, and more local, examples of projects that take decades to complete. These legacy projects are not typical, but also not uncommon. Pending the complexity, even newer, fast-tracked projects will take years to complete. While there are many valid reasons for these durations, one reason for delay that is not often discussed is staff turnover.

Pending the organization, there can be significant turnover in the PM positions. A challenge for new PMs is that it may take years for a project to progress from initiation to construction. During this time the new PM is still learning and experiencing tasks, milestones, and unexpected challenges for the first time. This learning curve can be accelerated if they manage multiple projects, but it will likely still take many years to tie it all together and find their own style. It will likely take many more years for their experiences to mature their abilities and perceptions. With staff turnover, the project often suffers. While the budget, scope, and schedule remain unchanged, most everything else may change. The focused direction. The team's culture, communications, and morale. Leadership and stakeholder's expectations. They can all change. Meanwhile, entropy is real. If left to their own devices, transportation projects will naturally evolve to higher states of disorder. Bureaucratic inertia can quickly take hold, particularly if the project is temporarily orphaned without an active PM or during the transition period to a new PM when they are working to get up to speed on project details, goals, expectations, risks, team members, and outstanding issues. The reality is, it is not uncommon for larger transportation projects to have multiple PMs over its lifespan.

Most DOTs and many localities have taken steps to minimize project disruption during staff turnover by creating an abundance of policies, Standard Operating Procedures (SOPs), and project management systems. These can be hugely helpful. But Transportation Project Management is not exactly a factory-like process where similar results can be produced with interchangeable parts. Employees have different strengths, weaknesses, interests, and limitations they will leverage in their jobs. Different PMs will run their projects differently. And that is okay. Different styles can be exceedingly successful.

The value in the policies, SOPs, and PM systems is they create an organizational and operational structure within which the project is developed and delivered. A well-defined PDP can make it much easier to pick up a project because once you discern where you are in the process, you should know what has been done and what is left to do. A detailed CPM schedule may be your best friend, and it should show you which tasks have been completed and by whom, which tasks are next, who is the responsible task manager, and when they are due. Established policies or PM-specific job aids can provide consistent and reliable guidance on how to accomplish that which remains. Scheduled project status updates may provide a detailed history of what has been done during these intervals, what is upcoming, project risks, and action items with responsible parties. A formal electronic documentation management system can be invaluable in quickly finding agreements, authorizations, permits, the scoping report, required forms, decision documents, meeting notes and minutes, public hearing transcripts, design exceptions or waivers, approved change management submissions and decisions, cost estimates, plans, and other essential project documentation. If your organization does not have these institutional assets, you are putting your projects at increased risk during your inevitable staff turnover.

On larger projects some organizations will assign a Deputy PM. This idea of an entrenched backup PM is sometimes also used by consultants in turn-key designs. But it is far more common that there is a single PM for each project that is responsible to know all the specifics and to develop and deliver the project on-time and on-budget.

So, what do you do when you are the PM taking over an ongoing project? Jump right in. There may be a honeymoon period, but if your name is on the project, then it is your responsibility. Leverage your organization's policies, SOPs, and project management systems. Move to quickly understand the project goals, its challenges, influential stakeholders, expectations, and risks. Get to know your team. Work to build relationships and learn their strengths and weaknesses. Keep the project momentum by focusing on the next immediate milestones. Unless a decision requires immediate action, don't rush to judgments. The notable exception to this is if the project is running late or over budget. If you are "righting the ship," then you need to establish clear command control early on. This does not necessarily translate into arbitrary assertions of positional power, but it does mean you may need to immediately address the culture and prioritization of project tasks. Pending project goals and expectations, you may need to do that which is necessary to put the project back on track. This can be difficult to do without alienating the team. In these instances, creating unified buy-in on shared goals and celebrating team successes can be powerful instruments to propel the project forward.

12.3 Managing Chaos

"The chaos doesn't end, you kinda' just become the calm."

– Nikki Rowe

As mentioned in Chapter 1, my favorite definition of a PM is an organizational leader dedicated to the imposition of order upon chaos, even if chaos is perfectly happy with the status quo. Transportation projects are inherently complex. Stakeholders are diverse and unpredictable. Applicable legislation and standards are ever evolving. Complicated solutions intertwine a myriad of engineering disciplines. Then, most often it is all constructed under traffic with the added challenge of maintaining acceptable traffic operations. Transportation projects can have an abundance of moving parts, many of which you do not directly control. Chaos can seem the order of the day. This can be especially true when a PM is balancing multiple projects. Each can be at a different stage of development. Each requiring different levels of attention in different ways with different stakeholders and team members.

A PM should bring clarity, order, and calmness to the project chaos. Some days, a bigger challenge is juggling conflicting priorities and bringing order to your daily schedule and responsibilities. The PM's daily to-do list can be varied, unpredictable, and ever changing. Some days are so busy may feel like you don't even have time to make a to-do list. On any given day you may review public-hearing schematics with the lead designer for one project, meet with a stakeholder to discuss potential noise wall impacts with a key stakeholder on another project, meet with your consultant and team members to solve a hydraulic design challenge on another project, talk with your programming folks to manage funding allocations on another project, review a draft interchange justification report on another project, work with the geotechnical engineer and surveyors to schedule additional borings on another project, do a forensic schedule analysis on a languishing project for your boss on another project, work with construction to determine if there really is enough room to place a crane in the shown easements on another project, provide webpage project updates on another project, attend an evening meeting with an influential stakeholder group on another project, and a million other things PMs do during their work to juggle many projects while striving to develop them all on time and on budget.

If before your work day is ended you make a daily to-do list for the next day (a practice I would highly encourage), it is not uncommon that halfway through the next day you will not have accomplished much on your prepared list. The work life of a PM is dynamic. Flexibility, focus, and perspective are keys to success. Remain flexible in approach, but maintain a laser-like focus on your objectives. It is not uncommon to, in real-time, prioritize urgent matters on different projects, all of which have to be done within a timeframe that dictates not all will get done. Effective PMs know how to manage chaos.

Time is a PM's most precious resource. Use it wisely. The temptation is to work to reduce the number of items on your list of action items, perhaps even starting with the easiest ones to expedite achievement. While this may be productive to a point, it is more important that you are strategic. Be intentional with your time. You control your day, not the other way around. As Coach John Wooden said, "*Never mistake activity for achievement.*"

Experience brings perspective, which is a wonderful tool to assist in prioritizing tasks. Wisdom is knowing which fires of the day are truly important. Don't do the most things you can do in a day; do the most impactful things you can do in a day. This means some days you may only work on one thing, but if it the most important, then that is what you should do. Filter out the static to concentrate on what really matters.

One important function of a PM is to make decisions. In addition to making a wise decision, a strong PM predicts and prepares for the intended and unintended consequences of the decision. They think about what other disciplines, team members, and stakeholders will be impacted. What will be the impacts to the project's budget, scope, and schedule? How will this impact resources availability?

While making wise decisions can be critical to success, indecision can kill it. As the saying goes, "*Be decisive. Right or wrong, make a decision. The road of life is paved with flat squirrels who couldn't make a decision.*" Sometimes you need to make a decision. Often this is in the midst of chaos where ambiguity is rampant. You may not know all the information you want or need to know. Especially in these circumstances, remember indecision is a decision. So be intentional and make a decision. Ask for advice. Defer to your organization's preferences or precedents. Consult key project documents. But make a decision. If you are unclear how to proceed, remember your first priority is to guide the project to success within the triple constraint of the established budget, scope, and schedule. If changes need to be made, then follow applicable change management procedures. Evaluate unexpected developments through risk management glasses, and follow established risk management procedures. Take care of your team. Do the right thing. And say *Thank You* to those who helped.

Chaos is a relative term, really more of a feeling. It is a perceived lack of control or order. A project can be progressing well, and then something changes. Momentum shifts. And like a rising tide the team, stakeholders, leadership, or even the PM can suddenly feel like the wheels are falling off even as you are racing downhill without trustworthy brakes. Silence feeds the fear and speculations. Particularly when things aren't going well and you don't know what to say, communication is imperative. There is a reason why citizens tune in to hear their leaders address a national emergency in real-time. It is not so much that we expect them to have all the answers, or even know all the questions. But it is reassuring to know there is someone in charge. It is comforting to know they are aware of the issue and working on a solution. Team members, executive leadership, and stakeholders are no different. The PM needs to exude reasonable and justified assurance that this too shall pass.

Make time out every day to think. Thinking is an undervalued and increasingly rare asset in the workplace. In our rush to do more with less, we often don't take the time to ask what it is we should be doing and why. As PM, you are captaining your projects' ships through the sea of development. It is imperative you take time to think about the looming storms, keep an eye on your outside threats, evaluate your crews' strengths and weaknesses, take inventory of your supplies and morale, review your shipping charter, and then chart a course for your fleet to success. You can't do this if you constantly reacting. You cannot project stability and instill confidence to your team, your stakeholders, or your leadership if you are constantly reacting. You need to think, plan, and execute.

When chaos rages as a tempest around you, be a steady hand. Be predictable in behavior. Avoid extreme highs and lows. Be approachable. Lead by example. The PM sets the tone for the team. Intentionally strive to convey a composed and calm demeanor. Be like a duck on a pond. Even though your legs may be paddling like crazy under the water, others see you calmly and gracefully gliding across the surface. Regardless of your schedule or remaining action items, give your full attention to the issue at the moment. Listen, think, make a decision, and move on. When things became more chaotic, you need to concentrate on your stillness. When arguments become emotional and frenetic, you need to focus on solutions. As Sun Tzu said, "*In the midst of chaos, there is also opportunity.*"

These same principles hold true when crisis occurs. In general, when crisis occurs, the following game plan seems to provide some reasonable guidance.

- Safety first – Identify and take whatever actions are necessary for the safety of those who may be immediately involved, and to ensure public safety
- Think – don't react immediately or overreact
- Define the problem
- Identify alternatives
- Don't access blame
- Take positive, authoritative action and pursue best solutions
- When the dust settles...evaluate the results and take corrective action, if necessary
- Communicate in all directions – up to leadership, down to staff and subject matter experts, and out to stakeholders

12.4 Surviving an Audit

"In this world nothing is certain but death and taxes."

– Benjamin Franklin

In public agencies, audits can seem similarly inevitable.

Audits can be conducted by federal, state, local, or professional entities investigating your finances, processes, protocols, and results. The focus of the audit can range from internal financials to civil rights compliance to environmental permits and program compliance to your business practices to your performance, and just about everything in between. Regardless of the specifics, there are a few commonalities of all audits that you should recognize that will help you survive the scrutiny and come out cleanly on the other side.

12.4.1 Auditors – Who are They and How Do They Think?

Those being audited often feel they are the target of a modern-day witch hunt. While there are certainly occasions where this is the case, in general that is not true. Auditors are simply doing their job. In the cases where audits are required for recertification or as a condition of oversight in administration of a program you are charged to administer, audits serve an important role that ultimately may benefit your organization. The more you can help the auditors complete their tasks, the better it will be, or at least the quicker it will be over.

With any audit it is important to set and maintain a proper perspective. Many auditors are bright individuals who are quite passionate about ensuring compliance and improving the world by improving your procedures. Some audits are extremely critical with results that can carry extraordinary and damaging consequences. Most are not. You should take every audit seriously, but as long as you are ethical, conscientious, and trying to do the right thing, most audit results will not be overly earth-shattering, regardless of their content. It is important to keep this perspective because no matter how prepared you may be, they will find something. Their job is to look over your shoulder, examine how you conduct a certain aspect of your business, and then write a report about it. They have to write something.

It is important to understand the mindset of an auditor. Generally, they are compliance minded regulators. Fundamentally, they think differently than an engineer or project manager. Their task is to produce reports on those that produce tangible, real-world progress. They are not driven to solve problems. Their job is to find problems. When they identify a problem, identifying a solution is often not their concern or interest. Many do not have experience, and may not even know how to solve the "real-world" problem. It is critical that you understand that to most auditors, the process is paramount, and far more important than the results. To many engineers and project managers, it can seem like common sense is secondary to the mantra of compliance accomplished with an arbitrary form and procedure that was surely created by someone who never actually built anything in their entire lives. Whether this is the case or not is irrelevant. Audits are inevitable and it is your job to safely guide your organization through them as painlessly and successfully as possible. Once you have acknowledged an auditor's mindset, you can more effectively create and execute your audit strategy. It is important to maintain this perspective when discussing the intended and unintended consequences of implementing any recommendations.

12.4.2 Preparing for the Audit

As with other aspects of life, there is no substitute for preparation. Auditors will generally be with you for a very short time, and they will make broad assumptions on what you do every day based upon the information presented and the impression they get from you during those short hours or days. Consequently, it is advantageous that you plan an audit strategy that includes having all needed documentation at your fingertips and choosing effective spokesman to lead the discussions.

General and President Dwight D. Eisenhower said, "*Plans are worthless. Planning is essential.*" Planning does not guarantee you favorable results, but lack of planning almost always guarantees you unfavorable results. No one accidentally has a great audit.

In most cases, you will have ample advance notice to prepare for an upcoming audit. If they do not provide one, request a list of suggested documents and records they will want to review. You will also want to know a schedule of their visit and how many people will be in their party. They may not provide you the list of names who be visiting, but they can likely give you a sense of the areas of expertise that will be represented and requested. This can be useful information in helping you determine the best space for the audit, and the personnel you will want to have present or available.

When formulating an audit strategy, you should start with the list of documents and records they will want to review. This will give you an indication of how wide a net they will cast into your data archives. It can also provide insight as to the focus. Pending the state of your records, their request can represent a significant effort. You need to plan so you have the requested documentation available and organized. It can be helpful to establish a number or color-coded filing system to expedite easy access. This responsiveness can go a long way to demonstrating you are taking the audit seriously, and is tangible evidence you are conscientiously trying to do the right thing.

Pending the audit, just as important as their requested records are your documented Standard Operating Procedures (SOPs). Auditors love SOPs. It allows them to easily examine your protocol and quickly have a feel for your workflow. SOPs allow them to easily check off the boxes on their forms that you have a documented procedure. SOPs also provide a vehicle that can be modified to easily incorporate their recommendations. If you do not have SOPs that you anticipate they will want, you can quickly create them or find another way to demonstrate your process is logical, consistent, and complies with the spirit and letter of the regulations.

As you prepare, it is appropriate to consider who will be involved in the audit. You need to make available those individuals who have specific roles or responsibilities that are critical to supplying accurate insight of your procedures. You want to carefully choose individuals who speak the auditors' language. The classic example is ensuring your accountant and controller are available to respond to financial questions. You also want to choose representatives who are skilled and deliberate communicators who say what they mean, mean what they say, and have the wisdom and discipline to practice discretion when appropriate. Charisma can help, but is not the most desired characteristic. Most auditors have little interest in establishing connections. The confidence and competency of your audit response team's lead spokesman can go a long way to providing a positive, forthright, and cooperative impression in the auditors' minds that can easily transfer to your program.

Many larger audits will have a multiple team approach. In this case, assign team leaders and support staff well ahead of the audit. Ensure everyone knows their role and authority during the audit. It is preferred that your organization speak *with* a single voice. The second-best option is to speak *as* a single voice. So go over with your staff what they should and should not discuss. Make sure each individual only speaks to their areas of direct responsibility. This will minimize conflicting reports. These reminders are especially important when auditors employ the method of attempting to separate staff on a team and ask them all the same question to see if consistent answers are provided. This is a logical tactic as they want to make sure the organization's SOPs are being carried out by those who actually do the work. So, if they ask the construction inspector about an accounting question, it is the inspector's responsibility to decline to answer and refer the auditor back to the team leader so he can make sure the right person is responding to their inquiries.

12.4.3 During an Audit

Rule #1 is never lie...ever. Even if you don't like the answer, it is always better to be truthful.

Do not guess or drift into hypothetical answers. Answer the question – no more – no less. Beware of open-ended questions, or inquiries that seem to come from way out in right field that may be intended to feel out areas of weakness. Don't feel compelled to fill silence; when you have answered the question, stop talking. Also be cautious if an auditing team asks the same question in different ways. They may be fishing for a slightly contradictory answer on a specific point of interest. So, if they ask you a question that sounds familiar, you may say, *"I think we covered that already. Was there an aspect of my previous answer you would like me to clarify?"*

As with most situations in life, your attitude can dramatically shape your experience. Regardless of your initial instincts, make the conscious decision to approach the audit with an open, honest, and positive attitude. Decide to view the audit as an opportunity to improve your processes through an independent appraisal by those who see many other organizations' operations. And remember, unless it is critical; this too shall pass.

12.4.4 After an Audit

Cultures within organizations vary widely, from absolutely no documented policies to a blind devotion to manuals. Each side of the spectrum has some benefits and potential pitfalls. Organizations with no documented policies often tout flexibility and the ability to nimbly address each situation on a case-by-case basis. In reality, each situation is unique and there can certainly be benefit to the flexibility of seeking common sense solutions. However, there is a high risk of making inconsistent and poor decisions, even as you open your organization to accusations of favoritism or discrimination, be they real or perceived. It is common that larger, government organizations are policy heavy. This eases onboarding of new hires, and better facilitates continued operations during staff transitions; however, policies should not restrict sound business decisions. It can also be easier to track and forecast workloads, expenses, and other managerial metrics for larger organizations. However, it is very easy for such organizations to go overboard. This is especially prevalent when those formulating the policies and protocol are detached from those actually doing the work, or have never actually done the work. Very quickly, the workflow can become a stiflingly documented process that cripples productivity while ensuring everyone has the correctly completed form in their own Cover Your own Ass (CYA) file. Most organizations are struggling to find their way somewhere in between. This is a constant battle as times, technology, regulations, expectations, organizational, professional and elected leadership, and personnel change. Most would agree that there are advantages to standardization provided the processes increase efficiency and productivity. The point at which these additional SOPs do not add value to the process is the point at which you may want to trend back on the flexibility side of the debate.

Audits can often bring this discussion and organizational struggle to the forefront as they tend to promote rigid documentation of workflow. During discussions leading up to and after the audit has concluded, it is important that a steady hand is guiding this ship who will take and keep a healthy perspective and resist expedient and compulsive reactions. Audits produce reports that typically include recommendations. Keeping an appropriate perspective is critical as you decide what to do with these recommendations. You may use your communication and political skills to shape the audit results and frame the recommendations to your executive and political leaders. If appropriate, do not underestimate the opportunities audit results and recommendations can present to be an impetus for needed or desired change.

12.5 Scaling Project Approaches

"You must learn a new way to think before you can master a new way to be."

– Marianne Williamson

Success breeds opportunity. The PMs on most larger projects were first granted that opportunity because of their success managing smaller projects. Not everyone can successfully make this transition. One common reason for failure is that may PMs naively believe they can scale up to larger projects by using the same processes, tools, and mindset that were successful on smaller projects. They don't fully appreciate the exponential increase in complexity. This can be the significance of the unintended consequences of poor decisions and missed deadlines, which can be crushing. The mounting avalanche of information and decisions can quickly snowball and become overwhelming without previous experience, strong procedures, and an experienced team. Successful large project PMs understand they cannot run a large project like they did a small project, or they will fail.

So, what defines a small vs. large project? Most rely on cost (budget) thresholds. Some rely on anticipated schedule durations. Others rely on the engineering complexity. More advanced organizations select or assign the PM based upon a collection of factors. This may include budget, schedule, and scope, but may also include considerations such as stakeholder involvement, anticipated public opposition, risk control, political sensitivity or indecision, team experience, utilities and right-of-way issues, presence of railroads, environmental concerns, specialized construction challenges, problem and solution complexity and communication demands.

Upscaling magnifies the bad and makes the good significantly harder to achieve. In order to combat this immutable law of scaling, a PM should focus on the following ten areas that should be intentionally addressed:

12.5.1 Budget

As a project budget grows, so do the financial complexities, oversight expectations, and accounting scrutiny. Routine tasks like processing invoices and tracking the budget can become far more complicated than expected. Additionally, as a project's budget and complexity increase, the estimation method may need to change. Be aware the level of effort to secure an accurate estimate on a larger project may be significant. Quantitatively converting qualitative risks to money, and then forecasting appropriate contingencies, reserves, and inflation rates can be as much art as science. Established financial systems, practiced procedures, proven policies, and experienced administrative support are critical to creating and maintaining the essential documentation needed to successfully execute the project and satisfy all contractual and funding obligations.

12.5.2 Scope

Fluidity and flexibility are essential to success of small projects. Larger projects require a very different administrative structure. The complexities and nuances of a large project scope can be significant. This can often result in a voluminous scoping document that the PM must vigilantly protect. Scope creep is real. On larger projects, it is imperative the PM rely on established change management procedures to document any changes, and integrate them into the workflow while minimizing impact to the critical path.

12.5.3 Schedule

Microsoft Excel can only carry you so far. Larger projects can become exceedingly complex. The number and interconnectivity of tasks require something more robust. Making, tracking, and maintaining schedules should be done in programs with dynamic, critical path method schedule engines such as MS Project, Project Web Application, or Primavera. These programs allow you to manage the project, and properly evaluate schedule impacts when performing risk analysis. These programs are also powerful enough to track and manage resources and costs, if desired.

12.5.4 Teams

PMs can have tremendous success managing small projects with streamlined, nimble teams that share common goals and a history of trust of productivity. There is a tipping point where the critical mass of larger projects requires a completely different paradigm. The size and the makeup of a large project team often includes hierarchical layers with new and changing faces. Building and maintaining a high performing team on a large project is a critical and significant accomplishment in its own right. A PM should expect turnover in staff and key team positions throughout the life of a larger project. Integrating new team members into the tasks, expectations, and culture of the project is critical to success.

12.5.5 Communications

It is simple math. A team of two people have one communication link. A team of three has three links. A team of four has six links. A team of five has ten possible communication links. As the size of the project team and number of stakeholders increases, so does the time a PM must dedicate to communication. Status reports, engineering summaries, meeting minutes, official letters, requests, approvals, press releases, emails, public involvement documentation, and so much more must all be carefully crafted, distributed, and stored. Perhaps nothing is more disruptive and demoralizing to a team than an open-ended Freedom of Information Act (FOIA) request on a large project where a comprehensive document management approach is nonexistent. A PM must control the message, and strive to keep communications among the team, and to the stakeholders, effective and accurate to efficiently forward progress while minimizing inaccurate assumptions, misconceptions rumors, and innuendos.

12.5.6 Project Complexities

PMs often rely on personal strengths and intuition to navigate ambiguity and successfully complete smaller projects. Larger projects can be exponentially more complex than smaller projects in intended and unintended ways. This often elevates the many Subject Matter Expert (SME) roles, and emphasizes the importance of their strategic, timely, and synergistic collaboration. Larger, more complex, projects must rely on established methodologies, appropriate tools, and proven procedures. Trying to scale up the flexibility of smaller projects can have disastrous results on larger projects.

12.5.7 Risks

Risks on smaller projects tend to be less in numbers, intensity, and complexity. Consequently, a small project PM can often very successfully identify, evaluate, and resolve risks as they arise in a rather informal manner. On larger projects, this is a recipe for disaster. Established risk tools (e.g., risk analysis, risk log, etc.) should be leveraged within a proven risk management environment. A PM must be vigilant in encouraging his team to identify, and appropriately elevate, risks so they may be properly evaluated and addressed. Risk resolution should follow established change management procedures that evaluate actions based upon resultant effects on budgets, scope, and schedule. Additionally, there may be management actions to accommodate risks (e.g., adjusting contingencies, updating schedule and then budget, etc.). The larger and more complex a project is, the more time the PM will spend on risk management.

12.5.8 Performance Metrics

Unless already in place on a program-wide basis, smaller projects rarely emphasize or utilize performance metrics. Conversely, performance metrics are fundamental to most larger projects. They can be an essential and authoritative tool to transparently communicate progress to stakeholders, the media, and the public. If properly chosen, performance metrics are also a powerful tool the PM should leverage to monitor progress, prioritize focuses, and drive productivity. Bottom line is what gets measured gets done.

12.5.9 PMs Project Organization

The day-to-day administrative challenges of a large project can be overwhelming. An organized document management system and communication plan are critical and foundational elements of success. Likewise, organizational policies and procedures are essential to direct and manage a larger project.

12.5.10 Stakeholder Involvement

Larger projects can galvanize a surprising number of local, regional, and even national stakeholder groups. The intensity, organization, persistence, creativity, and influence of an active Home Owner Association, local school's PTA, local business association, a church or synagogue, an environmental protection minded entity, a bike and pedestrian advocacy group, or a single-minded individual can be astounding. An experienced, large project PM proactively seeks out key stakeholders and involves them in the project and process. The time and effort to meet with them, listen to their evolving concerns, and keep them informed should not be underestimated.

12.6 Guidance Pyramid

"Good governance is less about structure and rules than being focused, effective, and accountable."

– Pearl Zhu

Managing a transportation project can be a delicate dance to satisfy a myriad of success criteria. In addition to the project's purpose and need, leadership and stakeholder expectations, engineering and economic realities, a PM must successfully navigate the project ship within its organization's applicable rules and regulations. Engineers often immediately revert to various federal, state, and local design standards. While these are tremendously important, standards only tell part of the story.

Governance is the structure by which roles and relationships between project team members and an organization's high-level decision makers are defined. The Guidance Hierarchy should consist of four levels, each conveying a critical aspect of the organization's governance strategy, as shown in Figure 12.1. Together they reflect and shape the organization's approach, culture, and expectations. Understanding the purpose of, and relationship between, Policies, Standards, Procedures, and Guidelines will better enable a PM to discern how to apply the various guidance documents to most efficiently manage and deliver their projects.

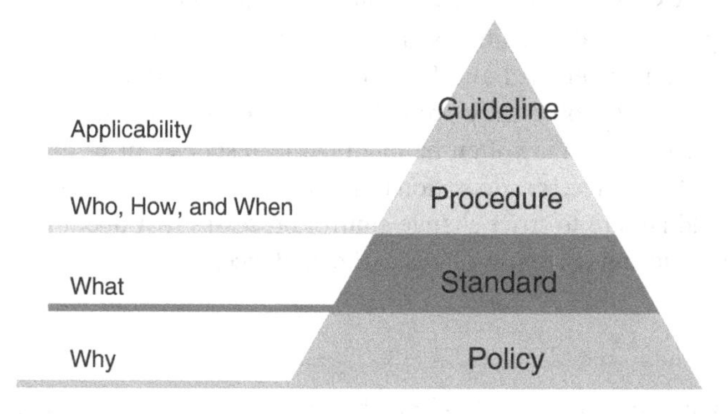

Figure 12.1 Guidance pyramid.

12.6.1 Policies

Successful organizations start with the *Why*. Policies should address the fundamental questions of Why an organization does what it is doing. They are typically formal statements that are written or approved by Senior Management. Driven by business objectives and consistent with mission and vision statements, policies convey the organization's principles and the risks leadership is willing to accept. They lay the foundation upon which all other aspects of the guidance hierarchy are built. They should be solid, and insulated from whimsy and hurried change.

12.6.2 Standards

After you know the *Why*, you can focus on the *What*. Standards detail the *What*. They describe what is required in quantifiable measures. In transportation projects these can be, and often are, established at Federal, State, and Local levels, with the higher levels trumping the lower levels. Standards are the rules that give formal policies shape and direction.

One critical nuance in standards is the vast intentional difference between the verbs "*shall*" and "*should*." "*Shall*" dictates a mandatory requirement that must be satisfied. "*Should*" opens the door for some flexibility in interpretation and application in the form of sound engineering judgment. In case it is needed or justified, every PM should know the avenues by which exceptions and waivers can be considered and granted for the various standards. Generally speaking, exceptions and waivers should not replace sound engineering design. They are intended to be a Plan B, and should be used as such.

12.6.3 Procedures

With the *Why* and *What* defined, an organization can effectively describe the *How*, which also includes the *Who* and *When*. Procedures provide detailed step-by-step implementation instructions on how to achieve the goal or mandate. The word *procedure* is derived from the word *process*. Procedures should build upon the *Why* of Policies and the *What* of Standards, and focus on the detailed guidance that directs individuals to successfully complete an established process. As such they should be readable and functional, finding the sweet spot of how much information to provide so all will understand and be able to comply.

While some procedures become entrenched and bear the formal or informal label of Standard Operating Procedures (SOPs), others are prone to change. Policies should stand the test of time. By contrast Procedures may often be modified in response, or to drive, changing business, industry, legislative, or operational preferences or requirements. Accordingly, procedures should adhere to strict change control processes that document changes and clearly tracks historic and current versions of the guidance.

12.6.4 Guidelines

Guidelines are less formal recommendations of action that can provide very effective guidance where other documents are silent. They typically take the form of general recommendations while inherently providing flexibility for the unknown and sound engineering judgment. Guidelines can prove particularly helpful when specific standards do not apply. These are the time proven advice of "General Rules of Thumb" and "Best Practices" that can help one navigate through a sea of ambiguity.

12.7 Planning Fallacy

> *"The planning fallacy is that you make a plan, which is usually a best-case scenario. Then you assume that the outcome will follow your plan, even when you should know better."*
>
> – Daniel Kahneman

Most transportation projects are delivered late and are over budget. How is this possible when they are typically designed and managed by intelligent, motivated, and well-intentioned engineers? Yet Boston's Big Dig that was supposed to be completed for $2.8 billion in 1997 was finally completed in 2007 for $14.8 billion. The Concorde supersonic airplane experienced cost overruns of 1,100%. The Sydney Opera House was over budget by 1,400%! How is this possible? Engineers are supposed to be good at math and logic games. And aren't schedules and cost estimates exactly that?

In 1979, Daniel Kahneman and Amos Tversky introduced the idea of the *Planning Fallacy*, the human phenomena in which we tend to grossly underestimate the time it takes to accomplish tasks. Kahneman later refined this principle in 2003 to extend beyond schedules and also include cost and risk.

Consider a survey by College Board of 1 million college students. 70% ranked their own leadership abilities above average compared to their peers, while just 2% rated themselves below average. In their abilities to get along with others, 25% ranked themselves in the top 1%. In another study, on average incoming college students expected to academically perform better than 84% of their peers. These opinions defy both logic and reality. You might think we get better at this with age and wisdom. But generally speaking, we are astonishingly terrible at estimating, be it in terms of time, cost, risk, or even our own abilities.

In addition to our own estimating limitations, there are other personal, professional, and institutional reasons that consciously or subconsciously promote the Planning Fallacy. This may be especially true within the transportation industry. Those that start a transportation project rarely finish it. Planners, politicians, and engineers need to sell the project so it is selected and funded. Planning predictions are naturally forward focused, which makes it far easier to disregard past performance and be overly optimistic with current situations and anticipated efficiencies from ongoing or anticipated leadership initiatives and process improvements. It is common to overestimate benefits and organizational efficiencies while underestimating schedule, costs, and risks. The reality is optimistic estimates boost the project's benefit/cost ration, and increases the chance of securing funding. Sell the idea and sort out the details later. Optimistic estimates make great political sound bites, resonate with the community, and can help advance individual careers. Optimistic estimates are consistent with an organization's aggressive objectives and stretch goals.

There are other practical reasons why the Planning Fallacy may occur. Successful people often assume others, and the organization itself, does or should have the same productivity as they do. The reality can be quite different. A better PM will deliver a project faster than a less skilled PM. Professional leadership is not an interchangeable commodity. Additionally, it is easy to create a best-case scenario for a high-priority project that does not take into account other projects and initiatives, that are known or may as yet be undefined, that will drain resources from an idealized schedule. Projects schedules are often created with the inherent assumption that they are and will always remain the top priority for all working on it from start to finish. Leveling resources and workloads across the program are rarely reflected in individual project schedules.

Further exacerbating the impact of the *Planning Fallacy* is the *Optimism Bias*, which is our natural tendency to underestimate the probability of experiencing adverse effects. This can go well beyond a glass half-full optimism. Our perception of risk can be in direct opposition to contrary evidence, and even our own personal experiences. We tend to misperceive causes of certain events, giving ourselves perhaps undue credit for positive outcomes, while we may readily attribute negative outcomes to external factors. We tend to overexaggerate our talents, wisdom, influence, and control over risk factors that we truly don't control, such as inflation, stakeholder and community sentiment, and so forth. We are also quick to discount the *Black Swan* event. Named after the historic belief that black swans did not exist until they were discovered in nature, this principle captures our inclination to disregard an unusual event or circumstance that we consider beyond the realm of normal expectations. In retrospect, these high-impact events happen more frequently than we acknowledge, such as a hurricane disrupting construction or supply chains in the southeast United States.

Complicating matters is the fact that in transportation projects, those making the optimistic estimates rarely bear the consequences of resulting budget and schedule busts. Underfunded projects with inadequate scopes and unrealistic schedules set unachievable expectations with leadership and stakeholders. There is also a real human impact on the PM and staff who find themselves in a very difficult situation that institutionally constrains their opportunities to succeed. Morale quickly erodes, accelerating with each request to justify the delays and cost overruns while providing yet another recovery action plan. If persistent and systemic, the discontent will ultimately produce staff ambivalence, turnover, and loss of agency credibility.

Once the original optimistic estimate is made, another tenant of the Planning Fallacy takes hold, *Anchoring*. This is when the original estimate intentionally or unintentionally becomes the default benchmark. Anchoring can be strengthened when a project schedule or budget is tied to other issues and leveraged for other uses. Consider the politician who promises to deliver a signature project by the next election, or a public speech by leadership committing to turning around a stagnant project by an arbitrary or meaningful date. The more entrenched that original anchored estimate becomes, the harder it is to revise it. This can be especially true in organizations that culturally discourage bad news that conflicts with the overall message and theme, regardless of its merit. Bearers of bad news may be labeled difficult, disloyal, or incapable. In extreme cases they may be professionally shunned, or even have their careers stalled or redirected.

Like so many other things in life, the first step toward a solution is acknowledging there is a problem. Recognition of the Planning Fallacy by Project Managers, Program Managers, Planners, and Leadership can go a long way to enabling an effective approach. Practically, it is paramount that schedule and budget estimates are prepared to appropriate baselines of similar projects. These reference class projects provide comparisons to actual past performance. This data should be adjusted to account for risk and confidence levels in the estimates. Mature organizations will evaluate their projects in order to create solid, program-wide schedule templates that balance past performance with department objectives. These templates should be regularly proofed against actual past performance.

Optimism is a wonderful attribute that enables us to look forward with eager anticipation. Optimism can motivate employees and inspire a community. Optimism can improve morale and create a vibrant and healthy culture. While there may be many reasons to be optimistic as we estimate project schedules, costs, and risks, it is our responsibility as engineers and public servants to also be realistic. Not only is it the better long-term business decision, it is the right thing to do. Credibility can take a lifetime to earn, and a moment to lose. Fiercely guard it.

12.8 Closing Thoughts

"To inspire starts with the clarity of WHY."

– Simon Sinek

In his book *Start with Why*, Simon Sinek details the critical importance of building your business foundation on the *WHY*. Most businesses focus on the *WHAT*. But businesses that build fanatical loyalty among their employees and customers have an inherent consistency of purpose and passion that permeates their operations at all levels. They start with the *WHY*.

Those of us fortunate enough to work in the transportation industry, never have to wonder about the *WHY*. We have a noble calling. We save lives. We ensure public safety. We fuel economic growth and development. We ensure the efficient movement of people and goods that drives, sustains, and enhances the economy. We make a difference. We make a direct, meaningful, and enduring impact on people's lives, improving the quality of life in the communities in which we live, work, worship, and play. As PMs, we are uniquely positioned to drive the development and delivery of transportation projects. This is a sober and honorable responsibility.

To all transportation project managers...Godspeed!

"We will either find a way, or make one."

– Hannibal

Index

Note: Page numbers followed by "*f*" refer to figures.

Transportation Project Management, First Edition. Rob Tieman.
© 2023 John Wiley & Sons Ltd. Published 2023 by John Wiley & Sons Ltd.

d